IET ELECTROMAGNETIC WAVES SERIES 47

Series Editors: Professor P.J.B. Clarricoats
Professor E.V. Jull

Electromagnetic Mixing Formulas and Applications

Other volumes in this series:

Electromagnetic Mixing Formulas and Applications

Ari Sihvola

The Institution of Engineering and Technology

Published by The Institution of Engineering and Technology, London, United Kingdom

First edition © 1999 The Institution of Electrical Engineers
Reprint with new cover © 2008 The Institution of Engineering and Technology

First published 1999
Reprinted 2008

The Institution of Engineering and Technology
Michael Faraday House
Six Hills Way, Stevenage
Herts, SG1 2AY, United Kingdom

www.theiet.org

British Library Cataloguing in Publication Data
A CIP catalogue record for this book is available from the British Library

ISBN (10 digit) 0 85296 772 1
ISBN (13 digit) 978-0-85296-772-0

Printed in the UK by TJ International, Padstow, Cornwall
Reprinted in the UK by Lightning Source UK Ltd, Milton Keynes

Contents

Preface

Throughout my working life I have been interested in heterogeneous materials and their properties. Mixtures are fascinating. It is impossible to give an exhaustive structural description of a sample of random medium. And yet in engineering applications problems have to be faced and solved which involve nondeterminately inhomogeneous media.

For these reasons and others, writing a book on electromagnetics of mixtures is not a straightforward task. Logic cannot uniquely predetermine the structure and direction of the text. What is my strategy? I have tried to begin the discussion with simple, idealised models and basic physics. Then I proceed to more complicated analysis. Finally, admitting that increasing the complexity in analysis is not sufficient, I present empirical mixing results and applications. Two groups of readers will be disappointed: on one hand those who would have wanted to find a rigorous full-wave analysis of the random medium problem, and on the other those who are looking for easy-to-use recipes for practical materials.

It has been difficult to decide, during the writing process, to what extent I should reference other literature. Most of the contents of the chapters have grown from my own studies of heterogeneous materials. Of course I have studied reports that my colleagues and rivals have published, and harvested their results. But I do not claim to have read all that has been written on a topic as vast as the modelling of electromagnetic materials. Therefore, I have included, for the most part, only citations to the latest scientific results. This book is primarily meant to be used as a textbook (I have included as exercises many problems that I have solved myself, and some that I would like to see someone else solve), and I have consciously set a limit to the number of references.

I feel grateful to many people with whom I have been working in electromagnetics and related fields, both in Finland and internationally. Much of this co-operation is reflected on the pages of the present book. In particular, I wish to thank Ismo Lindell. He encouraged me in the writing process.

In the preparation of the book I received practical help from Juha Avelin, Sami Ilvonen, Reena Sharma, and Sergei Tretyakov, my colleagues at the Electromagnetics Laboratory of the Helsinki University of Technology. Many library officers of our university assisted me in locating historical articles.

Christian Mätzler has kindly reviewed the manuscript to its considerable benefit. I would also like to thank the staff of IEE Publishing for their professionality and experience which greatly improved the text and its appearance.

The people of my childhood home and the members of my family have been a source of support. It is sad that my father is no longer here to share my joy for the completion of the manuscript. Let this book be dedicated to the memory of Heikki Sihvola.

Espoo, August 1999
Ari Sihvola

Chapter 1

Introduction

1.1 The philosophy of homogenisation of mixtures

To find out the density of a piece of matter, divide its mass by the volume. Homogenisation is an easy exercise in the fortunate case when one is looking for the average mass density. This also means that for a mixture that consists of components with different densities, the macrosopic density of the sample is the volume-based average of the densities of the components. The geometrical distribution of the components that compose the sample in the small-scale structure does not matter; the only thing that one is interested in is the total mass divided by the total volume.

Not all physical parameters, however, are as easy to homogenise as mass density. Physical media have a very great number of various properties that can be homogenised. In this book, a particular emphasis is given to the electrical properties of mixtures. Although the reader will find by the course of the present text that plain volume-average homogenisation along the lines of mass density is used in certain corners of the electromagnetics community, in general the averaging needs to be done in a manner where the laws of electromagnetics are respected and give their own perspective to the averaging process.

What is homogenisation? Certainly it is something more than averaging. We may think that homogenisation of a heterogeneous material is a process leading to its macroscopic characterisation with fewer parameters than those needed for a full description of the original object. For example, the whole microstructure with the dielectric variation of a sample is reduced to an effective permittivity.

A discussion about averaging and homogenisation has to be complemented with definitions of the length scales. Dimensions are of crucial importance and define the validity of the whole macroscopic description. As an example of the heterogeneity in different scales, we can think of snow. Snow can be considered homogeneous in many different everyday situations in the Winter. But if one happens to fall down during a leisurely stroll on a snow-covered field, the heterogeneity of the snow

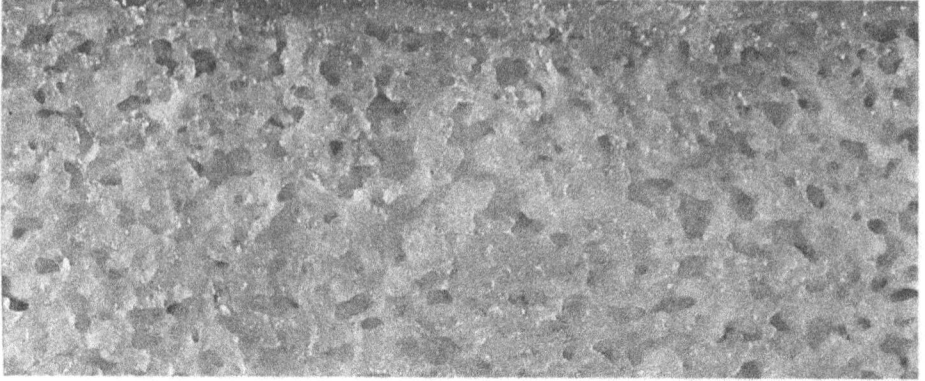

Figure 1.1: *A vertical cut of a seasonal snow layer. The width of the sample is 9 cm. Reproduced with permission of C. Mätzler (University of Bern).*

structure may be seen easily. A careful vertical cut from a snow layer is shown in Figure 1.1. The ice grains and air pores are distinct subregions in the snowpack, and in the scale of millimetres, this sample is clearly heterogeneous, and has a fine and seemingly random structure.

Now the salient question from the homogenising point of view is how the detailed ice grain structure has effect when we look at snow from a more distant point? On a postcard from the Alps, the white cover of the mountain tops and slopes is uniform, or perhaps it only changes colour because of sunlight falling at different angles on it. The microstructure is there in the large-scale image, but only in an averaged way.

Of course, the size of the sample to be averaged by our eyes needs to be large compared to the structure so that speaking of averages makes sense. It is so when we are sufficiently far away from the object. But in fact, more important than the mere distance of the observer from the observed is the wavelength of the light which carries the observation signals. It is well known that we cannot resolve the details of the atomic level of matter even with the best optical microscopes. This is not because of imperfect engineering achievements. It is a principle of physics that cannot be violated. Structures smaller than the wavelength of light are not separated in an image. They contribute a single pixel which is an average of everything under that region. If finer resolution is needed, electron microscopy has to be used. And as we look into smaller scales, new heterogeneities face us. Depending on the material, there may be several stages of various mesoscale structures to pass before we arrive at the molecular and atomic level, where the picture is again nonhomogeneous.

A relevant parameter in homogenisation problems is the ratio between the size of the inhomogeneities and the wavelength of the electromagnetic field that is used. If this ratio is much smaller than unity, the medium appears homogeneous to the wave. When the particles are large or of the order of wavelength they start to scatter radiation and then concepts of average parameters lose their usability. To use snow again as an example, the snow cover of land has an effect on the microwave signals that earth-orbiting satellites are measuring for remote sensing applications. Radiometers measure the radiation emission of the object and scatterometers measure their reflection properties. The wavelength of the signal is often much longer than the snow grain size, which means that, from the point of view of emission, reflection, and transmission characteristics, the snow cover acts as a homogeneous dielectric layer with effective permittivity.

In remote sensing applications, the user quite often seeks information about the physical properties of the object that are contained within the measurable dielectric properties but sometimes hidden in them. One may need, for example, the water equivalent of the snow cover in the Spring. This is certainly a problem where individual grain structures and their local distribution in snow are not important but rather averages over large areas. Here mixing formulas show their importance. If correct homogenisation is used, one can accurately relate the effective dielectric constant to the amount of water and ice in the snow cover.

The measure for the scale of inhomogeneity is obvious in mixtures where inclusions with clear boundaries are embedded in homogeneous environment. There are, however, also other types of structures in random media. Instead of discrete scatterers, the inhomogeneities may be continuous density fluctuations. Then the requirement of scale validity of homogenisation has to be determined from the relation of the wavelength to the correlation length of the spatial permittivity function.

All materials in nature may not be as simple mixtures as the snow cut in Figure 1.1. For example, the electron micrograph of the pancreatic cell shown in Figure 1.2 shows many different elements in the biological tissue. The various units that can be recognised are small compared to a broad range of wavelengths in the electromagnetic spectrum. And clearly separate inclusions, spherical and columnar, can be distinguished. Also in the microstructure of rocks, different shapes of the various components are often present, as can be observed in Figure 1.3, which shows a cut of dolerite (diabase) with ophitic structure.

The remote sensing problem where internal properties of heterogeneous media are sought using a measured macroscopic observable is an inversion of a mixing rule. Such a rule predicts the effective permittivity of a mixture as a function of its components. The material synthesis is a similar inverse problem. Indeed, homogenisation theories can be useful in the design of composite materials and other artificial dielectrics. Sometimes "real" dielectric materials may not be suitable for a given application, for example, because of weight limitations. In those cases one may try to synthesise a composite which would respond to electromagnetic excitation in the same way as the desired dielectric. In 1948 Winston E. Kock [1] suggested to make a dielectric lens lighter in weight by replacing the refractive material by a mixture of metal spheres in a matrix. Kock built lenses by spraying conducting paint on polystyrene foam and cellophane sheets. Of course, the number of metal inclusions needs to be sufficiently large compared to the wavelength at

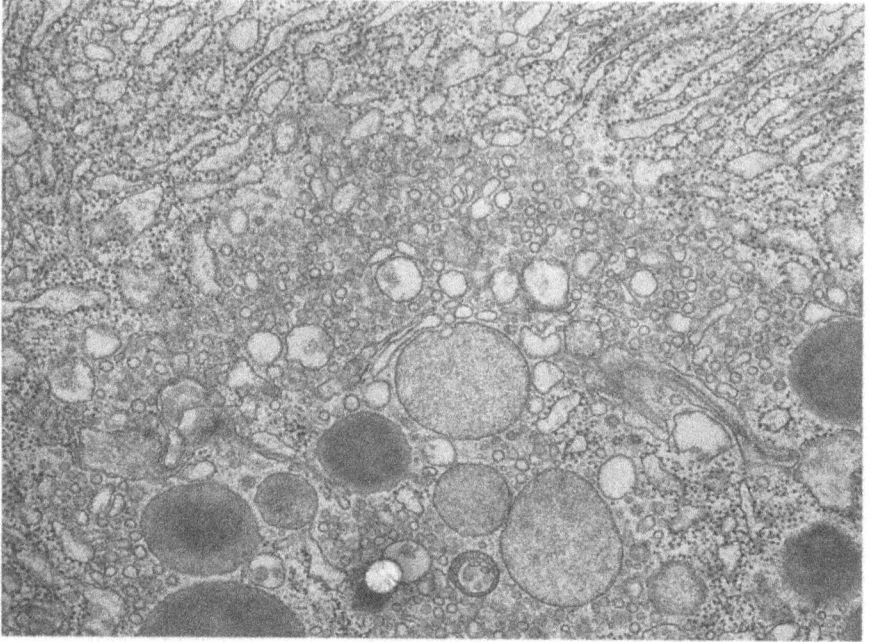

Figure 1.2: *A transmission electron micrograph from a thin (about 50 nm) section of a rat pancreatic exocrine cell. Reproduced with permission of J. Saraste (University of Bergen).*

the design frequency. The large polarisability effect of the metal inclusions adds to the electrical response of the neutral matrix and the result is an increase in the effective permittivity. Metal elements, however, also cause a small diamagnetic effect because of the induced currents, and hence the effective permeability of the composite is affected, too.

Weight is but one of the nonelectrical criteria that dictates the demands for composite materials design in the present-day technology. Other mechanical, elastic, and thermal properties restrict the choice of materials that can be used to manufacture the desired material. For example, in marine or aerospace systems the materials of which vessels and aircraft are made determine the extent to which they scatter electromagnetic radiation and are seen by radars. In such harsh environments the materials cannot be designed by purely electromagnetic formulas. This is something that an electrical engineer may admit, but he certainly adds that the composites that are to be used have to be very carefully electrically characterised in those cases when maximum (or minimum) radar visibility of the object is essential.

On the other hand, certain universality is contained in the dielectric homogenisation theories. In several fields of mathematical physics, equations similar to those used in the analysis of dielectrics appear. This being the case the resulting electri-

Figure 1.3: *Dolerite with ophitic structure; the length of the needle-shaped inclusions is about 1 mm. Reproduced with permission of B. Söderholm (Helsinki University of Technology).*

cal mixing principles are equally valid and applicable in those other problems. Such problems in which the analogies and exact correspondences to the dielectric problem can be found are, for example, the magnetic permeability, electric conductivity, thermal conductivity, particle diffusion, irrotational fluid flow, and even certain elastic properties of matter. Some of these problems will be addressed later in the present book.

Before going into electromagnetic analysis of materials, let us take a short historical look at the work done in the field.

1.2 Historical background

The known history of electrostatics has its origins some 2600 years back in Miletus, in the Presocratean Greece, where Thales the philosopher made observations on the peculiar properties of amber. The Greek word for this precious fossil resin, amber, is indeed *electron*. As centuries passed, people learned to distinguish between resinous and vitreous electricity and invented ways to generate and accumulate small amounts of charge, enough to be released through a sparking discharge. Scientific studies, however, of the dielectric properties of materials are much more recent. Stephen Gray showed in 1729 that electricity could pass through several hundred feet along packthread, which discovery made the distinction between conducting and insulating materials possible. An important invention somewhat later was the Leyden jar, which gave people the possibility to store electric energy in much larger quantities and for longer times than before. The Leyden jar, or phial, as it also was called, may have been instrumental in bringing forth the concept of permittivity, the dielectric constant.

It is known that the great Henry Cavendish was interested in electrical corollaries of chemical differences in materials. He examined minutely the specific electrical capacity of various materials in the 1770s. His measurements of the electrical properties of substances remained unpublished for a hundred years, and Michael Faraday repeated similar capacitor experiments in the 1830s. Faraday called insulators "dielectrics" and by measuring the change of charge accumulation properties of two conductors when various dielectrics replace air between them, he determined the "specific inductive capacity" of these substances.[1] This term was the natural choice in Faraday's philosophy in which he identified the response due to electric force as "curved lines of inductive action." This corresponds to our concept of the electric flux density. In the latter half of the nineteenth century the term "dielectric constant" appeared in the literature; for example, James Clerk Maxwell in his Treatise used that word instead of specific inductive capacity.[2] A possible natural reason to call an insulator dielectric is the following. If we consider a charged capacitor with no material between the plates, the plates experience an electric action at a distance, but when the space between the plates is filled by an insulator the action is carried through this matter even though the insulator cannot conduct electricity.[3]

It may be worth noting that the magnetic constant of materials corresponding to the electric capacity that became under investigation after Faraday's studies on the diamagnetic properties of bismuth was called "magnetic inductive capacity" by Maxwell. Lord Kelvin preferred to characterise this quantity with the term "magnetic permeability," and realising the mathematical analogy between the problems regarding various properties of matter, he spoke as well about "thermal permeability" and "electric permeability." The term "permittivity" for the dielectric permeability seems to be of later origin.[4]

What, then, can we say about early studies of electrical properties of heterogeneous media? Speculations about the effective dielectric characterisation of mixtures

---

[1] Credit for the discovery of the dielectric polarisation and the specific inductive capacity has been offered in the literature to many other investigators who were active before Faraday. At least the following names can be suggested to have influenced Faraday [2]: Amedeo Avogadro, Tiberio Cavallo, John Canton, Thomas Milner, Abraham Bennet, and George Adams. See also the discussion on the term *diëlektrisch* by Riess [3].

[2] The original reports on the early history of dielectrics are sometimes difficult to find or understand, and a scholar with interest in the history of science needs often resort to secondary studies. [4–9] list some of the valuable collections and resources for a science detective to start with.

[3] The prefix "di" in the word di-electric comes probably from *dia*, which is Greek for "through."

[4] The linguistic roots of these terms can be recognised in the Latin words: *per* (through), *meare* (to glide, flow, or pass), and *mittere* (to send).

can be traced already to the works by Amedeo Avogadro and Faraday. But serious quantitative studies regarding dielectric properties of mixtures or conglomerates of different materials began to emerge around the mid-1800s. Siméon-Denis Poisson's theory of magnetism helped Octavio F. Mossotti [10] to formulate equations for the effect of a dielectric inclusion on its environment. Rudolf Clausius, more known by his work on the study of heat, was also interested in electrodynamics, and he studied the relative effective dielectric constant ϵ_r of a collection of molecules. His book [11] gives a clear derivation of the effective dielectric constant, showing that the ratio

$$\frac{\epsilon_r - 1}{\epsilon_r + 2}$$

is proportional to the number of molecules in the unit volume. Later literature uses extensively the name Clausius–Mossotti formula for a relation that contains this ratio. The Clausius–Mossotti relation will be discussed very much on the pages to follow in this book.

After J.C. Maxwell's unification of electricity and magnetism in 1864 and the discovery of the electromagnetic nature of light, a possibility opened to connect between the optical and dielectric properties of matter. This meant that the effective optical properties of heterogeneous matter could be derived from polarisability considerations of the inclusions in the mixture. Ludvig Valentin Lorenz [12, 13] developed an extensive theory of the refractive index of matter assuming that the density of matter is determined by the density of rigid molecules that are located in free space. He was able to show that the refractive index A of matter is determined by its volume v in such a way that the quantity

$$v\frac{A^2 - 1}{A^2 + 2}$$

is practically constant.[5] The new Maxwellian theory also inspired the young Hendrik Antoon Lorentz into studies of the velocity of light in a medium which he assumed to consist of molecules separated by aether. For the refractive index, his finding was the same as Lorenz's [14], and the result is frequently cited as the Lorenz–Lorentz formula.

A more formal treatment of the problem was undertaken by Lord Rayleigh [15] who calculated the effective material parameter of a mixture where spherical or cylindrical inclusions are ordered in a rectangular lattice. He saw the generality of his own analysis that applied not only for the refractive index but also for the conductivity of heat or electricity. His result gave a connection between the properties of the inclusions and the macroscopic medium. Not much later [16] he generalised

---

[5]For a modern reader, Lorenz's conclusion is perhaps not transparently obvious but can be intuitively appreciated by combining the rigidity assumption of the molecules with the fact that increased volume means less dense composite matter. This presumably displays a lower refractive index.

the theory for ellipsoidal inclusions, a problem which was treated in the Clausius–Mossotti spirit by Anton Lampa in 1895 [17].

A name that appears very commonly in present-day literature on mixing formulas is that of J.C. Maxwell Garnett. He derived the famous relation between the effective dielectric constant of a medium where metal spheres, having given optical properties, occupy randomly positioned a given volume fraction in the host medium. His result is known as the Maxwell Garnett formula[6] [19]. Garnett focused his attention to the effect which silver, gold, potassium, and sodium particles and films can create in glasses. This was not a new problem: Faraday had performed 50 years earlier [20] an extensive series of qualitative experiments on how the colours of light are affected by the "great power of action" of small metal particles. Garnett's study was analytical. His aim was to describe the optical properties of the conglomerate for a very wide range of volume fractions of the metal; from zero to unity, which is an outrageously broad variation from the point of view of today's assumed applicability range of the Maxwell Garnett formula. Around the same time, Gustav Mie was interested to explain the light-colour problem in terms of electromagnetic scattering [21].

The analyses of effective properties of mixtures became more well-defined and exact by the early 20th century. Otto Wiener presented [22] a lucid theory of the refraction constant in 1910. He extended, with half-heuristic reasonings, the Clausius ratio into the form

$$\frac{\epsilon_r - 1}{\epsilon_r + u}$$

and used that extensively for new mixing predictions. The included parameter u Wiener called the *Formzahl*, the form coefficient. The formzahl value can be easily connected to the depolarisation factor of the inclusion, and is, of course, $u = 2$ for spherical inclusions. Mixing rules for the effective conductivity of mixtures, where also ellipsoidal inclusions shapes were allowed, were derived slightly later by H.C. Burger [23].

Other scientists working with homogenisation of mixtures during these and earlier times included E. Ketteler, T.H. Havelock, and K. Lichtenecker. A rigorous analysis leading to the Lorenz–Lorentz formula was provided by the "extinction theorem" which was derived for crystalline media by P.P. Ewald [24] and for isotropic media by C.W. Oseen [25]. This "Auslöschungssatz" [26] allows the dipoles in the medium to radiate with the effect of extinguishing the incident wave and replacing it with another wave that has a different propagation velocity. An obvious inter-

---

[6]Perhaps "Garnett formula" would be the name proper, since the three given names of J.C.M. Garnett were James Clerk Maxwell (see [18] for interesting biographical details of the Garnett family). This person must not be mixed with his senior fellow countryman J. Clerk Maxwell, the father of Maxwell equations. The risk of confusion increases when we observe the fact that Maxwell's Treatise [6, Articles 314 and 430] already contains relations—for the conductivity problem—corresponding to the (dielectric) Garnett formula.

pretation of this state of affairs is that the continuum possesses effectively another refractive index.

A great step forward in the phenomenological mixing theories was taken in the 1930s when many studies of Dirk Anton George Bruggeman were published (for example, [27]). Bruggeman proposed new mixing approaches that led to mixing rules that were qualitatively different compared to the earlier homogenisation principles. In the Netherlands, very extensive research work on the dielectric and magnetic properties of materials was conducted in universities and the research laboratories of the Philips company before and also after the war [28]. This was in connection with the ferrite development work with pulverised magnetic materials, and because of the analogy between the electric and magnetic materials problems, the results can be carried over into dielectric modelling.

In addition to electrical descriptions of solids and gases, models for the dielectric behaviour of liquids were suggested. Peter Debye published studies of the behaviour of polar molecules [29] already in 1912. However, the classical Clausius–Mossotti way to calculate the local field that is supposed to excite a given molecule, when applied to dipolar liquids, did not satisfy experiments. Lars Onsager in 1936 refined the calculation of the exciting field by adding a "reaction field" [30] which is in fact an effect of the permanent dipole moment on itself through the surrounding polarisation. Onsager's theory, and its further developments by John Kirkwood are in good agreement with the observed dielectric behaviour of water.

The monograph of Böttcher [31] reviews the mixing principles in use till the mid-1900s. Since then, the explosive growth of the importance of materials science has created an enormous amount of literature dealing with homogenisation problems in recent decades. Solid-state studies and, later, analyses on disordered media have provided us with many different effective-medium theories which can be used— with varying degrees of success—to predict the macroscopic properties of dielectric mixtures. The technological significance of various colloids, cermets, photographic emulsions, foams, metal sols, and other suspensions that are materials with small but not microsopic inclusions has kept the research particularly strong in the field of dielectric mixtures. It has been stated that from the studies of colloid science, thin discontinuous films, and colour centres, a totally new discipline has formed, *cluster science*. A presentation of all new theories would require efforts from someone having read and absorbed more than I have. So, instead of an attempt to give a review of those theories, the next section provides a list of texts that may partly console the reader who would have preferred one.

1.3   Literature

The following articles and monographs concern dielectric and other properties of matter from the homogenisation point of view that reflect the profile of my personal library. The list is by no means exhaustive.

A classic book on the dielectric responses of materials is the monograph by J.H. Van Vleck [32]. It gives the quantum-mechanical theory of dielectric permittivity. Perhaps easier to read by an electrical engineer is the text written by Arthur von Hippel, which discusses macroscopic and microscopic dielectric modelling and also emphasises practical applications. The book is nowadays available as a reprint [33]. Good insight into dielectric models can be gained from Feynman's teachings [34] but there are also other books that describe enjoyably the macroscopic polarisation phenomena in matter, as examples [31], [35–38]. Much tabular data on dielectric properties of materials has been collected into the volume [39], as with the handbook by Neelakanta [40]. See also [41] for data on the optical properties of materials, especially metals. The corresponding properties for various rocks and minerals are tabulated and illustrated in [42].

The review article by Kranendonk ja Sipe [43] discusses thoroughly the foundations of macroscopic description of dielectric systems both for crystalline and amorphous media. The collection [44] serves the same purpose from a slightly different point of view. A detailed and informative description of dielectric modelling of materials is the review by Fuller Brown [45], and more of a historical rarity is the summary by Karl Lichtenecker of the permittivity studies from the early part of the 20th century [46]. For another treatment of mixing models with emphasis on losses, the text by Van Beek [47] can be consulted. Materials containing water have been a challenge in dielectric modelling as mixtures; see de Loor's work [48] on this topic with a historical flavour. A reader in need of a thorough treatment of the foundations of material electromagnetics should find for herself the book by De Groot and Suttorp [49].

For nonmagnetic media, the frequency-dependent dielectric properties give information also about how the materials respond to optical excitation. Discussion about the optical constants of materials can be found, for example, in [50,51]. See also the collection [52] where many historical papers relevant to the optical and dielectric properties of composite materials are republished. When a liquid is treated as a molecular mixture, a careful analysis needs to be performed to correctly determine the local field that acts on polar molecules. See the classics [53–55] for a discussion.

The extensive study by Bruggeman concerning dielectrical heterogeneities [27] was supplemented by a similar research on the elastic constants of materials [56]. For a general overview of mixing relations of transport phenomena applicable to many various fields of physics, see the text by Batchelor [57]. A universal look on disordered and related systems is given in [58], and a later monograph on such systems from the localisation point of view has been written by Sheng [59]. The reader is urged to try to acquire a copy of the oft-cited review by Landauer [18], which is a good reference to the history and state of the field in the 1970s. Slightly later appeared Bergman's view of the dielectric modelling problem [60]. Of other article collections along these lines [61] must be mentioned. Homogenisation in the spirit of the present book is treated in [62]. In [63], a mathematical introduction to the homogenisation method for several physical problems is given, as well as [64], which suits a more experimentally-oriented reader. A cluster scientists' view on dielectric mixtures can be found in the comprehensive treatment by Kreibig and Vollmer [41].

In addition to analysing dielectric mixtures, the homogenisation problem can be approached with a synthetic purpose to design a material with desired properties. A concise treatment of artificial dielectrics in the microwave-engineering sense of the term can be found in Chapter 12 of Collin's book on guided waves [65].

1.4 Outline of the book

To begin discussion about the dielectric properties of mixtures, Chapter 2 introduces the various phenomena that cause the dielectric response in materials. The concept of permittivity is defined and discussed. Also the varieties of the material response to electric and magnetic fields are discussed.

Thereafter the focus is on mixtures and heterogeneities. The remainder of the book is divided into two parts. The first part (Chapters 3–7) presents the classical mixing principles. The idea is to start with the most simple case: isotropic two-phase mixture with spherical geometries. The classical Maxwell Garnett rule for this mixture is derived and then gradually generalised into more complex geometries and materials responses. In separate chapters, anisotropic, magnetoelectric, and nonlinear mixtures are discussed.

The analysis in Chapters 5–7 makes use of dyadic and six-vector algebra. Although dyadic and six-vector algebras are introduced and a basic toolbox is provided in separate sections, a reader with no previous familiarity with dyadics may prefer to bypass these chapters. I give my permission to such a reading strategy: after Chapter 7, the book takes another direction, with perhaps a less dogmatic view on the effective properties.

Chapters 8–13 form the second part of the discussion on mixtures. There we accept the fact that the dielectric properties of real-life materials do not always follow the classical mixing principles that are introduced in the chapters of the first part. The assumptions inherent in the derivation of the classical mixing principles are discussed. In one of the chapters, the great variety of alternative mixing formulas for heterogeneous materials is presented, as well as the bounds and limitations to which the predictions of a mixing rule are subject. The frequency dependence of the dielectric properties of bulk matter may be quite different from the particulate medium and the related special effects and phenomena are given also attention. Towards the end of the book, attention is given to several engineering and geophysical applications of mixing rules.

After a concluding chapter, two appendices follow. Appendix A collects basic results of dyadic algebra, and Appendix B the most important mixing rules.

References

[1] KOCK, W.E.: 'Metallic delay lenses', *Bell System Technical Journal*, 1948, **27**, pp. 58-82

[2] HEILBRON, J.L.: 'Electricity in the 17th and 18th centuries' (University of California Press, Berkeley, 1979)

[3] RIESS, P.: 'Ueber die Wirkung nichtleitender Körper bei der elektrischen Influenz', *Annalen der Physik und Chemie*, 1854, **XCII**, (7), pp. 337-347

[4] THOMSON, W. (Lord Kelvin): 'Reprint of papers on Electrostatics and Magnetism' (Macmillan, London, 1872)

[5] FÖPPL, A.: 'Einführung in die Maxwell'sche Theorie der Elektricität' (Teubner, Leipzig, 1894)

[6] MAXWELL, J.C.: 'A treatise on electricity and magnetism', Volumes I & II, third edition, (Clarendon, Oxford, 1904)

[7] GILLESPIE, C.C. (Ed.): 'Dictionary of scientific biography' (Scribner's, New York, 1974)

[8] FARADAY, M.: 'Experimental researches in electricity' *Phil. Trans. of the Royal Society of London*, 1838, Series 11

[9] EKELÖF, S.: 'Catalogue of books and papers in electricity and magnetism belonging to the institute of history of electricity' (Chalmers University of Technology, Göteborg, Sweden, 1991)

[10] MOSSOTTI, O.F.: 'Discussione analitica sull'influenza che l'azione di un mezzo dielettrico ha sulla distribuzione dell'elettricità alla superficie di più corpi elettrici disseminati in esso', *Memorie di Matematica e di Fisica della Societa Italiana delle Scienze*, (Modena), 1850, **XXIV**, Parte seconda, pp. 49-74

[11] CLAUSIUS, R.J.E.: 'Die mechanische Behandlung der Electricität', Abschnitt III (F. Vieweg, Braunschweig, 1879)

[12] LORENZ, L.: 'Experimentale og theoretiske Undersøegelser over Legemernes Brydningsforhold', *Det Kongelige Danske Videnskabernes Selskabs Skrifter*, Naturvidenskabelig og mathematisk afdeling, 1869, Femte Række, Ottonde Bind, (8), pp. 205-248

[13] LORENZ, L.: 'Ueber die Refractionsconstante', *Annalen der Physik und Chemie*, 1880, **IX**, (9), pp. 70-103

[14] LORENTZ, H.A.: 'Ueber die Beziehung zwischen der Fortpflanzungsgeschwindigkeit des Lichtes und der Körperdichte', *Annalen der Physik und Chemie*, 1880, **IX**, (4), pp. 641-665

[15] RAYLEIGH, Lord: 'On the influence of obstacles arranged in rectangular order upon the properties of the medium', *Philosophical Magazine*, 1892, **34**, pp. 481-502

[16] RAYLEIGH, Lord: 'On the incidence of aerial and electric waves upon small obstacles in the form of ellipsoids or elliptic cylinders, and on the passage of electric waves through a circular aperture in a conducting screen', *Philosophical Magazine*, 1897, **44**, pp. 28-52

[17] LAMPA, A.: 'Zur Theorie der Dielektrica' *Sitzungsber. d. mathem.-naturw. Classe d. Akad. d. Wiss. Wien*, 1895, **104**, Abth. II.a., pp. 681-723

[18] LANDAUER, R.: 'Electrical conductivity in inhomogeneous media', *in* GARLAND, J.C., and TANNER, D.B. (Eds.): 'Electrical transport and optical properties of inhomogeneous media' (American Institute of Physics, Conference Proc. 40, 1978), pp. 2-45

[19] MAXWELL GARNETT, J.C.: 'Colours in metal glasses and metal films', *Trans. of the Royal Society,* (London), **CCIII**, 1904, pp. 385-420

[20] FARADAY, M.: 'Experimental researches in chemistry and physics' (Taylor and Francis, London, 1991), pp. 391-443

[21] MIE, G.: 'Beiträge zur Optik trüber Medien, speziell kolloidaler Metallösungen', *Annalen der Physik*, 1908, **25**, (3), pp. 377-445

[22] WIENER, O.: 'Zur Theorie der refraktionskonstanten', *Berichte über die Verhandlungen der Königlich-Sächsischen Gesellschaft der Wisseschaften zu Leipzig*, Math.–phys. Klasse, 1910, **62**, pp. 256-277

[23] BURGER, H.C.: 'Das Leitvermögen verdünnter mischkristallfreier Legierungen', *Physikalische Zeitschrift*, 1919, **20**, (4), pp. 73-75

[24] EWALD, P.P: 'Zur Begründung der Kristalloptik', *Annalen der Physik*, 1916, **49**, (1), pp. 1-38; **49**, (2), pp. 117-143

[25] OSEEN, C.W.: 'Über die Wechselwirkung zwischen zwei elektrischen Dipolen und über die Drehung der Polarisationsebene in Kristallen und Flüssigkeiten', *Annalen der Physik*, 1915, **48**, (17), pp. 1-56

[26] FAXÉN, H.: 'Der Zusammenhang zwischen Maxwellschen Gleichungen für Dielektrika und den atomistischen Ansätzen von H.A. Lorentz u.a.', *Zeitschrift für Physik*, 1920, **2**, (3), pp. 218-229

[27] BRUGGEMAN, D.A.G.: 'Berechnung verschiedener physikalischer Konstanten von heterogenen Substanzen, I. Dielektrizitätskonstanten und Leitfähigkeiten der Mischkörper aus isotropen Substanzen', *Annalen der Physik*, 1935, Ser. 5, **24**, pp. 636-664

[28] CASIMIR, H.B.G., and GRADSTEIN, S. (Eds): 'An anthology of Philips research' (N.V. Philips Gloeilampenfabrieken; Centrex Publishing Co., Eindhoven, 1966)

[29] DEBYE, P.: 'Einige Resultate einer kinetischen Theorie der Isolatoren', *Physikalische Zeitschrift*, 1912, **13**, (3), pp. 97-100

[30] ONSAGER, L.: 'Electric moments of molecules in liquids', *Journal of the American Chemical Society*, 1936, **58**, (8), pp. 1486-1493

[31] BÖTTCHER, C.J.F.: 'Theory of electric polarisation' (Elsevier, Amsterdam, 1952)

[32] VAN VLECK, J.H.: 'The theory of electric and magnetic susceptibilities' (Clarendon, Oxford, 1932)

[33] HIPPEL, A. VON: 'Dielectrics and waves' (Artech House, Boston, 1995)

[34] FEYNMAN, R.P., LEIGHTON, R.B., and SANDS, M.: 'The Feynman lectures on physics' Volume II: Mainly electromagnetism and matter (Addison–Wesley, Reading, Massachusetts, 1964)

[35] ROBINSON, F.N.H.: 'Macroscopic electromagnetism' (Pergamon Press, Oxford, 1973)

[36] ROBERT, P.: 'Electrical and magnetic properties of materials' (Artech House, Norwood, Mass., 1988)

[37] SCAIFE, B.K.P.: 'Principles of dielectrics' (Clarendon Press, Oxford, 1989)

[38] SOLYMAR, L., and WALSH, D.: 'Electrical properties of materials', Sixth Edition (Oxford Science Publications, OUP, 1998)

[39] HIPPEL, A. VON. (Ed): 'Dielectric materials and applications' (Artech House, Boston, 1995)

[40] NEELAKANTA, P.S.: 'Handbook of electromagnetic materials' (CRC Press, Boca Raton, Florida, 1995)

[41] KREIBIG, U., and VOLLMER, M.: 'Optical properties of metal clusters', Materials Science Series, **25**, (Springer, Berlin, 1995)

[42] EGAN, W.G., and HILGEMAN, T.W.: 'Optical properties of inhomogeneous materials. Applications to geology, astronomy, chemistry, and engineering' (New York, Academic Press, 1979)

[43] VAN KRANENDONK, J., and SIPE, J.E.: 'Foundations of the macroscopic electromagnetic theory of dielectric media', *Progress in Optics* (WOLF, E., Ed.), 1977, **XV**, pp. 245-350

[44] KELDYSH, L.V., KIRZHNITZ, D.A., and MARADUDIN, A.A. (Eds): 'The dielectric function of condensed systems', Modern problems in condensed matter sciences, **24**, (North Holland, Amsterdam, 1989)

[45] FULLER BROWN, W.: 'Dielectrics', *in* FLÜGGE, S. (Ed): 'Handbuch der Physik' (Encyclopedia of physics), **XVII**, Dielektrika, (Springer, Berlin, 1956), pp. 1-154

[46] LICHTENECKER, K.: 'Die Dielektrizitätskonstante natürlicher und künstlicher Mischkörper', *Physikalische Zeitschrift*, 1926, **27**, (4/5), pp. 115-158

[47] VAN BEEK, L.K.H.: 'Dielectric behaviour of heterogeneous systems', *in* BIRKS, J.B. (Ed.), *Progress in Dielectrics*, **7** (Heywood Books, London, 1967), pp. 69-114

[48] DE LOOR, G.P.: 'Dielectric properties of heterogeneous mixtures' (Thesis, University of Leiden, 1956). See also 'Dielectric properties of heterogeneous mixtures with a polar constituent', *Appl. Sci. Res.*, B, 1964, **11**, pp. 310-320

[49] DE GROOT, S.R., and SUTTORP, L.G.: 'Foundations of electrodynamics' (Amsterdam, North Holland, 1972)

[50] WARD, L.: 'The optical constants of bulk materials and films' (Adam Hilger, Bristol, 1995)

[51] BOHREN, C.F., and HUFFMAN, D.R.: 'Absorption and scattering of light by small particles' (John Wiley, New York, 1983)

[52] LAKHTAKIA, A. (Ed): 'Selected papers on linear optical composite materials' SPIE Milestone Series, **MS120**, (SPIE Optical Engineering Press, Bellingham, Washington, 1996)

[53] DEBYE, P.: 'Polar molecules' (Chemical Catalog Company, New York, 1929)

[54] FRENKEL, J.: 'Kinetic theory of liquids' (Dover, New York, 1955)

[55] FRÖHLICH, H.: 'Theory of dielectrics', Second Edition (Clarendon Press, Oxford, 1986)

[56] BRUGGEMAN, D.A.G.: 'Berechnung verschiedener physikalischer Konstanten von heterogenen Substanzen, III. Die elastischen Konstanten der quasiisotropen Mischkörper aus isotropen Substanzen', *Annalen der Physik*, 1937, Ser. 5, **29**, pp. 160-178

[57] BATCHELOR, G.K.: 'Transport properties of two-phase materials with random structure', *Annual Review of Fluid Mechanics*, 1974, **6**, pp. 227-255

[58] ELLIOTT, R.J., KRUMHANSL, J.A., and LEATH, P.L.: 'The theory and properties of randomly disordered crystals and related physical systems', *Reviews of Modern Physics*, July 1974, **46**, (3), pp. 465-543

[59] SHENG, P.: 'Introduction to wave scattering, localization, and mesoscopic phenomena' (Academic Press, San Diego, 1995)

[60] BERGMAN, D.J.: 'The dielectric constant of a composite material—a problem in classical physics', *Physics Reports*, July 1978, **43**, (9), pp. 377-407

[61] ERICKSEN, J.L., KINDERLEHRER, D., KOHN, R., and LIONS, J.-L. (Eds): 'Homogenization and effective moduli of materials and media' (Springer, New York, 1986)

[62] LAKHTAKIA, A.: 'Frequency-dependent continuum electromagnetic properties of a gas of scattering centers', *in* EVANS, M., and KIELICH, S. (Eds.), 'Modern nonlinear optics', Part 2, *Advances in Chemical Physics*, 1993, **85**, pp. 311-359

[63] PERSSON, L.E., PERSSON, L., SVANSTEDT, N., and WYLLER, J.: 'The homogenization method: An introduction' (Studentlitteratur and Chartwell Bratt, Lund, 1993)

[64] GRIMVALL, G.: 'Thermophysical properties of materials', Selected topics in solid state physics, Vol. XVIII, WOHLFARTH, E.P., (Ed.), (North-Holland, Amsterdam, 1986)

[65] COLLIN, R.E.: 'Field theory of guided waves', Second Edition (IEEE Press, New York, 1991)

Part I

To observe the pattern:
Classical and neoclassical mixing

Chapter 2

Physics behind the dielectric constant

Dielectric materials—or, as they are also called, *dielectrics*—are such media that do not conduct electricity. In the present text a definition so narrow is not strictly followed but certainly the most important property of dielectric materials in our discussion is their ability to store, not conduct, electrical energy. A measure for this property is the permittivity or dielectric constant of the material. In fact, permittivity is only a higher-level invention to calculate approximatively the electric response of matter. Underneath it a great amount of detailed physics is hidden. Let us try to start the discussion of dielectric materials with a look at the various polarisation mechanisms.

2.1 Polarisation phenomena in matter

Matter—although electrically neutral in average—is composed of charged elements. On the atomic scale electrons, with negative electric charge, surround a positively charged nucleus. In solids, the atomic configurations include complicated bonding mechanisms where electrons share their orbitals between neighbouring atoms and molecules.

The ideal dielectric does not allow its electrons to be carried around by the electric field. Instead, the force that an applied field exerts on charges displaces these from their equilibrium positions. A classical picture of this situation includes a restoring force that tries to bind the electrons to their undisturbed locations. The balance between the two forces determines the final static situation, in which case there is a net displacement of positive charges into the direction of the electric field and electrons into the opposite direction. No new charges are created and the net charge is zero. A separation of charges is equivalent to a dipole moment,

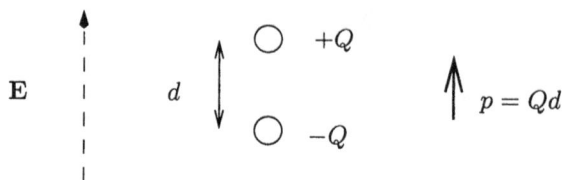

Figure 2.1: *The dipole moment is proportional to the electric field.*

and polarisability is a measure for the relation between the dipole moment and the electric field. Figure 2.1 shows the situation.

But because matter can be composed of charge distributions in many different ways, its response to the electric excitation may also be classified into different types of polarisation. The literature often separates the polarisation mechanisms into electronic, atomic, ionic, orientational, and interfacial polarisations. These different mechanisms of polarisation have different response times. Therefore the effects of the mechanisms can be separated, especially if the response of the material is analysed in the frequency domain.

Electronic polarisation

The phenomenon of electronic polarisation is caused by the displacement of the charge centre of the electron cloud with respect to the nucleus, and it can be observed, undisturbed by other effects, in noble gases. A quantitative description of the polarisability can be calculated by using a classical model of electrons that are elastically bound to their undisturbed positions and respond to an electric field like harmonic oscillators.[1] But certainly, a proper description of the electronic polarisation requires quantum-mechanical treatment [2].

The response of such a system means a resonant behaviour. The polarisation amplitude that is created by the electric field depends strongly on the time-variation of the field. For a certain frequency, especially around the resonant frequency of the oscillator, the induced dipole moment may be very large. The properties of the resonance are also determined by the friction that is associated with the electronic response mechanism. Friction lowers the amplitude and broadens the resonance curve.

The electronic polarisation manifests itself at optical and ultraviolet frequencies because the light mass of the electron gives it the possibility to respond easily to such a fast-varying excitation.

---

[1]See, for example, [1] for a comparison of the classical and quantum-mechanical descriptions of the electronic and other polarisability mechanisms.

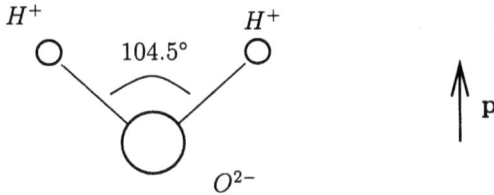

Figure 2.2: *The water molecule has a permanent dipole moment.*

Atomic and ionic polarisation

More complicated polarisation mechanisms arise when we treat molecules. Molecules are formed of atoms, and when atoms are bound to each other with various types of forces they will not always share their electrons symmetrically. It often happens that the eccentric distribution of the electron clouds leads to a situation that the atoms in the molecule acquire charges. And when an electric field is present its effect is to displace atoms or groups of atoms thus creating dipole moments. Carbon dioxide gas is an example of such a medium.

The ionic polarisation is a similar mechanism but the molecules that are capable of displaying it are formed by atoms which are bound together by ionic bonds. Examples of such materials are sodium chloride, NaCl, and other crystals of salt. Atomic and ionic polarisations are lower-frequency phenomena than electronic polarisation and show their maximum effect in the optical and infra-red wavelengths. The absorption lines associated with these effects can be quite sharp in gases but are broadened in liquids and solids.

Orientational polarisation

The bonding of atoms that form a molecule may lead to a situation in which the molecule has a dipole moment even in the absence of an electric field. The water molecule, Figure 2.2, is an example of a familiar material having such a property. A molecule with a permanent dipole moment suffers a torque in an electric field that tries to align the dipole with itself. Unlike in the previous cases of electronic, atomic, and ionic polarisations, now the whole molecule as such is experiencing a force. Hence it is to be expected that the time constants in the process are longer than in those other polarisation mechanisms. This orientational polarisation is indeed a low-frequency phenomenon, and optical waves do not contribute in polarising a medium through orientation of the molecules.

A quantitative model for the orientation polarisation can be calculated by assuming the molecules to be randomly oriented. An external electric field tends to orient the permanent dipoles, but on the other hand their thermal motion is a force that tries to preserve the randomness. This situation is formally similar to the magnetic field and response to it in paramagnetic materials. The resulting polarisability is linear for small field amplitudes and inversely proportional to the temperature, as was shown by Langevin and Debye.

In the condensed state, the model is complicated by the interaction between neighbouring molecules. An orientation due to the external field can be modelled as a relaxation process with a characteristic time constant. The resulting relation between the electric field and the average polarisation is an important relation for dipolar molecules, the Debye equation, which will be studied in more detail later in this book. Characteristic of this response is the frequency dependence in the polarisation of the fluid which may be considerably stronger for low frequencies than for optical frequencies. The transition between these two regimes happens around the relaxation frequency.

Regarding terminology, it is instructive to note the classification by Fröhlich [3]. His view of the polarisation mechanisms emphasises the division of materials by their electrical response into *nonpolar*, *polar*, and *dipolar* substances. Nonpolar substances show optical polarisation only, polar substances have infra-red polarisation as well, and dipolar substances display in addition orientational polarisation. Hence Fröhlich's classes fall quite closely to the three classes of electronic, atomic and ionic, and orientational polarisation that were discussed above.

Interfacial polarisation

Although the preceding polarisation mechanisms are the most important for homogeneous materials quite often another polarisation type is added to a dielectric description of matter in practical situations. Even if a medium as a whole is nonconducting, sometimes it can be composed of mesoscale structures of isolated inclusions that can be conducting. These conducting regions need not be of electronic conductivity of metals; also other charge transport mechanisms may happen in liquids and solids. In ionic liquids charge may be transferred by electrolytic convection, and in crystals, lattice defects may propagate. When an electric field acts on this kind of material, these conducting regions act as polarisable "macromolecules." Charge is accumulated at the interfaces between the conducting and nonconducting regions, thus creating macroscopic polarisation. This causes a large increase of the polarisability at low frequencies. The phenomenon is known as the Maxwell–Wagner effect and was first discussed in J.C. Maxwell's Treatise [4, Art. 328–330], and later more thoroughly studied by Wagner [5].

In bioelectromagnetism, the Maxwell–Wagner effect is very important when the electrical properties of tissues are measured. Cell membranes that separate the cytoplasm and extracellular space are poorly conducting and therefore prevent global conductivity from taking place. However, because of the interfacial polarisation effect, considerable permittivity variations can be observed around the kHz–MHz region. In the biomedical engineering literature, these variations are often referred to as β-dispersion, and are distinguished from α-dispersion (which is observed at audio frequencies) and γ-dispersion (due to the free-water relaxation at microwave frequencies) [6]. Later in Section 13.6, the Maxwell–Wagner effect is treated quantitatively.

2.2   Conduction and complex permittivity

For readers with a background in electrical engineering, the discussion about the frequency dependence of the polarisation mechanisms may gain in clarity if field equations and macroscopic constitutive relations are taken into the analysis. Let us therefore take another look at dielectric materials using equations for the macroscopic electromagnetic fields.

2.2.1   Field relations

Maxwell equations are partial differential equations that give the relations between the electric and magnetic field quantities:

$$\nabla \times \mathbf{E} = -\frac{\partial \mathbf{B}}{\partial t} \tag{2.1}$$

$$\nabla \times \mathbf{H} = \frac{\partial \mathbf{D}}{\partial t} + \mathbf{J} \tag{2.2}$$

$$\nabla \cdot \mathbf{D} = \varrho \tag{2.3}$$

$$\nabla \cdot \mathbf{B} = 0 \tag{2.4}$$

Here all quantities can be functions of space and time as is implicit in the spatial (∇) and temporal ($\partial/\partial t$) differentiations. The quantities and their units in the SI system are the following:[2]

E   the electric field vector [V/m]

H   the magnetic field vector [A/m]

D   the electric flux density vector [As/m^2]

B   the magnetic flux density vector [Vs/m^2]

J   the current density vector [A/m^2]

ϱ the charge density scalar [As/m^3]

Often the flux densities are called displacements.

The Maxwell equations have to be supplemented by so-called constitutive relations which relate the flux densities to the fields:

$$\mathbf{D} = \mathbf{D}(\mathbf{E}, \mathbf{H}) \tag{2.5}$$

$$\mathbf{B} = \mathbf{B}(\mathbf{E}, \mathbf{H}) \tag{2.6}$$

---

[2]Vectors are marked by boldface characters, and scalars by ordinary italics.

Here the functional dependence is determined by the nature of the material in which the fields exist. In a vacuum, the relations become trivial

$$\mathbf{D} = \epsilon_0 \mathbf{E} \tag{2.7}$$
$$\mathbf{B} = \mu_0 \mathbf{H} \tag{2.8}$$

with the free-space permittivity ϵ_0 and permeability μ_0 which have the values

$$\epsilon_0 = \frac{1}{c^2 \mu_0} \approx 8.854 \times 10^{-12} \frac{\text{As}}{\text{Vm}} \tag{2.9}$$

$$\mu_0 = 4\pi \times 10^{-7} \frac{\text{Vs}}{\text{Am}} \tag{2.10}$$

and the parameter c appearing above is the speed of light in a vacuum, which is (by a definition) $c = 299\,792\,458$ m/s exactly.

The constitutive relation in a material is not as simple as the vacuum relation (2.7). The electric displacement is larger, and reads

$$\mathbf{D} = \epsilon_0 \mathbf{E} + \mathbf{P} \tag{2.11}$$

where $\mathbf{P}$ is the average polarisation,[3] the electric dipole moment density. In its plain form for a linear isotropic dielectric material, the polarisation at each point is proportional to the electric field

$$\mathbf{P} = \chi_e \epsilon_0 \mathbf{E} \tag{2.12}$$

and here χ_e is called the electric susceptibility. The permittivity, or dielectric constant, ϵ, is the relation between the flux density and the electric field:

$$\mathbf{D} = \epsilon \mathbf{E} \tag{2.13}$$

Here the relation is a multiplication by a scalar. But in more complicated materials, the response cannot be written so easily. Some of these complex materials will be treated later in this book.[4]

Quite often a relative-to-vacuum permittivity ϵ_r is used which is dimensionless:

$$\epsilon = \epsilon_r \epsilon_0 \tag{2.14}$$

and obviously $\epsilon_r = 1 + \chi_e$.

---

[3]Electromagnetics literature uses the term "polarisation" with two different meanings. In this book we are mostly interested in the polarisation that is denoted by $\mathbf{P}$, the response of matter to electric excitation (or in the case of magnetic response, the magnetic polarisation). But for radar engineers and optical physicists the first thing "polarisation" brings into mind is the way the electric field of the plane wave is behaving, and they talk about linear and circular polarisations. Readers belonging to the latter group need to adapt to the present terminology.

[4]One has to bear in mind that in spite of its omnipresence in electrical engineering literature, an equation like (2.13) cannot be a very deep and fundamental description concerning matter.

2.2.2   Time-harmonic fields

The differential character of Maxwell equations becomes much simpler if the time dependence of the sources and fields is assumed to be sinusoidal. Then a time-derivative means a multiplication by the angular frequency and a 90-degree phase shift. The habit of electrical engineers is to use complex vectors because then the whole differentiation with respect to time can be replaced by algebraic multiplication by a complex scalar. The convention of the time variation according to $\exp(j\omega t)$ with ω being the angular frequency means that the time derivative is equal to multiplication by $j\omega$.

With the introduction of complex fields also the material parameters may become complex. This is not necessarily an uncomfortable complication of the analysis. In fact, the complex permittivity of a material is a very useful concept. Consider fields in a homogeneous material that is conducting. If an electric field is acting in this medium, charges start to move, which is tantamount to electric current. This relation between the current density $\mathbf{J}$ and the electric field $\mathbf{E}$ is given in linear Ohm's law

$$\mathbf{J} = \sigma\mathbf{E} \tag{2.15}$$

where σ is the conductivity of the medium.

Now we can take one of the Maxwell equations, (2.2)

$$\nabla \times \mathbf{H} = \mathbf{J} + \frac{\partial \mathbf{D}}{\partial t} \tag{2.16}$$

Here, two current terms are juxtaposed: the conduction current of the original Ampère law and the Maxwell displacement current as the time-derivative of the flux density. Writing the currents for time-harmonic fields and in terms of the electric field $\mathbf{E}$, we have

$$\mathbf{J} + \frac{\partial \mathbf{D}}{\partial t} = \sigma\mathbf{E} + j\omega\epsilon\mathbf{E} = j\omega\left(\epsilon - \frac{j\sigma}{\omega}\right)\mathbf{E} \tag{2.17}$$

The conductivity term is now combined with the permittivity, and we can generalise the permittivity in such a way that conduction contributes to its imaginary part.

Conduction current is in phase with the electric field, and this means that energy of the field is lost. In general, if the permittivity is complex, the medium is dissipative. But conduction current is not the only loss mechanism in dielectrics. As we saw in the previous section, various molecular polarisation mechanisms involved friction which means that electrical energy is wasted into thermal form. These contributions add to the imaginary part of the permittivity on the same footing as the conduction loss. Once the complex permittivity is known, it can be calculated how much the material attenuates electromagnetic waves that propagate through it.

Because the plane wave, propagating in the z direction, has a functional dependence $\exp(-jkz)$, and the dispersion equation for isotropic nonmagnetic materials is

$$k^2 = \omega^2 \mu_0 \epsilon \tag{2.18}$$

where now the permittivity is a complex scalar, the wave attenuates exponentially. The dependence is

$$\exp\left(-|\text{Im}\{k\}|z\right)$$

which means that the amplitude of the field is decreased into 37% of its original value after a propagation distance $1/(\omega\sqrt{\mu_0\epsilon_0}|\text{Im}\sqrt{\epsilon_r}|)$. This distance is also known as the penetration depth.

Note that the imaginary part of the permittivity ϵ and the wave number k are negative for dissipative materials. This inconvenience is due to the traditional convention of the time-dependence $\exp(j\omega t)$.[5] The sign is compensated in practice by the separation of the real and imaginary parts of the permittivity in the following way:

$$\epsilon = \epsilon' - j\epsilon'' = \epsilon_0(\epsilon'_r - j\epsilon''_r) \tag{2.19}$$

where ϵ' and ϵ'' are both real. Now ϵ'' is positive (or at least non-negative) for passive materials.

A measure for the dissipation quite often used in microwave studies is the loss tangent $\tan \delta$. This quantity is defined by

$$\tan \delta = \frac{\epsilon''}{\epsilon'} \tag{2.20}$$

2.2.3   Dispersion

Because the physical mechanisms that are responsible for causing the electric polarisation in matter depend strongly on the time-variation of the excitation the permittivity of matter is dependent on the frequency of the field variation. This property is called dispersion.[6] Dispersion is seen in the functional dependence on the angular frequency of the real and imaginary parts of the permittivity. A typical dispersion curve of a material displaying the various kinds of polarisability mechanisms above is shown in Figure 2.3.

In Figure 2.3 the electronic and atomic polarisation mechanisms cause a resonance behaviour in the $\epsilon'(\omega)$ curve, associated with a peak in the $\epsilon''(\omega)$ curve. A very typical characteristic of the dispersion curve above and below the resonances is that the real part of the permittivity increases over the major part of the spectrum.

---

[5]The physics community (and also some electrical engineers) prefer to use a complex notation with an opposite sign; their assumed (and omitted) time-dependence in all field and source quantities is $\exp(-i\omega t)$. This convention changes the sign of the imaginary parts of the permittivity of lossy materials, and also means that the reactance of a capacitor is positive unlike in the conventional electrical engineering. However, that notation is not followed here.

[6]To be exact, the frequency dependence of the permittivity is a sign of *temporal* dispersion. This should be distinguished from another important characteristic in certain materials, the *spatial* dispersion, to be discussed later.

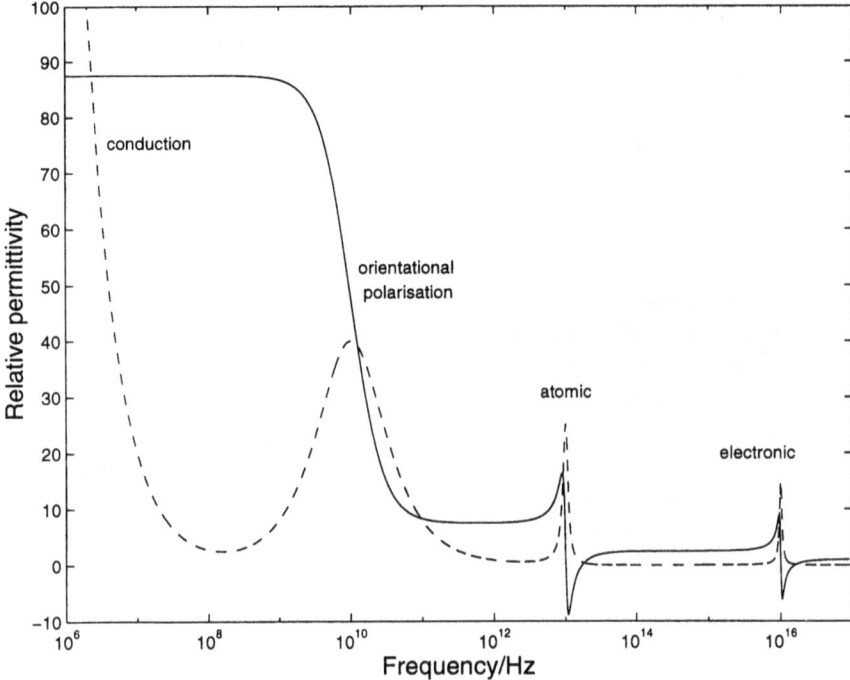

Figure 2.3: *The frequency dependence $\epsilon'(\omega)$ (solid curve) and $\epsilon''(\omega)$ (dashed curve) of the real and imaginary parts of a material in which various polarisation mechanisms are present. Note that the variations in the two curves are connected, which is a property that has a mathematical form in the Kramers–Kronig relations.*

The increase is indeed termed as normal dispersion. However, at a narrow band close to the resonance (and above it), the real part decreases strongly. This is the region of anomalous dispersion.

The connection between the frequency behaviours of the real and imaginary parts can be rigorously formulated. Starting from the basic physical principle of causality, which means that the polarisation response of the matter to an electric excitation cannot precede the cause, the following Kramers–Kronig relations have to be fulfilled for the real and imaginary parts of the permittivity function [7]

$$\mathrm{Re}\{\epsilon(\omega)\} = \epsilon_\infty + \frac{2}{\pi}\,\mathrm{PV}\int_0^\infty \frac{\omega'\mathrm{Im}\{\epsilon(\omega')\}}{\omega'^2 - \omega^2}d\omega' \tag{2.21}$$

$$\mathrm{Im}\{\epsilon(\omega)\} = -\frac{2\omega}{\pi}\,\mathrm{PV}\int_0^\infty \frac{\mathrm{Re}\{\epsilon(\omega')\} - \epsilon_\infty}{\omega'^2 - \omega^2}d\omega' \tag{2.22}$$

Here PV stands for the principal value part, which means that in the integration, the singular point $\omega' = \omega$ is symmetrically excluded. These relations, connected to Hilbert transforms, are valuable in the analysis of the dispersive properties of materials.[7] For example, a wide-band measurement of the real part of the permittivity can be used to gather information about the imaginary part and its frequency dependence.

2.2.4 Complex resistivity

The idea of connecting conductivity with the imaginary part of the permittivity of a material can be reversed. It is not uncommon to find literature where the ohmic character is more emphasised than the dielectric capacity, and the electrical properties are looked at mostly from the conduction point of view. If a medium is more like a conductor than an insulator, the conduction current term dominates over the displacement current in (2.2). Then it is natural to generalise the conductivity in such a way that the permittivity contributes a frequency-dependent imaginary part to it, as is seen in Equation (2.17): $\sigma + j\omega\epsilon$. In geo-electromagnetics applications, the quantity of *complex resistivity* ρ is defined [9]:

$$\rho = \frac{1}{\sigma + j\omega\epsilon} \tag{2.23}$$

which simplifies into the "ordinary" resistivity $1/\sigma$ for ideal conductors or DC current.[8] The unit of resistivity is Ωm. For good conductors or low frequencies when the displacement term is much smaller than the conduction term, the complex resistivity can be approximatively written as

$$\rho \approx \frac{1}{\sigma} - j\frac{\omega\epsilon}{\sigma^2} \tag{2.24}$$

The capacitive character of this lossy material is reflected in the negative imaginary part of the complex resistivity.

The continuation of the permittivity or resistivity into the complex domain—although done here only half-rigorously—is quite helpful in the analysis of dielectric properties of materials, in particular heterogeneous materials. As will be later shown in the present book, the mixing rules for the macroscopic permittivity of mixtures can be used for lossy materials. Then the permittivity contains information about the losses in its imaginary part. But if the losses start to dominate over the dielectric character, the mixing rules can be "turned around" and be thought of as telling something about the effective conductivity. The displacement part is only a perturbational correction to the resistive mixing formula which retains the form of the original dielectric formula.

---

[7] For Hilbert transforms, see [8].

[8] To avoid ambiguity caused by the Greek "rho" character, in this book the charge density is denoted by ϱ and the resistivity by ρ.

2.3   Higher-order polarisation mechanisms

The response of the matter to electric excitation was until now assumed to be very simple: linear, spatially local, and isotropic. This kind of behaviour occurs when a polarisable unit in matter (atom, molecule, conducting region) is exposed to an electric field $\mathbf{E}$ and an electric dipole moment $\mathbf{p}$ is created in such a way that it is linearly dependent on the field

$$\mathbf{p} = \alpha\mathbf{E} \qquad (2.25)$$

through a scalar polarisability α. But such a response of a material to electromagnetic field is very simple, and real materials have many ways of behaving more complicatedly. Several of those characteristics will be treated in forthcoming chapters from the point of view of mixtures; let us here shortly present the various complex polarisation mechanisms in homogeneous matter.

2.3.1   Anisotropy and multipole moments

Anisotropy means that the dielectric response of matter is dependent on the direction of the electric field. This means that the relation in Equation (2.25) cannot be given as a multiplication by a scalar polarisability α. However, still in the anisotropic case the linear character of the relation can be retained if the polarisability is allowed to be a dyadic or a tensor. This means that when the dipole moment components are calculated from the field a matrix expansion has to be used of the polarisability dyadic. Many materials and dielectric objects are anisotropic. This property can be understood to arise for example from geometrical effects in the microstructure which make it easy for charge to accumulate in certain directions while the creation of polarisation may be difficult if the electric field vector points to another direction.

When matter is macroscopically neutral, in other words the net charge over a representative volume vanishes, the main contribution to the electrostatic energy comes from the dipole moments in the medium. In the expansion for electrostatic energy (see, for example, [10]), the charge is multiplied by the electric potential and the dipole moment is multiplied by the electric field. Therefore in neutral materials and under homogeneous electric field, the dipole moments account for the energy density. However, presenting the dipole moment distribution is not an exhaustive description of the polarisation state of matter. Charge distributions in real materials may be so complicated that higher-order multipoles are created in the polarisable units, like molecules or larger regions. The electrostatic energy of a quadrupole or any higher multipole vanishes in a uniform field but if the electric field is not spatially constant the multipoles contribute to the energy. For example, the electric quadrupole components have to be taken into account by weighing them with the components of the electric field gradient.

Should we treat matter that consists of elements that polarise themselves in such a manner where the dipole moment is not an exhaustive description, the macroscopic averaging also needs special consideration. Because of possible quadrupole corrections, the macroscopic flux density **D** contains contributions from the higher-order multipole densities, in addition to the polarisation due to the average value of the dipole moments.

The inclusion of multipoles of higher order than the dipole requires on the level of equations that the response of matter be described with relations where dyadics and matrices are not sufficiently general. For quadrupole effects, triadics are needed, or alternatively third-rank tensors. The various symmetries that the structure of matter may possess are reflected in the properties of the medium tensors of all orders [11, 12].

2.3.2   Magnetic polarisation

Electric fields cause electric polarisation in matter. But electrostatic fields are only a part of a more general setting of electromagnetism the laws of which are governed by Maxwell equations. Electric and magnetic fields and their effects are coupled. Magnetic effects in materials are sometimes so strong that they cannot be neglected. In many microwave engineering applications, magnetic polarisation in materials is exploited in the design of devices in such ways that their response is qualitatively different from their nonmagnetic counterparts. Magnetic polarisation in materials is due to the concentration of magnetic moments in them. A macroscopic magnetic moment emerges for example from a loop of current, and this picture helps in understanding magnetic properties of matter, although the origins of magnetic effects are primarily quantum-mechanical.

Matter is composed of atoms. Atoms have electrons, protons, and neutrons as constituents. Electrons can be quasi-classically thought to be circling the nucleus, so that this movement creates an orbital magnetic moment because the electron carries a charge. But in creating magnetisation this "movement" is not necessary as the spins of these particles cause additional magnetic moments. In fact, the magnetic effects of the atom and matter in general are not limited to electronic properties; an important phenomenon is the magnetic moment configuration of the protons in the nucleus, which is taken advantage of in biomagnetism and medical engineering in MRI devices.[9]

A measure for the magnetic response to a material is its magnetic susceptibility χ_m which is the relation of the average magnetisation to the magnetic field that causes it. In electrical engineering, a useful concept is the (magnetic) permeability μ, which is defined by the susceptibility:

$$\mu = \mu_0(1 + \chi_m) \tag{2.26}$$

where $\mu_0 = 4\pi \times 10^{-7}$ Vs/Am is the permeability of a vacuum.

---

[9]MRI (magnetic resonance imaging) is the modern replacement for the acronym NMR (nuclear magnetic resonance).

But the magnetic effects—although always present in matter—are very small for large classes of materials. The two important "nonmagnetic" material groups are the diamagnetic and paramagnetic classes. Diamagnetism is due to the orbital part of the electronic magnetic polarisation. In response to a magnetic field, the induced magnetisation is oppositely directed as predicted by Lenz's law in electrodynamics. Therefore the magnetic susceptibility is negative for diamagnetic materials. Examples of these materials are copper, silver, and water, but their susceptibilities are of the order $\chi_m \approx -1 \times 10^{-6}$. But also macroscopical conducting loops in heterogeneous materials contribute to a similar effect; see [13] for the analysis of diamagnetism by eddy currents in the liquid water phase in wet snow.

About as small in magnitude, but positive, is the magnetic susceptibility of magnesium, which is an example of paramagnetic material. For paramagnetic materials, the induced magnetic dipole moments are, on the average, oriented in the same direction as the incident magnetic field. Paramagnetism appears in materials where independent permanent magnetic moments respond to magnetic excitation. However, due to thermal motion which randomises the dipole orientations, the resulting magnetisation remains very small: only in low temperatures, like for example in the case of liquid helium, a paramagnetic susceptibility of the order of $+10^{-3}$ can be observed.

Magnetically more active are materials in which there exists a coupling between the adjacent permanent magnetic moments in the structure of the material. If the magnetic moments are all aligned and parallel to each other the material is ferromagnetic. Due to this ferromagnetic coupling very large magnetic polarisation can be created. Ferromagnetic materials may also possess a spontaneous polarisation in the absence of the external magnetic field. The magnetic susceptibility of such materials can be very large, of the order of thousands or even a million in certain expensively engineered soft magnetic materials. However, because of the strongly nonlinear nature of the ferromagnetism, the polarisabilities are dependent on the field amplitude and its history. This special character, and the anisotropy and loss mechanisms of such magnetism, are due to the domain structure of the materials: the ferromagnetic coupling is limited within a certain volume in the medium that is bounded by so-called Bloch walls. Behind the walls there are other domains with their own magnetisation directions.

Antiferromagnetic materials also have coupled moments, but they are antiparallel and therefore no large polarisation is formed. For ferrimagnetic materials—of which ferrites with their nonconducting properties are very important in radio engineering—the net density of magnetic moments does not vanish but the coupling still causes a spontaneous magnetisation. Magnetic materials are very sensitive to temperature, and ferro-, ferri-, and antiferromagnetism is lost in sufficiently high temperatures.

In so-called magnetoelectric materials it may happen that magnetic polarisation is caused by electric field and vice versa. Therefore, the general description of materials with magnetic and dielectric responses requires that both electric and magnetic polarisations are allowed to depend on all the possible electric and magnetic excitations: not only on the electric and magnetic fields but also on all of their temporal and spatial derivatives. For the high-order terms in these expansions, the relations contain tensors of increasing rank. The magnetic effects are, however, characterised by the fact that time reversal inverts the magnetic field and magnetic moment, which is not the case for electric field and electric moment. This fact can be used if we need to consider which types of solid state crystals can theoretically exhibit such effects, because the symmetry of their structure is reflected in their property tensors [12].

2.3.3   Other polarisation effects

Electric polarisation phenomena, in addition to magnetic effects, can also be more complicated than linear isotropic or anisotropic responses. The cause of an electric dipole is not necessarily only the electric and magnetic field or their derivatives. Also, nonlinear phenomena analogous to the coupling effects in the magnetic polarisability are allowed. Often the polarisation effects due to nonelectrical causes are much smaller in magnitude than the direct effects. However, in the following some of these effects are briefly discussed.

Ferroelectricity

Ferroelectricity has received its name by its analogy to ferromagnetism, although iron and other ferromagnetic materials are not ferroelectric, and indeed, few ferroelectric materials contain any iron at all.[10] Ferroelectricity is the spontaneous alignment of electric dipoles by mutual interactions [15]. Ferroelectric materials are characterised by very high relative permittivities, which may reach orders of 10^3 or even 10^4. Not unlike ferromagnetic materials, ferroelectricity disappears above a temperature called the (ferroelectric) Curie temperature. Examples of ferroelectric materials are Rochelle salt[11] and barium titanate, of which the latter material has recently been very applicable in microwave engineering as a ceramic, sintered from powder. The electric material most analogous to a permanent magnet is *electret*, an electrified substance carrying charges of opposite polarity at its extremities. Its usefulness is, however, diminished by free charges that are attracted from air to its surfaces and which tend to cancel the polarisation, a disadvantage from which its counterpart, the permanent magnet, does not suffer.

The irreversibility of the domain structure of a ferroelectric material makes its electric response history-dependent and nonlinear. The average electric polarisation plotted as a function of an alternating electric field is not a single curve. Rather, it is

---

[10] In fact, some authors do not like to use the term "ferroelectric." For example, Böttcher [14] prefers to talk about "Seignette electricity" instead.

[11] Tetrahydrate of potassium-sodium tartrate.

a hysteresis loop. The nonlinear and saturating character of ferroelectric materials means that the permittivity is much larger at low field strengths than for high values of the field.

Nonlinearity

Nonlinearity along the ferromagnetic and ferroelectric vein is typically a static and low-frequency phenomenon. But nonlinearity in materials is also very important at the other end of the electromagnetic spectrum, in optical frequencies. The harmonic generation of light is due to the nonlinear character of a medium which is exposed to an electric field with strong amplitude. Also the parametric amplification of microwave signals is a beneficial consequence of nonlinear response of matter. The progress in solid state physics and laser technology in the latter part of the 20th century has increased our understanding about the nonlinear properties of matter. Unlike the ferroelectric materials which may exhibit nonlinear responses for relatively weak fields, the optical applications of nonlinearity deal with very strong field amplitudes.

The constitutive relation for a nonlinear material is of course again more complicated than for linear materials. The electric polarisation created by an electric field has terms which depend on the second- and higher-order powers of the field [16, 17]. And this relation can also be anisotropic and depend differently on the various combinations of the components of the electric field. Therefore, for example, the second-order nonlinear susceptibility is a third-order tensor. Of course, symmetry of the medium can again be used to reduce the complexity of the property tensors of a nonlinear material. For example, the third-order susceptibility tensor has to vanish in centro-symmetric media, from which it follows, for example, that the so-called Pockels effect is forbidden. However, the next-order nonlinearity with a tetradic constitutive relation—the Kerr effect—is allowed. From the point of view of the present book it is important to note that the nonlinear character of materials may be strongly enhanced by the heterogeneous structure of mixtures. This is due to the fact that the field amplitudes may increase locally to a great extent due to geometrical boundary effects near gaps and other sharp corners. In Chapter 7 nonlinear materials are discussed in more detail.

Piezoelectricity

Piezoelectricity is a phenomenon already discovered by the Curie brothers in 1880. They found that electric fields could be created in certain asymmetrical crystals by mechanical compression in some directions. An example of such a medium is tourmaline, which is also strongly anisotropic and displays dichroism: it is of different colour when viewed in different axes. This effect is often labelled as the *direct* piezoelectric effect. By thermodynamic grounds, one can expect that for such materials also the converse effect exists, which is indeed the case: the application of a potential difference in the material creates a mechanical distortion. A natural way to make use of the piezoelectric phenomenon is in electro-mechanical resonators and transducers, like microphones and pressure sensors. The constitutive relations for piezoelectric materials are more complicated than for ordinary anisotropic materials because the mechanical stresses and strains themselves are second-order tensors, and hence their relations with electric field and flux density have to be presented by third-order tensors [18].

The relation between the electric field as a cause and the mechanical strain as an effect may also contain nonlinear terms. The name for this component of the distortion is *electrostriction*. Of course, a compression of matter increases its density and may also cause the increase of the permittivity; therefore piezoelectricity makes the material nonlinear.

Pyroelectricity

Pyroelectricity is a property of such a material which develops electric polarisation when it is heated. Hence, the material may show polarisation even in the absence of an external electric field. Thermal expansion changes the permanent moments of the sample through piezoelectric-type effects. Crystals that have a polar axis are pyroelectric in addition to being piezoelectric, and these mechanisms can be used as a basis for classification of the various crystallographic point groups [19]. Pyroelectricity was first observed in quartz in 1824 and it can be obviously used as a sensor in thermometers.

With a closer look one has to distinguish between the *primary* and *secondary* pyroelectricity [20]. The electric polarisation can be a direct consequence of a change in temperature (primary pyroelectricity). But also the secondary effect (so-called false pyroelectricity) is possible: thermal expansion causes a strain, and then electric polarisation is created if the material is piezoelectric.

Nonlocality

Another important concept in the treatment of the response of materials is the distinction between locality and nonlocality. In one of the previous sections the memory effects of the dielectric polarisation were discussed. That meant that because there is always slowness in the response of matter, the induced dipole moments were not only dependent on the electric field at the same moment but also on previous field values. This "temporally nonlocal" character also made the permittivity function dependent on frequency. In fact, the properties of the material in the time-domain description have their counterparts in the frequency-domain [21].

Electromagnetic fields depend not only on time but also on the three space coordinates. Therefore the common engineering approach which involves Fourier-transformed functions often plays not only with frequency ω of the wave but often also with a spectrum of plane waves. Then the dependence on the space variables is hidden in functions on the wave vector **k**. The description of the material response means that the permittivity has to be written as a function of both Fourier variables: $\epsilon = \epsilon(\omega, \mathbf{k})$. And because frequency and wavevector are two independent variables, we have to distinguish between temporal and spatial dispersion. Spatial dispersion means that the dielectric polarisation in the medium is dependent on the field variations in space in addition to the pointwise field values. Spatial dispersion is an important phenomenon in the design of artificial materials with a complex response where one can deliberately create a forced polarisation at a neighbouring point in the medium by using, for example, small transmission-line elements that are mixed in sufficient quantities into the microstructure of an otherwise locally responding material.

This list of different types of polarisation mechanisms is not exhaustive. Magneto-striction, piezoresistivity, electro-optical mechanism, and galvano/thermomagnetic effects like Hall, Ettingshausen, and Nernst effects are examples of other phenomena that may be observed in human-made and other natural materials. Although great engineering efforts have been achieved in making practical use of these phenomena, we shall not elaborate on those. The emphasis to follow in this book will be put mostly on the electrical polarisation caused by electric and, to a certain extent, magnetic fields.

Problems

2.1 Show that the Kramers–Kronig relations (2.21)–(2.22) can be written in the form

$$\mathrm{Re}\{\epsilon(\omega)\} = \epsilon_\infty + \frac{1}{\pi} \mathrm{PV} \int_0^\infty \frac{\mathrm{Im}\{\epsilon(\omega')\}}{\omega' - \omega} d\omega' \tag{2.27}$$

$$\mathrm{Im}\{\epsilon(\omega)\} = -\frac{1}{\pi} \mathrm{PV} \int_0^\infty \frac{\mathrm{Re}\{\epsilon(\omega')\} - \epsilon_\infty}{\omega' - \omega} d\omega' \tag{2.28}$$

2.2 A crude model for the imaginary part of the permittivity of a material is that it is constant ϵ'' within a frequency range $\omega_1 < \omega < \omega_2$ and vanishes otherwise. Use Kramers–Kronig relations to calculate the real part of the permittivity of this material. It is obviously more dispersive than the assumed imaginary part. Plot the curves $\epsilon'(\omega)$, $\epsilon''(\omega)$ as a function of frequency.

2.3 Assume in this problem that now the real part of the permittivity is piecewise constant:

$$
\begin{aligned}
\epsilon'(\omega) &= \epsilon_0, & \omega < \omega_1 \\
\epsilon'(\omega) &= \epsilon', & \omega_1 < \omega < \omega_2 \\
\epsilon'(\omega) &= \epsilon_0, & \omega_1 < \omega
\end{aligned}
$$

Use Kronig–Kramers relations to calculate the imaginary part $\epsilon''(\omega)$ of the permittivity of this material. Plot the curves $\epsilon'(\omega)$, $\epsilon''(\omega)$ as a function of frequency.

2.4 In Section 2.2.2, the conduction current density term was connected with the displacement current in such a manner that the permittivity could be generalised to be a complex quantity. The simple model for the permittivity of plasma,

$$\epsilon = \epsilon_0 \left(1 - \frac{\omega_p^2}{\omega^2} \right)$$

although clearly real-valued, can be derived using a similar reasoning.

Derive this relation. Assume that, in a vacuum, charged particles (with charge Q and number density n) are forced to move according to a sinusoidally varying electric field. Let the angular frequency of the field variation be ω . Make use of the nonrelativistic Newton law for the charged particle: $\mathbf{F} = m\mathbf{a}$, where $\mathbf{F}$ is the force, and m and $\mathbf{a}$ are the mass and acceleration vector of the particle. What is the plasma frequency ω_p?

References

[1] HIPPEL, A. VON: 'Dielectrics and waves' (Artech House, Boston, 1995)

[2] VAN VLECK, J.H.: 'The theory of electric and magnetic susceptibilities' (Clarendon Press, Oxford, 1932)

[3] FRÖHLICH, H.: 'Theory of dielectrics', Second Edition (Clarendon Press, Oxford, 1986)

[4] MAXWELL, J.C.: 'A treatise on electricity and magnetism', Volumes I&II, Third edition (Clarendon, Oxford, 1904)

[5] WAGNER, K.W.: 'Erklärung der dielektrischen Nachwirkungsvorgänge auf Grund Maxwellscher Vorstellungen', *Archiv für Elektrotechnik*, 1914, **2**, (9), pp. 371-387

[6] FOSTER, K.R., and SCHWAN, H.P.: 'Dielectric properties of tissues and biological materials: a critical review', *CRC Critical Reviews in Biomedical Engineering*, 1989, **17**, (1), pp. 25-104

[7] JACKSON, J.D.: 'Classical electrodynamics', Second Edition (John Wiley & Sons, New York, 1975), (Third Edition, 1999)

[8] BRACEWELL, R.: 'The Fourier transform and its applications', Second Edition (McGraw-Hill, New York, 1986)

[9] WAIT, J.R.: 'Geo-electromagnetism' (Academic Press, New York, 1982)

[10] ROBINSON, F.N.H.: 'Macroscopic electromagnetism' (Pergamon Press, Oxford, 1973)

[11] GRAHAM, E.B., PIERRUS, J., and RAAB, R.E.: 'Multipole moments and Maxwell's equations', *Journal of Physics B: At. Mol. Opt. Phys.*, 1992, **25**, pp. 4673-4684

[12] BIRSS, R.R.: 'Symmetry and magnetism' (North Holland, Amsterdam, 1966)

[13] MÄTZLER, C.: 'Eddy currents in heterogeneous mixtures', *Journal of Electromagnetic Waves and Applications*, 1988, **2**, (5/6), pp. 473-479

[14] BÖTTCHER, C.J.F.: 'Theory of electric polarisation' (Elsevier, Amsterdam, 1952)

[15] STRUKOV, B.A., and LEVANYUK, A.P.: 'Ferroelectric phenomena in crystals' (Springer, Berlin, 1998)

[16] BLOEMBERGEN, N.: 'Nonlinear optics', 4th edition (World Scientific, Singapore, 1996)

[17] SHEN, Y.R.: 'Principles of nonlinear optics' (Wiley, New York, 1984)

[18] LANDAU, L.D., and LIFSHITZ, E.M.: 'Theory of elasticity', Third edition (Butterworth-Heinemann, Oxford, 1995)

[19] LANDAU, L.D., and LIFSHITZ, E.M.: 'Electrodynamics of continuous media', Second Edition (Pergamon Press, Oxford, 1984)

[20] NYE, J.F.: 'Physical properties of crystals. Their representation by tensors and matrices' (Clarendon Press, Oxford, 1985)

[21] KARLSSON, A., and KRISTENSSON, G.: 'Constitutive relations, dissipation and reciprocity for the Maxwell equations in the time domain', *J. Electromagnetic Waves and Applications*, 1992, **6**, (5/6), pp. 537-551

Chapter 3

Classical mixing approach

After the introduction of the basic types of dielectric behaviour of matter, let us shift the attention to heterogeneous media. To begin with, in this chapter we shall analyse the simplest model for a dielectric mixture. Isotropic dielectric spheres are embedded as inclusions in an isotropic dielectric environment. The two material components of the mixture are known by various names: the inclusion phase is often called *guest*, and the environment *host*, or *matrix*. If we look at this geometry from such a distant position that only averages matter, it may feel natural to associate a macroscopic permittivity with the mixture, which can be calculated if the permittivities of the two components are known.

The basic mixing rule, the so-called Maxwell Garnett formula, will be derived. In later chapters this simple mixing rule will be generalised for cases where the geometry or materials of the problem are more complicated. However, even in these more complex cases the mathematical appearance of the mixing formula retains much of its form. This is one of the motivations for starting the analysis with the introduction of the basic mixing rule. Later on, in Part II, the discussion will be extended to more complicated mixing principles which involve interaction of inclusions in dense materials.

The order of presentation in this chapter is to first reveal the main result, the Maxwell Garnett mixing rule. After that, the theoretical validity and consistency of this result is supported in various ways. An extremely important concept to be introduced is the polarisability of a spherical dielectric sphere. The static problem of the dielectric sphere in uniform electric field is given special attention. This is important for two reasons; first, the polarisability arises from its solution, and second, the principle and the structure of the solution can be transferred, mutatis mutandis, to inclusions that are made of more complicated materials or shapes.

But let us start by showing where to aim.

3.1   Average fields and Maxwell Garnett rule

Let us consider the mixture shown in Figure 3.1 where spherical inclusions with permittivity ϵ_i occupy random positions in the environment of permittivity ϵ_e. Let the fraction f of the total volume be occupied by the inclusion phase, and then the volume fraction $1 - f$ is left for the host.

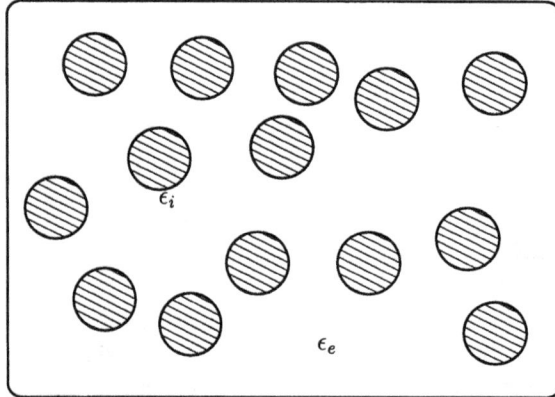

Figure 3.1: *Dielectric spheres are guests in the dielectric background host. This geometry makes the simplest type of dielectric mixture. Note the randomness in the position of the spheres.*

What is the effective macroscopic permittivity ϵ_{eff} of this mixture? Let us define ϵ_{eff} as the relation between the (volume)-average field and flux density:

$$< \mathbf{D} > = \epsilon_{\text{eff}} < \mathbf{E} > \tag{3.1}$$

The average field and flux density can be written by weighing the fields with the corresponding volume fractions:

$$< \mathbf{D} > = f\epsilon_i\mathbf{E}_i + (1 - f)\epsilon_e\mathbf{E}_e \tag{3.2}$$
$$< \mathbf{E} > = f\mathbf{E}_i + (1 - f)\mathbf{E}_e \tag{3.3}$$

where we assume the fields $\mathbf{E}_e$ and ϵ_i to be constants. Then we can write for the effective permittivity

$$\epsilon_{\text{eff}} = \frac{f\epsilon_i A + \epsilon_e(1 - f)}{fA + (1 - f)} \tag{3.4}$$

where A is the field ratio between the internal field and the external field: $\mathbf{E}_i = A\mathbf{E}_e$. Suppose now that we are aware of the fact[1] that the field ratio is $A = 3\epsilon_e/(\epsilon_i + 2\epsilon_e)$. Then the effective permittivity can be written as

---

[1]See for example [1, Sect. 4.4].

$$\epsilon_{\text{eff}} = \epsilon_e + 3f\epsilon_e \frac{\epsilon_i - \epsilon_e}{\epsilon_i + 2\epsilon_e - f(\epsilon_i - \epsilon_e)} \tag{3.5}$$

This result (3.5) is known as the Maxwell Garnett mixing formula. Most of the rest of this chapter goes to persuade the reader that although we jumped over details in the derivation of this result, it has a solid electrostatic foundation. To show that the result is correct requires us to analyse the behaviour of a single spherical dielectric particle. Let us start by looking at a dielectric sphere and its response to an electric field.

3.2   Polarisability of dielectric sphere

3.2.1   Polarisability and dipole moment

Assume that a spatially uniform electric field $\mathbf{E}_e$ permeates a homogeneous space with permittivity ϵ_e. Let us then introduce a foreign particle in this infinite background. When the original situation is deformed by this inhomogeneity, the field in the neighbourhood of the inhomogeneity will be perturbed. In particular, inside the particle the electric field may be quite different from the unperturbed value that would be there at that point in the absence of the particle.

If one is interested merely in fields outside the particle, this problem can be replaced by an equivalent problem: the whole space is again filled with permittivity ϵ_e but now instead of the inhomogeneity there is an electric dipole source.[2] The dipole moment $\mathbf{p}$ is in linear relation with the external field. The coefficient that carries this relation is called *polarisability*, and denoted by the symbol α:

$$\mathbf{p} = \alpha \mathbf{E}_e \tag{3.6}$$

The polarisability of an inclusion is therefore the simplest measure of its response to an incident electric field. Polarisability is an important concept that is very much used in the present and following chapters.

For a homogeneous sphere the polarisability is easy to calculate. The dipole moment is proportional to the internal field within the inclusion, its volume, and the dielectric contrast between the inclusion and the environment:

$$\mathbf{p} = \int (\epsilon_i - \epsilon_e)\mathbf{E}_i \, dV \tag{3.7}$$

Note that the polarisation is referred to the environment with permittivity ϵ_e. This does not need to be vacuum. The advantage of such a definition is that it is easy to treat mixtures where the host medium has arbitrary permittivity.

---

[2] A more accurate description includes additional multipole terms into the equivalent problem [2, Sect. 6.3]. The electric dipole is the leading term in the multipole series, and for some basic geometries, like a sphere, the only existing one.

Because the dielectric contrast between the inclusion and the environment is non-zero only within the volume of the sphere the integration is limited to this volume. And let us reuse the result for the internal field $\mathbf{E}_i$ which is induced in a sphere in a uniform and static external field $\mathbf{E}_e$: it is also uniform, static, and parallel to the external field:

$$\mathbf{E}_i = \frac{3\epsilon_e}{\epsilon_i + 2\epsilon_e} \mathbf{E}_e \tag{3.8}$$

Then the dipole moment can be written as

$$\mathbf{p} = (\epsilon_i - \epsilon_e) \frac{3\epsilon_e}{\epsilon_i + 2\epsilon_e} \mathbf{E}_e V = 4\pi a^3 \epsilon_e \frac{\epsilon_i - \epsilon_e}{\epsilon_i + 2\epsilon_e} \mathbf{E}_e \tag{3.9}$$

Now the polarisability can be written immediately:

$$\alpha = V(\epsilon_i - \epsilon_e) \frac{3\epsilon_e}{\epsilon_i + 2\epsilon_e} \tag{3.10}$$

where the permittivities of the inclusion and its environment are denoted by ϵ_i and ϵ_e, respectively. The volume of the sphere is V. Note that the polarisability is a scalar. This is because the inclusion material is isotropic and its shape is spherically symmetric. The dielectric isotropy means that the response of the medium does not depend on the direction of the electric field, only on its amplitude.

Let us take a closer look at the theoretical backgrounds of the polarisability expression (3.10).

3.2.2   Consistency of the field solutions

Why is the field solution (3.8) correct? It is because we can show that it satisfies the requirements of the problem. To analyse the solution in detail, let the external field $\mathbf{E}_e$ be directed along the z-axis according to Figure 3.2. For simplicity in the analysis, the sphere with radius a is located in the origin of the spherical co-ordinate system.

The total field system can be considered to be composed of three fields, the original external uniform field $\mathbf{E}_e$, the internal field $\mathbf{E}_i$ which is the total field inside the sphere, and the perturbational field $\mathbf{E}_d$ which is the effect of the sphere on its surroundings. Then the total field outside the inclusion ($r > a$) is $\mathbf{E}_e + \mathbf{E}_d$.

The fields are

$$\mathbf{E}_e = \mathbf{u}_z E_e \tag{3.11}$$

$$\mathbf{E}_i = \mathbf{u}_z \frac{3\epsilon_e}{\epsilon_i + 2\epsilon_e} E_e \tag{3.12}$$

$$\mathbf{E}_d = \frac{\epsilon_i - \epsilon_e}{\epsilon_i + 2\epsilon_e} \left(\frac{a}{r}\right)^3 E_e \left(2\cos\theta\, \mathbf{u}_r + \sin\theta\, \mathbf{u}_\theta\right) \tag{3.13}$$

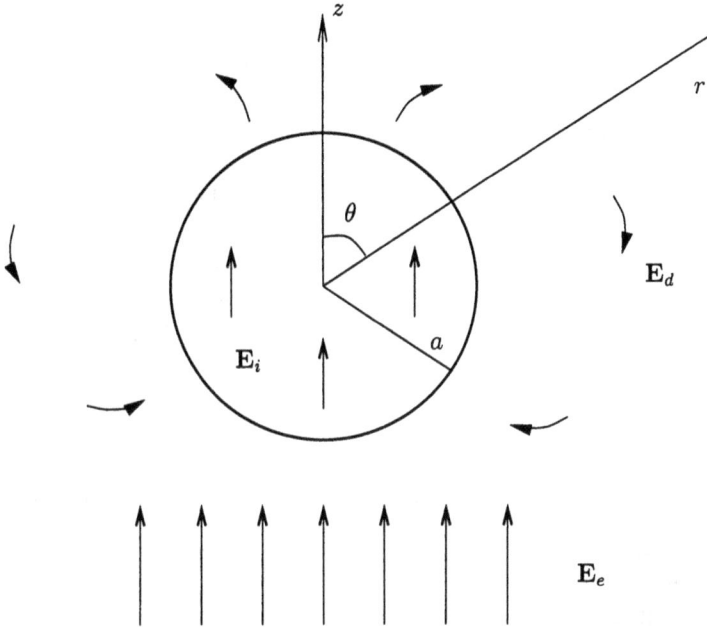

Figure 3.2: *A homogeneous dielectric sphere in a uniform electric field* $\mathbf{E}_e$ *causes a dipole field perturbation* $\mathbf{E}_d$ *in its surroundings.*

where $\mathbf{u}_z$ is the unit vector along the z-axis, and similarly $\mathbf{u}_r$ and $\mathbf{u}_\theta$ are unit vectors along the r and θ directions, respectively. The field $\mathbf{E}_d$ can also be called the dipole field because of its spatial dependence. The field from a dipole decays as the inverse of the cube of the distance from the dipole.

Now we can check that the solution (3.11)–(3.13) is acceptable. A static field solution should be curl-free: $\nabla \times \mathbf{E} = 0$, and it is easy to see that all the three field functions satisfy this condition. Also, far away from the inclusion ($r \gg a$) the perturbation of the sphere becomes negligible, whence only the field $\mathbf{E}_e$ remains. And finally, the solution set is consistent in the respect that the boundary conditions [3, Sect. 1.5] over the interface at $r = a$ are satisfied.

Firstly, the tangential component (the component along the interface) of the total electric field has to be the same on both sides of the interface:

$$\mathbf{u}_\theta \cdot \mathbf{E}_i = \mathbf{u}_\theta \cdot (\mathbf{E}_e + \mathbf{E}_d) \tag{3.14}$$

Likewise, as the second boundary consition, the normal component (the component perpendicular to the interface) of the electric flux density has to be continuous across the interface. Because the flux density is the permittivity times the electric field, the following must be satisfied:

$$\mathbf{u}_r \cdot \epsilon_i \mathbf{E}_i = \mathbf{u}_r \cdot \epsilon_e (\mathbf{E}_e + \mathbf{E}_d) \tag{3.15}$$

Both boundary conditions hold for $r = a$, as can be seen through the observations $\mathbf{u}_\theta \cdot \mathbf{u}_z = -\sin\theta$ and $\mathbf{u}_r \cdot \mathbf{u}_z = \cos\theta$.

Concluding from this check, we may accept the field solutions (3.11)–(3.13), and the already anticipated result for the field ratio

$$A = \frac{|\mathbf{E}_i|}{|\mathbf{E}_e|} = \frac{3\epsilon_e}{\epsilon_i + 2\epsilon_e} \tag{3.16}$$

was legal.

3.2.3   Dipole moment as solution for the external problem

The dipole moment and polarisability (3.10) of the dielectric sphere were calculated by integrating the polarisation density relative to the environment over the whole region. This method of solving α could be labelled "the internal method."

On the other hand, an "external method" is available. The result (3.9) can also be derived by studying the field outside the scatterer. A static dipole with moment $\mathbf{p} = \mathbf{u}_z p$ in a homogeneous and isotropic environment with permittivity ϵ_e generates a dipole field (see, for example, [4, Sect. 3.3])

$$\mathbf{E}_d = -\nabla\left(\frac{\mathbf{p}\cdot\mathbf{r}}{4\pi\epsilon_e r^2}\right) = -\nabla\left(\frac{p\cos\theta}{4\pi\epsilon_e r^2}\right) \tag{3.17}$$

where the expression in brackets is the electric scalar potential of a static dipole. Note, again, that this dipole produces its field in the environment material (not necessarily a vacuum) whence the permittivity in the denominator is ϵ_e. A comparison between (3.13) and (3.17) gives us the dipole moment exactly of the amplitude of (3.9).

It may look redundant to present two methods, internal and external, for deriving the dipole moment of a dielectric sphere, and in the present simple isotropic case admittedly the algebra is not complicated. However, it is useful to distinguish conceptually the external and internal methods. The internal method suits for the present problem, and in fact the dipole moment can be written down by intuition that is tuned to the internal approach. But when the problem involves more complicated inclusions, like, for example, anisotropic or heterogeneous particles, the internal field may become nonuniform and the integral no longer reduces to a multiplication with the volume of the particle. Then it is much more advantageous to resort to the external method and study the spatial dependence and the amplitude of the dipole field that surrounds the inclusion.

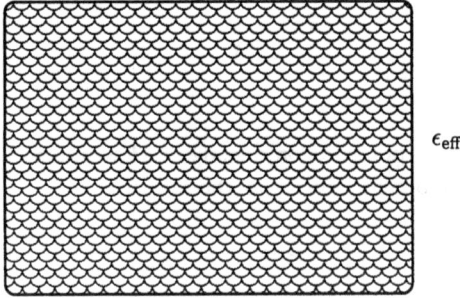

ϵ_{eff}

Figure 3.3: *An effective-medium description for the heterogeneous dielectric sample of Figure 3.1.*

3.3   Mixture with spherical inclusions

Equipped with the polarisability of a single scatterer it is now possible to attack the problem of a mixture which contains a distribution of inclusions. The idea of the homogenisation of a mixture is to replace the scatterers by dipole moments which average into the electric polarisation. The Clausius–Mossotti relation gives the effective permittivity of the mixture as a function of the polarisabilities, and the Maxwell Garnett formula replaces the polarisabilities with explicit material parameters of the inclusions.

3.3.1   Clausius–Mossotti formula

Now that the polarisability of a single sphere is known (3.10), the effective permittivity can be calculated for a mixture where many spheres of this kind are embedded in a background medium. Figure 3.1 could be one sample of such a heterogeneous material. A measure for the guest phase presence is the number density of the spheres n, with unit m^{-3}.

We need to know how to define the effective, or macroscopic permittivity for this heterogeneous sample, in other words how to replace the medium in Figure 3.1 by the one in Figure 3.3.

The classical way to determine the effective permittivity of a mixture is to follow the constitutive relation for a dielectric material. This is a simple relation between the electric field $\mathbf{E}$ and the electric flux density $\mathbf{D}$. Taking the averages of the two we can define the effective permittivity according to (cf. Equation (3.1))

$$< \mathbf{D} > = \epsilon_{\text{eff}} < \mathbf{E} > = \epsilon_e < \mathbf{E} > + < \mathbf{P} > \qquad (3.18)$$

where the spatial average of any function $\mathbf{f}(\mathbf{r})$ over a representative volume[3] of the

---

[3]What volume can be considered as representative is difficult to state in exact terms. It has to be large enough. Here the volume V_{mix} refers to the volume of the mixture but a smaller volume is sufficient if the value of the integral does not change essentially with increase in the integration volume.

sample is noted by

$$< \mathbf{f} >= \frac{1}{V_{\text{mix}}} \int\limits_{V_{\text{mix}}} \mathbf{f}(\mathbf{r}) \, dV \qquad (3.19)$$

The average electric polarisation density $< \mathbf{P} >$ induced in the mixture is connected to the dipole moment density in the mixture:

$$< \mathbf{P} > = n\mathbf{p}_{\text{mix}} \qquad (3.20)$$

where now $\mathbf{p}_{\text{mix}}$ is the dipole moment of a single inclusion in the mixture, which is in general different from the previously calculated dipole moment $\mathbf{p}$ in free-environment. Here it is also assumed that all dipole moments are of equal strength, and n is the number density of the dipoles. As a reminder of the dimensions of the quantities, the units are $[\mathbf{D}] = [\mathbf{P}] = \text{As/m}^2$, $[\mathbf{p}_{\text{mix}}] = [\mathbf{p}] = \text{Asm}$, $[\mathbf{E}_e] = \text{V/m}$, and $[n] = \text{m}^{-3}$. The unit of polarisability is $[\alpha] = \text{Asm}^2/\text{V}$.

The next step is to establish the relation how the average polarisation density depends on the average electric field. This is the point where hand-waving arguments are often used in the literature. Although we know the relation between the dipole moment and the external field of a single particle, the situation is now complicated by the neighbouring particles and the randomness of the mixture. Therefore in a mixture, especially when it is dense, one cannot assume the field exciting one inclusion to be the same as in the analysis leading to the result (3.10) where the inclusion was free from the effect of any disturbing neighbours.

In the classical approach to calculate the field that "excites" a given inclusion in the mixture, all other inclusions are replaced by the average polarisation which uniformly surrounds the mixture. The exciting electric field is then taken to be as the local electric field $\mathbf{E}_L$ which exists within a fictitious cavity that has the shape of the inclusion. The situation felt from the inclusion's point of view is depicted in Figure 3.4.

Now one can calculate that the surrounding polarisation gives an additional contribution to the field, thus increasing the amplitude of the external average field. It is dependent on the shape of the cavity, and for a spherical shape it is [5]

$$\mathbf{E}_L = < \mathbf{E} > + \frac{1}{3\epsilon_e} < \mathbf{P} > \qquad (3.21)$$

where $1/3$ is the so-called depolarisation factor of a sphere.[4]

Because the exciting field is larger than the average field, also the in-mixture dipole moment $\mathbf{p}_{\text{mix}}$ is larger than a free-environment dipole moment $\mathbf{p}$, because the ratio between the dipole moment and the field remains the same polarisability α:

$$\mathbf{p}_{\text{mix}} = \alpha\mathbf{E}_L \qquad (3.22)$$

---

[4]The additional field contribution $\frac{1}{3\epsilon_e} < \mathbf{P} >$ is often called the "Lorentz field" [6].

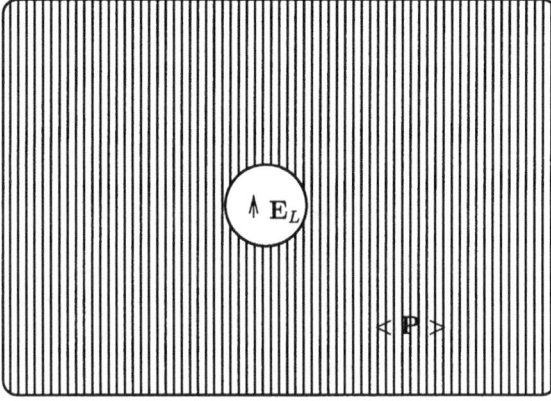

Figure 3.4: *To calculate the field $\mathbf{E}_L$ exciting a single inclusion, the neighbourhood is replaced by the average polarisation $< \mathbf{P} >$. The polarisation is taken away from the volume of the inclusion in question; only a spherical cavity is left.*

Combining the equations leaves us with the average polarisation $< \mathbf{P} > = n\alpha\mathbf{E}_L$, and then the effective permittivity can be written (see Equation (3.18)):

$$\epsilon_{\text{eff}} = \epsilon_e + \frac{n\alpha}{1 - \dfrac{n\alpha}{3\epsilon_e}} \tag{3.23}$$

This famous equation can often be seen in the form

$$\frac{\epsilon_{\text{eff}} - \epsilon_e}{\epsilon_{\text{eff}} + 2\epsilon_e} = \frac{n\alpha}{3\epsilon_e} \tag{3.24}$$

The relation (3.24) carries the name *Clausius–Mossotti formula*, although it deserves the label *Lorenz–Lorentz formula* as well.[5]

If the density of the inclusions is small, the mixture is dilute. Then the Clausius–Mossotti formula can be written by taking the limit of small n:

$$\epsilon_{\text{eff}} \approx \epsilon_e + n\alpha \tag{3.25}$$

Note that this approximation (3.25) is tantamount to using the average field $< \mathbf{E} >$ instead of the local field $\mathbf{E}_L$ to calculate the induced dipole moment, leading to $\mathbf{p}_{\text{mix}} = \mathbf{p} = \alpha < \mathbf{E} >$.

3.3.2 Maxwell Garnett mixing rule

The Clausius–Mossotti/Lorenz–Lorentz-relation contains microscopic quantities like polarisabilities and scatterer densities which are relevant in the molecular description

---

[5]See the historical references in Chapter 1 for the original articles.

of matter. However, for macroscopic engineering use these are not always those most convenient to play with. It may be preferable to perform the calculations with the permittivities of the components of the mixture. To achieve this, a combination of the Clausius–Mossotti formula (3.24) with the polarisability expression (3.10) gives us

$$\frac{\epsilon_{\text{eff}} - \epsilon_e}{\epsilon_{\text{eff}} + 2\epsilon_e} = f \frac{\epsilon_i - \epsilon_e}{\epsilon_i + 2\epsilon_e} \tag{3.26}$$

where $f = nV$ is a dimensionless quantity, the volume fraction of the inclusions in the mixture. The relation (3.26) can be found in the literature under the name *Rayleigh mixing formula* [7].

A comparison to the Clausius–Mossotti formula shows that the Rayleigh equation does not contain information about an individual scatterer. Only the volume fraction of the guest phase and the permittivities appear in the mixing rule. In fact, the assumption that the spheres in the mixture be of the same size can be relaxed when using this result, as long as all of those, including the largest ones, are small compared to the wavelength of the operating field.

Simple algebra applied upon (3.26) leaves us with the Maxwell Garnett mixing formula [8]. According to this famous mixing equation, the effective permittivity of a mixture ϵ_{eff} obeys the following rule:

$$\epsilon_{\text{eff}} = \epsilon_e + 3f\epsilon_e \frac{\epsilon_i - \epsilon_e}{\epsilon_i + 2\epsilon_e - f(\epsilon_i - \epsilon_e)} \tag{3.27}$$

which is the same that we met in the beginning of Section 3.1. Formula (3.27) is in wide use in very diverse fields of application. The beauty of the Maxwell Garnett formula[6] is in its simple appearance combined with its broad applicability. The effective permittivity relative to ϵ_e is determined only by two parameters, the inclusion permittivity relative to background ϵ_i/ϵ_e and the volume fraction of inclusions f:

$$\frac{\epsilon_{\text{eff}}}{\epsilon_e} = 1 + 3f \frac{\epsilon_i/\epsilon_e - 1}{\epsilon_i/\epsilon_e + 2 - f(\epsilon_i/\epsilon_e - 1)} \tag{3.28}$$

The Maxwell Garnett formula satisfies the limiting processes for vanishing inclusion phase

$$f \to 0 \qquad \Rightarrow \qquad \epsilon_{\text{eff}} \to \epsilon_e \tag{3.29}$$

and vanishing host medium

$$f \to 1 \qquad \Rightarrow \qquad \epsilon_{\text{eff}} \to \epsilon_i \tag{3.30}$$

---

[6]The name may betray: J.C. Maxwell Garnett, who published this formula in 1904, was a different man than the father of electrodymamics, J.C. Maxwell. See the historical introduction in Section 1.2.

The latter limit case ($f \to 1$) is a rather academic one because high volume fractions may be difficult to connect with the model of Figure 3.1. Space cannot be filled with monodisperse spheres. The regime $f \approx 1$ is admittedly a problematic one for the Maxwell Garnett rule and there its applicability is generally doubted. However, purely geometrically a total filling of the space by the guest phase can be achieved by using smaller spheres to fill the pores that remain between large spheres. A continuous fractal filling process leads to vanishing host phase without limit.

A perturbation expansion for the Maxwell Garnett rule (3.27) gives the mixing equation for dilute mixtures ($f \ll 1$):

$$\epsilon_{\text{eff}} \approx \epsilon_e + 3 f \epsilon_e \frac{\epsilon_i - \epsilon_e}{\epsilon_i + 2\epsilon_e} \tag{3.31}$$

Figure 3.5 shows the prediction of the Maxwell Garnett formula for different values of the dielectric contrast ϵ_i / ϵ_e. Shown is the susceptibility ratio

$$\frac{\epsilon_{\text{eff}} - \epsilon_e}{\epsilon_i - \epsilon_e}$$

which vanishes for $f = 0$ and is unity for $f = 1$, independently of the inclusion-to-background contrast. The figure shows clearly the fact that the effective permittivity becomes a very nonlinear function of the volume fraction for large dielectric contrasts. This means that the *relative* effect in the increase of the permittivity of a mixture from its background value is better for "weak" dielectric spheres than for "strong" ones.

But the dielectric contrast between the phases can also be opposite: if the inclusion spheres are dielectrically softer than the environment ($\epsilon_i < \epsilon_e$), the effective permittivity of the mixture is smaller than the background value. Figure 3.6 compares two mixtures where the component phases are interchanged. One of the phases is air (ϵ_0) and the other has a permittivity $20\epsilon_0$.

The message of the figure is that the Maxwell Garnett model is not symmetric. If it were, the two curves in Figure 3.6 should intersect at a point that corresponds to a volume fraction of 50%. Host and guest are not contributing on an equal basis to the effective permittivity. This is of course not to be required of the Maxwell Garnett model as the geometric structure of the mixture is nonsymmetric in the first place.

3.3.3 Q_2 function for mixture analysis

The foregoing treatment and analysis of mixtures has resulted in formulas where a simple quotient of differences and sums appears in a similar form, as is succintly shown by Equation (3.26). It is advantageous to define a function which contains this rational relation because then the use of the function helps to shorten mixing relations. Let us call the following relation the Q_2 function:[7]

---

[7]This is an extension of the Q function, the properties and applications of which have been discussed in detail in [9].

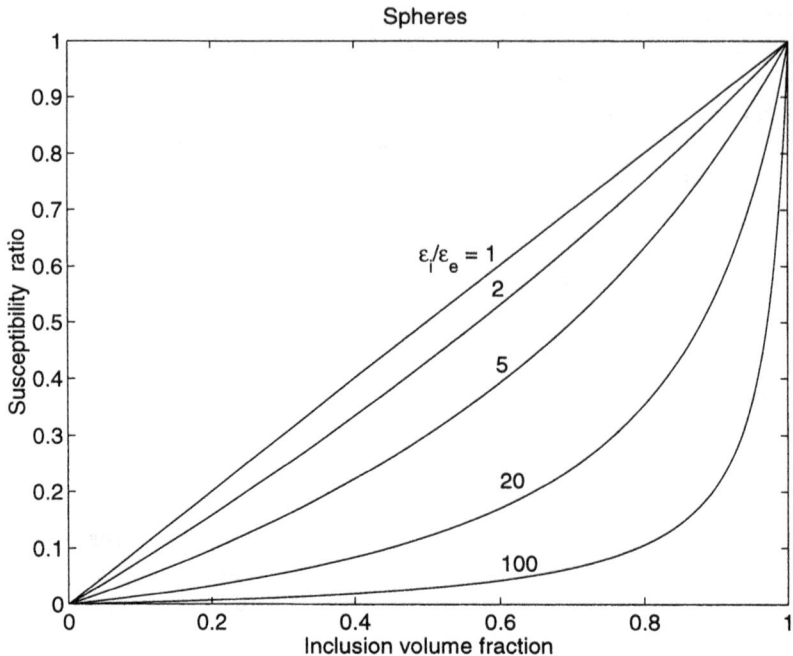

Figure 3.5: *The susceptibility ratio $(\epsilon_{\text{eff}} - \epsilon_e)/(\epsilon_i - \epsilon_e)$ for the Maxwell Garnett prediction of the effective permittivity of a mixture with spherical inclusions of permittivity ϵ_i in a background medium of permittivity ϵ_e.*

$$Q_2(x) = \frac{1-x}{1+2x} \tag{3.32}$$

This expression is a bilinear form which has the property that when it operates on a complex argument, it conserves cirles. The interesting property of the Q_2 function is that it is the inverse of itself:

$$Q_2(Q_2(x)) = x \quad \rightarrow \quad Q_2^{-1}(x) = Q_2(x) \tag{3.33}$$

This property is of great help in solving problems that involve Maxwell Garnett and other mixing rules. Other properties of the Q_2 function include: $Q_2(0) = 1$, $Q_2(1) = 0$, and $Q_2(\infty) = Q_2(-\infty) = -1/2$. Figure 3.7 shows the behaviour of the $Q_2(x)$ function.

With this function, the Maxwell Garnett mixing rule can be written in the form

$$Q_2(\epsilon_e/\epsilon_{\text{eff}}) = f Q_2(\epsilon_e/\epsilon_i) \tag{3.34}$$

which can be seen from (3.26). Therefore, using the inverse property of the Q_2 function, the effective permittivity can be seen to follow the rule

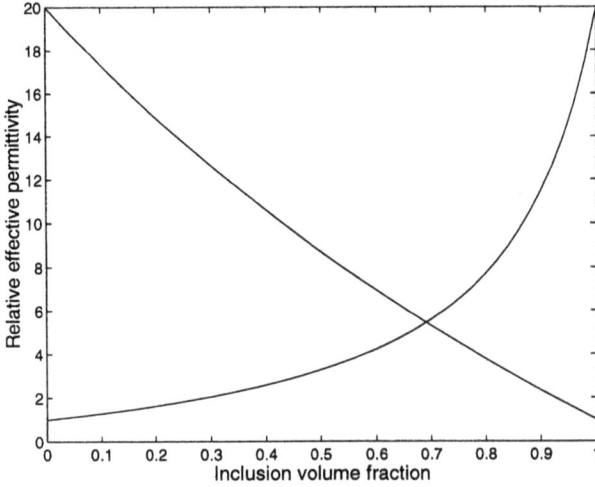

Figure 3.6: *A comparison of complementary mixtures: the effective permittivity ϵ_{eff} is shown for two mixtures where the contrast between the phases is 20. The roles of inclusion and environment have been interchanged in the two models: $\epsilon_i = 20\epsilon_e$ in the rising curve and $\epsilon_e = 20\epsilon_i$ in the falling curve.*

$$\epsilon_{\text{eff}} = \frac{\epsilon_e}{Q_2[fQ_2(\epsilon_e/\epsilon_i)]} \tag{3.35}$$

But it is clear that the inclusion permittivity obeys a similar-looking rule:

$$\epsilon_i = \frac{\epsilon_e}{Q_2[(1/f)Q_2(\epsilon_e/\epsilon_{\text{eff}})]} \tag{3.36}$$

This relation is useful in the inverse problem, where the macroscopic properties are a priori known (like in the composite material design problem) or measured (in remote sensing applications, for example) and one needs to know the permittivities of the components.

3.4 Discussion on basic field concepts

The mixture analysis presented in the previous sections stayed at a simple phenomenological level. That was intentional, and was done with the objective that the reader might have the opportunity to observe the salient steps in the derivation of the final mixing rule. The approach contained approximations which shall be discussed and partly corrected for in the following chapters of the book.

The existing literature that discusses homogenisation principles uses a broad variety of concepts and terms that enter the analysis. Let us discuss some of the

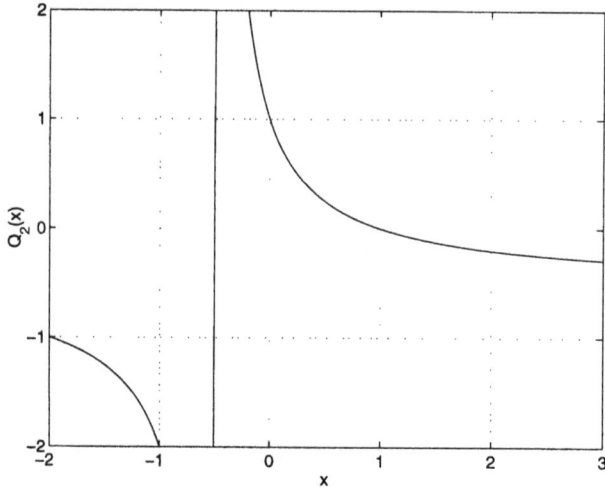

Figure 3.7: *The Q_2-function.*

most common ones of these and relate them to the analysis leading to the Maxwell Garnett mixing rule (3.27).

3.4.1 Macroscopic and microscopic fields

In classical electromagnetism we use Maxwell equations in which the fields that appear are macroscopic field quantities. But, of course, matter is not a continuum. When we look at materials within scales of the molecular and atomic dimensions, the continuous description breaks down, and along with it the macroscopic and smooth picture of the fields. Atom sizes are around 10^{-10} m and the nucleus is a tiny fraction of the atom, 10^{-15} m. If we were able to plot the electric field with the scale of these distances, the behaviour would certainly be wildly varying. The field is surely large, close to the nucleus and much smaller between the atoms. Fortunately, for macroscopic applications the detailed analysis of these variations is not necessary. In our treatment we want to talk mostly about larger-scale electric and magnetic fields which are not bothered by the atomic and molecular details. This requires that the size of the particles (for example the spheres in Figure 3.1) should be large compared to the atomic dimensions. For an illuminating discussion about the spatial averaging process of fields, see [1, Sect. 6.7], [10, Chapters 5–6].

Although the scope of the present book is within macroscopic electromagnetics it is sometimes instructive to make use of field terms from the molecular description of matter. A certain analogy exists between molecules on one hand, and macroscopic particles in dielectric mixtures on the other, as far as the local fields are concerned. The local field that was calculated in the previous section was used as the field that excites the dipole moment, through the known polarisability relation. In the studies

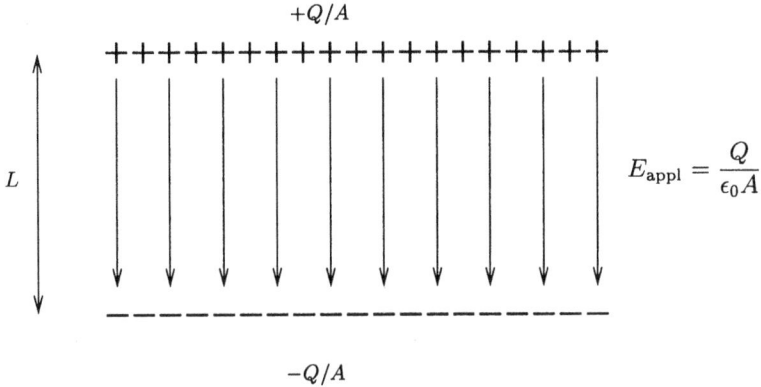

Figure 3.8: *In the simple planar case, the applied field due to external charges is proportional to the charge density on the plates.*

of the molecular response of solid state lattices, the same local field is often assumed for cubic crystals.

Regarding the various fields acting in the surroundings of a particular inclusion in the mixture it is to be emphasised that in the literature one may find also other field concepts than those that appeared in the analysis of the present section. One rather commonly appearing distinction is the one between the *applied* electric field and the *average* electric field. It is perhaps useful at this point to analyse the roles of these fields in more detail, before moving on to homogenise more complicated mixtures.

The applied field upon a given scatterer, inclusion, molecule, or body in general is defined as the field produced by fixed charges external to the body. On the other hand, the average field is the real microscopic field averaged over a representative volume in the mixture, for example over the volume of a crystal cell. These two field amplitudes are different at a given point in matter because of the very dipole moments and polarisation that are induced by the external charges. Why this is so can perhaps be made easier to understand using the following explanation.

Suppose that the external charges are distributed uniformly over two parallel planar surfaces that are separated in free space by a distance L according to Figure 3.8. The plane surfaces have equal areas A, and opposite charges $+Q$ and $-Q$ are distributed on them such that the surface charge density is constant: $\rho_s = \pm Q/A$. If the plates are very wide compared to their separation, the fringing fields due to open sides of the plates can be neglected. Then the electric field strength between the planes is uniform and of magnitude

$$E_{\mathrm{appl}} = \frac{\rho_s}{\epsilon_0} = \frac{Q}{\epsilon_0 A} \tag{3.37}$$

where ϵ_0 is the vacuum permittivity.

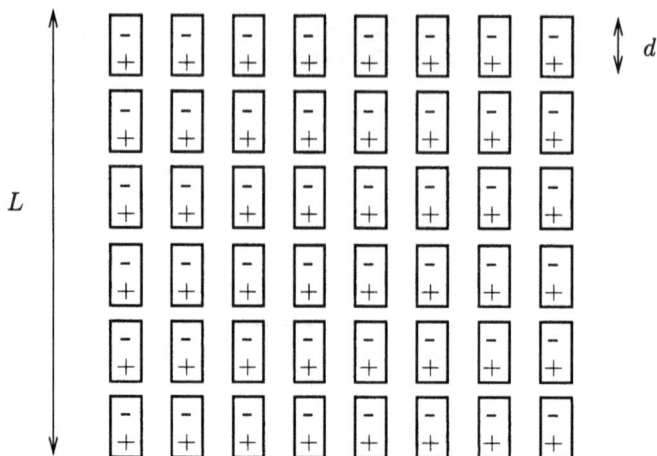

Figure 3.9: *The induced dipoles are assumed to be touching each other in the vertical direction, and hence the adjacent + and − ends cancel each others' effect.*

If now a homogeneous dielectric body is brought between the applied charges, it will be polarised. An ordered array of electric dipoles will be created. Let the number of these dipoles be $N = N_x N_y N_z$ where N_z is the number of dipoles there are along a line between the plates, and N_x, N_y are the corresponding numbers in the two transverse directions. Given also that a single dipole consists of two opposite charges $\pm q$, separated by a distance d in the z-direction, the average field between the plates can be calculated.

A heuristic way to arrive at the relation between the applied and average fields is to assume that the dipoles "almost touch" each other along the direction of the electric field as shown in Figure 3.9. Then the connection between the dipole length d and the distance between the free charges L is $L = N_z d$ (we assume that the dielectric body fills the whole space between the planar charges). Now the opposite charges of the ends of adjacent dipoles in the z-direction cancel each other's effect. What remains is an additional amount of negative charges at the plane of the positive external charge $+Q$, and an equal-magnitude positive charge density at the other end. The charge density is naturally the number of charges divided by the area: $N_x N_y q/A$. This means that the average field within the matter is the field caused by the primary charge density the amplitude of which is diminished by this equivalent charge density:

$$
\begin{aligned}
E_{\text{av}} &= \frac{\rho_s - N_x N_y q/A}{\epsilon_0} \\
&= E_{\text{appl}} - \frac{N_x N_y N_z q d}{\epsilon_0 A d N_z}
\end{aligned}
$$

$$= E_{\text{appl}} - \frac{Np}{\epsilon_0 AL}$$

$$= E_{\text{appl}} - \frac{P}{\epsilon_0} \tag{3.38}$$

where $P = Np/V$ is the dipole moment density in the volume V, which volume is AL in this geometry. This result shows a clear difference between the applied and the average fields which is discussed in many treatments of dielectric mixtures.

However, in the analysis of the present chapter, the distinction between the applied and average fields is unnecessary as the external charges are not of primary importance. One can start by introducing a potential difference between capacitor plates which creates a uniform field over a sufficiently large volume. The electric field is obviously equal to the voltage divided by the distance between the plates. Now if matter—homogeneous or inhomogeneous—is brought between the plates, the average field does not change, given that the voltage remains the same. Therefore we do not need to bother about the applied electric field.

3.4.2 Shape of the cavity in a crystal

The derivation of the Maxwell Garnett rule as it was done in the present chapter regarded the mixture as a fluid. No regular lattice was assumed but the inclusions were randomly positioned in the environment. The shape of a given inclusion affected the analysis in two ways. First, the polarisability was determined by the internal field, which of course depends on the shape of the inclusion. But also the local field acting on an inclusion (3.21) depends on the fictitious cavity in which the dipole moment is created, and the most natural choice for the shape of this cavity is to use the inclusion itself.

When the local electric field is calculated within a solid state crystal, the shape of the cavity in which the field has to be estimated is not necessarily a sphere. A natural building block in a lattice is a parallelepiped, and in special cases a cube. Because of the symmetry of a cube, the local field is often calculated according to a similar formula (3.21) as for a spherical cavity. However, even for different symmetrical cavity forms, the local fields may be different.

If an external electric field $\mathbf{E}$ excites a uniform polarisation $\mathbf{P}$ into matter, and a cavity is carved within the polarised matter, the cavity field, which also can be termed now the local field, can be calculated using the so-called depolarisation dyadic[8] $\overline{\overline{L}}$ [5]:

$$\mathbf{E}_L = \mathbf{E} + \frac{1}{\epsilon_0}\overline{\overline{L}} \cdot \mathbf{P} \tag{3.39}$$

---

[8]In later chapters, the depolarisation dyadic (the dyadic character of $\overline{\overline{L}}$ is distinguished by its double overbar) will be under consideration in more detail. For information about dyadic algebra, see Appendix A.

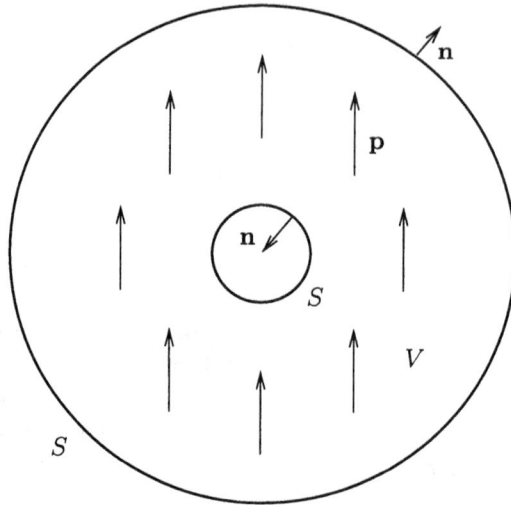

Figure 3.10: *The effect of the dipoles inside the exclusion volume has to be integrated over the volume limited by the larger spherical surface (the exclusion boundary of Figure 3.4) but not containing the point where the local field is calculated.*

The field within a cube is in fact rather difficult to evaluate in general. Numerical solutions are needed. However, if we only treat the field at the centre point, the depolarisation dyadic is the usual one-third of the unit dyadic, and the local field is the same as for a sphere. This gives credibility to the studies where cubic lattices are analysed with a spherical exclusion volume. One must, however, bear in mind that in the cubic case we must fix our local field point in the very centre of the inclusion whereas, for spherical cavities, the field is the same everywhere due to the fact that the depolarisation dyadic is independent of the position of the field point.

3.4.3   The internal dipoles

In the derivation of the local electric field acting on a given molecule, the effect is split into many different components, some of which have been analysed above. One may ask what approximation is made when we carve the cavity into the material. And why, for then we miss the effect of part of the polarisation in the medium.

However, it turns out that in the previous symmetric case the effect of the removed polarisation vanishes. The perturbational field due to dipoles that are located around the point of interest but limited to the exclusion volume can be calculated by integrating the dipole fields. The geometry is shown in Figure 3.10, where the volume V contains the dipoles that we excluded in our earlier analysis.

What were we calculating? We needed the electric field that in some sense "excites" the given inclusion. This field we called the local field $\mathbf{E}_L$, and it is calculated in the centre point of Figure 3.10. The sources are distributed over the volume of the exclusion sphere from which the field point is punctured away (volume V in the figure). Because the field due to dipoles is a negative gradient of the dipole potential ϕ_d, the total field effect due to these "internal dipoles" is

$$\mathbf{E}_{\text{int}} = -\int_V \nabla \phi_d \, dV \qquad (3.40)$$

Note that the result is here calculated by considering the equivalent situation where the source would be at the centre point and the field distributed over the volume. Then, use of Gauss' law brings this integral equal to

$$\mathbf{E}_{\text{int}} = -\oint_S \phi_d \, \mathbf{n} dS \qquad (3.41)$$

where the integration is now over the two surfaces bounding the punctured volume, with the unit normal $\mathbf{n}$ pointing outwards from V. The dipole potential is (cf. Equation (3.17)) proportional to $r^{-2} \cos \theta$, where θ is the angle of the field point from the polarisation direction. And because the surface element is proportional to r^2, the contributions from the inner and outer surface are equal and opposite to each other. Hence $\mathbf{E}_{\text{int}} = 0$.

3.4.4   Alternative routes to Maxwell Garnett formula

The canonical character of the Maxwell Garnett relation and its classical status warrant it a large amount of discussion in the electromagnetics literature. But the argumentation that leads to it is not always very clear. The combination of the microscopic and macroscopic pictures may lead to obscurities, and it is important to define carefully the concepts with which one is operating.[9] Certainly, other ways are available to derive the result (3.27) than the one shown previously. Let us briefly mention other argumentations.

Average field and average flux density

In textbooks very often the Maxwell Garnett formula is derived by basing it on the polarisabilities and the static field solution. But the relation between the average field and the average flux density which started the present chapter is indeed also very appealing. Only two results were needed, of which the first one was the field ratio of the internal field to the constant external field. But the other assumption which was tacitly made in Section 3.1 was that, in the averaging of the fields, the dipole field contribution was neglected. Fortunately, this field component (Equation (3.13) for a single inclusion) vanishes in the calculation of the averages: the volume

---

[9]In this connection it is good to point out the study by Aspnes [11], where he separates, interestingly, the approaches leading to the Clausius–Mossotti formula into "average/solve" and "solve/average" principles.

integral of this field vanishes in the region external to the inclusion. This conclusion follows from the same reasoning as was done in the previous section 3.4.3.

Energy balance

Another way to the Maxwell Garnett result (3.27) has been shown by Hashin and Shtrikman [12]. Consider a homogeneous medium with the effective permittivity ϵ_{eff}. Suppose we wish to replace a spherical volume of this medium by a composite sphere consisting of an inner spherical part with permittivity ϵ_i and the remaining spherical shell with permittivity ϵ_e. Now let us pose the following question: if we wish to perform this replacement in such a way that there is no change in the total electric energy stored in the body, what conditions follow from this for the composition of the replacing sphere?

This requirement of self-consistency leads to the choice of volume fractions of the two permittivities in exactly those proportions that are predicted by the Maxwell Garnett mixing formula.

Problems

3.1 Prove that the internal and external field vectors (3.11)–(3.13) are curl-free as they should be in order to be static electric fields.

3.2 Show that the two types of potentials

(a) $\phi_c(\mathbf{r}) = Cr \cos \theta$

(b) $\phi_d(\mathbf{r}) = Dr^{-2} \cos \theta$

satisfy Laplace equation $\nabla^2 \phi(\mathbf{r}) = 0$, where r and θ are two of the three co-ordinates in the spherical co-ordinate system. The Laplacian reads

$$\nabla^2 \phi(r, \theta, \varphi) = \frac{1}{r^2} \frac{\partial}{\partial r} \left(r^2 \frac{\partial \phi}{\partial r} \right) + \frac{1}{r^2 \sin \theta} \frac{\partial}{\partial \theta} \left(\sin \theta \frac{\partial \phi}{\partial \theta} \right) + \frac{1}{r^2 \sin^2 \theta} \frac{\partial^2 \phi}{\partial \varphi^2}$$

3.3 Consider a dielectric sphere of permittivity ϵ_i and radius a that is brought into a uniform electric field $\mathbf{E}_e = \mathbf{u}_z E_e$. Put the origin of the co-ordinate system at the centre of the sphere. The polarisation within the sphere exerts a perturbation in the field distribution such that the total field remains uniform inside the sphere but outside the dipole field disturbs the uniformity of the external field.

Plot the magnitude of the total electric field (sum of the external field and the perturbational field) along the z-axis. Plot also the scalar electric potential as a function of z, as the permittivity constrast is

(a) $\epsilon_i / \epsilon_e = 2$

(b) $\epsilon_i/\epsilon_e = 10$

3.4 Convince yourself that the "external method" (Subsection 3.2.3) gives the same result for the polarisability as the "internal method." In other words, calculate explicitly the dipole field $\mathbf{E}_d$ of Equation (3.17), and compare it with (3.13).

3.5 For the geometry and excitation of the previous problem, draw the

   (a) total electric field function $\mathbf{E}(z, \theta)$
   (b) total electric flux density $\mathbf{D}(z, \theta)$

   in a plane that halves the sphere. Represent the vectors as arrows. Can you justify the satisfaction of boundary conditions from your figure?

3.6 In Figure 3.5, the effective properties of such a mixture were shown which could be termed "raisin pudding" mixtures. This means that dielectrically hard inclusions were embedded in a soft environment ($\epsilon_i > \epsilon_e$). Draw a similar plot for the complementary mixture, for a "Swiss cheese" mixture, in which the inclusions have smaller permittivity than the environment. Use a broad range of dielectric contrasts, anticipating that the upper-left region of the figure be filled.

   Is the resulting illustration symmetric to Figure 3.5 with respect to the diagonal? If not, why?

3.7 In Figure 3.6, the curves of the complementary mixtures do not predict the same effective permittivity for equal amounts of the two materials. Rather, the same effective permittivity is predicted for around 70% volume of the inclusion. Let us call this crossover volume fraction the "complementary threshold fraction" f_t. Plot f_t as a function of the ratio between permittivities of the two components.

3.8 The Q_2 function was seen to have the property that its inverse is the same function Q_2. Study the more general rational function

$$Q(x) = \frac{a + bx}{c + dx} \tag{3.42}$$

What are the requirements for the parameters $a, b, c,$ and d such that the inverse of this Q function is the very same Q function?

References

[1] JACKSON, J.D.: 'Classical electrodynamics' (Second Edition, John Wiley & Sons, New York, 1975), (Third Edition, 1999)

[2] LINDELL, I.V.: 'Methods for electromagnetic field analysis' (IEEE Press and Oxford University Press, 1995)

[3] KONG, J.A.: 'Electromagnetic wave theory' (John Wiley & Sons, New York, 1986)

[4] VAN BLADEL, J.: 'Electromagnetic fields' (Hemisphere Publishing Co., Washington, 1985)

[5] YAGHJIAN, A.D.: 'Electric dyadic Green's function in the source region', *Proceedings of the IEEE*, 1980, **68**, (2), pp. 248-263

[6] KITTEL, C.: 'Introduction to solid state physics' (Wiley, New York, 1986, 6th Edition)

[7] RAYLEIGH, Lord: 'On the influence of obstacles arranged in rectangular order upon the properties of the medium', *Philosophical Magazine*, 1892, **34**, pp. 481-502

[8] MAXWELL GARNETT, J.C.: 'Colours in metal glasses and metal films', *Trans. of the Royal Society*, (London), Vol. CCIII, 1904, pp. 385-420

[9] LINDELL, I.V., and SIHVOLA, A.: 'The quotient function and its applications', *American Journal of Physics*, 1998, **66**, (3), pp. 197-202

[10] ROBINSON, F.N.H.: 'Macroscopic electromagnetism' (Pergamon Press, Oxford, 1973)

[11] ASPNES, D.E.: 'Local-field effects and effective-medium theory: a microscopic perspective', *American Journal of Physics*, August 1982, **50**, (8), pp. 704-709

[12] HASHIN, Z., and SHTRIKMAN, S.: 'A variational approach to the theory of the effective magnetic permeability of multiphase materials', *Journal of Applied Physics*, 1962, **33**, (10), pp. 3125-3131

Chapter 4

Advanced mixing principles

A whole chapter has now been devoted to the introduction of the basic Maxwell Garnett mixing formula and various analytical exercises to illustrate the theory. The mixing analysis was in many respects idealised. The mixture consisted of two phases and the inclusions were assumed to be spherical in shape. It is perhaps time to relax these assumptions and try to see whether it is possible to say more about mixtures. In the present section, let us generalise the Maxwell Garnett theory to allow variation in the structure and materials of the mixture.

4.1   Multiphase mixtures

Multiphase mixtures contain more than two homogeneous material constituents. If one of the components can be treated as host the generalisation of the effective permittivity calculation of the previous chapter is easy. Such a situation is depicted in Figure 4.1. A natural way to approach the problem of multiphase mixtures is to repeat the polarisability calculations which were done in Chapter 3 for the inclusion phase, now for each of the guest phases. It turns out that the previous analysis for the two-phase mixture can be followed very closely.

The averaging relations interconnect the field terms and dipole moments within inclusions. But from the point of view of a single inclusion, all the neighbouring polarisation density is replaced by plain average polarisation, although it is "coloured" by the variety of surrounding dipoles. Then the local field expression remains the same in form as in (3.21):

$$\mathbf{E}_L = <\mathbf{E}> + \frac{1}{3\epsilon_e} <\mathbf{P}> \qquad (4.1)$$

Here again, only the shape matters through the depolarisation factor. As the spherical shape has still been taken for the inclusion, the coefficient $1/3$ appears. But the

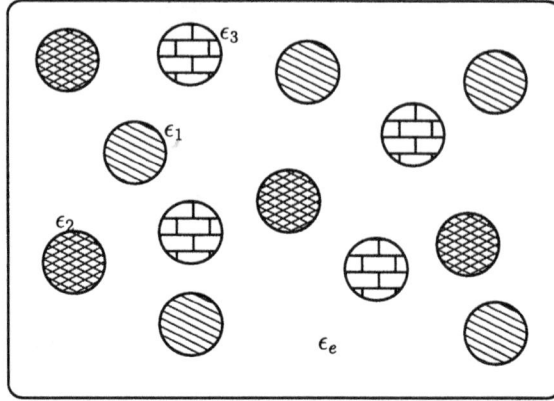

Figure 4.1: *In a simple model for a multiphase mixture, one of the components forms the host matrix (the "environment" ϵ_e), and the other constituents are embedded as inclusions in it.*

average polarisation density $< \mathbf{P} >$ has to account for all the different components of the mixture. Therefore, instead of Equation (3.20), where the total polarisation density came from the one guest phase, now each guest phase contributes one such term to a sum of many:

$$< \mathbf{P} > = \sum_{k=1}^{K} n_k \mathbf{p}_{k,\text{mix}} \tag{4.2}$$

where now n_k is the number density of the kth inclusion phase, and $\mathbf{p}_{k,\text{mix}}$ is the dipole moment of a single inclusion of that phase.[1] Note that the number of different materials forming the mixture is $K+1$ because the environment material is not counted in the sum.

The final result for the effective permittivity is

$$\frac{\epsilon_{\text{eff}} - \epsilon_e}{\epsilon_{\text{eff}} + 2\epsilon_e} = \sum_{k=1}^{K} \frac{n_k \alpha_k}{3\epsilon_e} = \sum_{k=1}^{K} f_k \frac{\epsilon_k - \epsilon_e}{\epsilon_k + 2\epsilon_e} \tag{4.3}$$

where f_k is the volume fraction of the inclusions of the kth phase in the mixture, α_k is the polarisability of such a sphere, and ϵ_k is its permittivity. And of course this relation can be solved for the effective permittivity:

$$\epsilon_{\text{eff}} = \epsilon_e + 3\epsilon_e \frac{\displaystyle\sum_{k=1}^{K} f_k \frac{\epsilon_k - \epsilon_e}{\epsilon_k + 2\epsilon_e}}{1 - \displaystyle\sum_{k=1}^{K} f_k \frac{\epsilon_k - \epsilon_e}{\epsilon_k + 2\epsilon_e}} \tag{4.4}$$

---

[1]The subscript "mix" refers to the fact that the dipole moments are created by the local field in the mixture. Cf. Section 3.3.

Here again, all inclusions of all phases were assumed to be spherical, although not necessarily of the same size.

4.2 Ellipsoidal inclusions

The assumption of spherical shape for the inclusions needs to be relaxed because many natural media possess inclusions of other forms. The polarisability of small particles can of course be calculated for any shape but in general this requires numerical effort. The only shapes for which simple analytical solutions can be found are ellipsoids. Fortunately, ellipsoids allow many practical special cases, like discs and needles, for example.

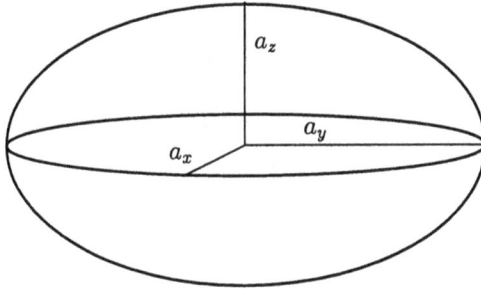

Figure 4.2: *The geometry of an ellipsoid. The semi-axes a_x, a_y, and a_z fix the Cartesian co-ordinate system.*

4.2.1 Depolarisation factors

The important parameters in the geometry of an ellipsoid are its depolarisation factors. If the semi-axes of an ellipsoid in the three orthogonal directions are a_x, a_y, and a_z, according to Figure 4.2, the depolarisation factor N_x (the factor in the a_x-direction) is

$$N_x = \frac{a_x a_y a_z}{2} \int_0^\infty \frac{ds}{(s + a_x^2)\sqrt{(s + a_x^2)(s + a_y^2)(s + a_z^2)}} \qquad (4.5)$$

For the other depolarisation factor N_y (N_z), interchange a_y and a_x (a_z and a_x) in the above integral.

The three depolarisation factors for any ellipsoid satisfy

$$N_x + N_y + N_z = 1 \qquad (4.6)$$

A sphere has three equal depolarisation factors of $1/3$. The other two special cases are a disc (depolarisation factors $1, 0, 0$) and a needle ($0, 1/2, 1/2$). For ellipsoids of revolution, prolate and oblate ellipsoids, various closed-form expressions for the integral (4.5) can be found in [1,2].

Prolate spheroids $(a_x > a_y = a_z)$ have

$$N_x = \frac{1-e^2}{2e^3}\left(\ln\frac{1+e}{1-e} - 2e\right) \tag{4.7}$$

and

$$N_y = N_z = \frac{1}{2}(1 - N_x) \tag{4.8}$$

where the eccentricity is $e = \sqrt{1 - a_y^2/a_x^2}$. For nearly spherical prolate spheroids, which have small eccentricity, the following holds:

$$N_x \simeq \frac{1}{3} - \frac{2}{15}e^2 \tag{4.9}$$

$$N_y = N_z \simeq \frac{1}{3} + \frac{1}{15}e^2 \tag{4.10}$$

For oblate spheroids $(a_x = a_y > a_z)$,

$$N_z = \frac{1+e^2}{e^3}(e - \tan^{-1}e) \tag{4.11}$$

$$N_x = N_y = \frac{1}{2}(1 - N_z) \tag{4.12}$$

where $e = \sqrt{a_x^2/a_z^2 - 1}$. And again, for nearly spherical oblate spheroids,

$$N_z \simeq \frac{1}{3} + \frac{2}{15}e^2 \tag{4.13}$$

$$N_x = N_y \simeq \frac{1}{3} - \frac{1}{15}e^2 \tag{4.14}$$

Figures 4.3 and 4.4 show the behaviour of the depolarisation factors for prolate and oblate spheroids of revolution as functions of the axis ratio.

For a general ellipsoid with three different axes, the depolarisation factors have to be calculated from the integral (4.5). The integral can be written in terms of tabulated elliptic integrals [3, Chapter 17] and can be enumerated if one has the energy to write subroutine codes for a software package, like Matlab or Mathematica, for example. A student of ellipsoids who is able to appreciate numerical calculations must find the work by the authors Osborn and Stoner especially praiseworthy. They have given the depolarisation factors of a general ellipsoid in forms of tables and graphics [4,5].[2]

Choose the co-ordinates in a way that the semiaxes of the ellipsoid fall into the order

$$a_x > a_y > a_z$$

---

[2]See also the early study by Fricke [6] and a recent work by Weiglhofer [7].

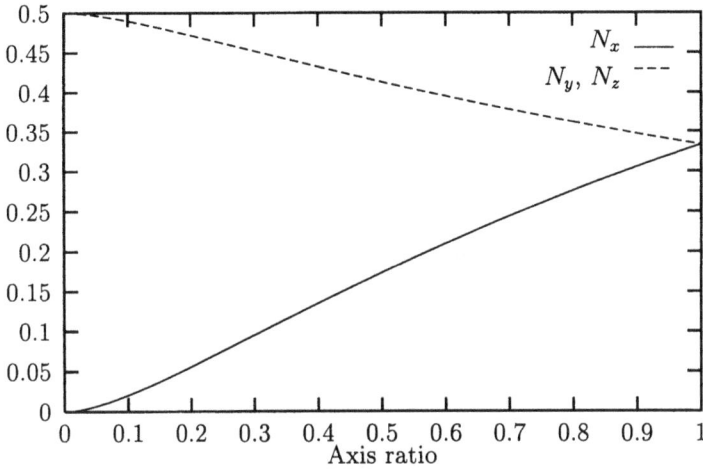

Figure 4.3: *The depolarisation factors of a prolate ellipsoid as functions of the axis ratio a_z/a_x $(= a_y/a_x)$. Note the fact that the three factors for a given ellipsoid sum to unity.*

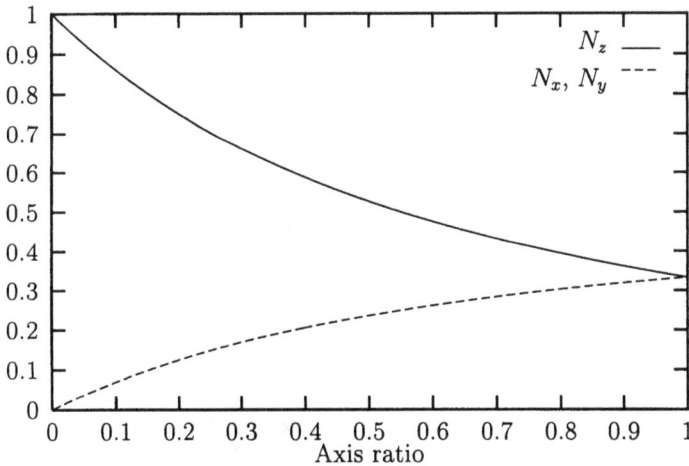

Figure 4.4: *The depolarisation factors of a oblate ellipsoid as functions of the axis ratio a_z/a_x $(= a_z/a_y)$. Note the fact that the three factors for a given ellipsoid sum to unity.*

Then the depolarisation factors are

$$N_x = \frac{a_x a_y a_z}{(a_x^2 - a_y^2)\sqrt{a_x^2 - a_z^2}} [\mathrm{F}(\phi, k) - \mathrm{E}(\phi, k)] \tag{4.15}$$

$$N_y = 1 - N_x - N_z \tag{4.16}$$

$$N_z = \frac{a_y}{a_y^2 - a_z^2} \left[a_y - \frac{a_x a_z}{\sqrt{a_x^2 - a_z^2}} \mathrm{E}(\phi, k) \right] \tag{4.17}$$

where the incomplete elliptic integrals are defined as

$$\mathrm{F}(\phi, k) = \int_0^\phi \frac{d\theta}{\sqrt{1 - k^2 \sin^2 \theta}} \tag{4.18}$$

$$\mathrm{E}(\phi, k) = \int_0^\phi \sqrt{1 - k^2 \sin^2 \theta} \, d\theta \tag{4.19}$$

4.2.2   Polarisability components of an ellipsoid

Because of the broken geometrical symmetry of an ellipsoid it is to be expected that the dipole moment induced in an ellipsoid is dependent on the direction of the electric field that excites it. In general, the dipole moment vector has a different direction than the field. It is only in the three principal axis directions that the field creates a dipole moment in alignment with the field.

Consider the ellipsoid in Figure 4.2, and assume that it is exposed to an external uniform x-directed electric field $\mathbf{E}_e$. The internal field $\mathbf{E}_i$ is also uniform and x-directed, and the field ratio is

$$\mathbf{E}_i = \frac{\epsilon_e}{\epsilon_e + N_x(\epsilon_i - \epsilon_e)} \mathbf{E}_e \tag{4.20}$$

from which the natural special case of a sphere (Equation (3.8)) emerges by the choice $N_x = 1/3$.

The polarisability component of the ellipsoid for x-directed field is then

$$\alpha_x = \frac{4\pi a_x a_y a_z}{3} (\epsilon_i - \epsilon_e) \frac{\epsilon_e}{\epsilon_e + N_x(\epsilon_i - \epsilon_e)} \tag{4.21}$$

Likewise the y- and z-directed components of the total polarisability can be written, by replacing N_x with N_y and N_z. The polarisability can be presented in matrix form, which in the present case is diagonal in the xyz co-ordinate system:

$$\alpha = \begin{pmatrix} \alpha_x & 0 & 0 \\ 0 & \alpha_y & 0 \\ 0 & 0 & \alpha_z \end{pmatrix} \tag{4.22}$$

The components of this polarisability matrix α depend on the choice of the co-ordinate axes. In this book, another notation, the so-called *dyadic* notation is preferred [8]. The dyadic notation is explained in detail in Section 5.2 (see also Appendix A).

Using this co-ordinate independent dyadic notation, the polarisability dyadic reads

$$\overline{\overline{\alpha}} = \frac{4\pi a_x a_y a_z}{3}(\epsilon_i - \epsilon_e) \sum_{j=x,y,z} \frac{\epsilon_e}{\epsilon_e + N_j(\epsilon_i - \epsilon_e)} \mathbf{u}_j \mathbf{u}_j \tag{4.23}$$

where the double overbar denotes that the quantity is a dyadic.[3]

The polarisability dyadic can be thought of as an operator acting on the external field vector to produce another vector, the dipole moment:

$$\mathbf{p} = \overline{\overline{\alpha}} \cdot \mathbf{E}_e \tag{4.24}$$

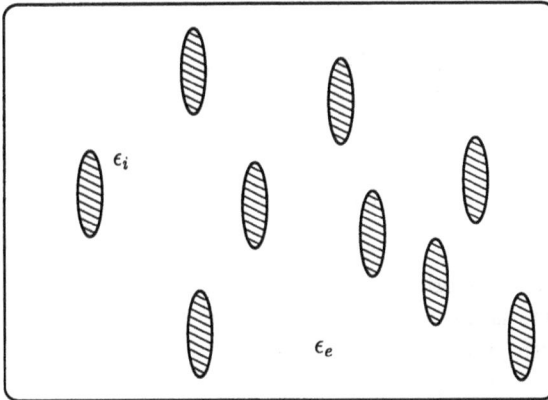

Figure 4.5: *A mixture with aligned ellipsoidal inclusions.*

4.2.3 Aligned orientation

Now consider a mixture where ellipsoids of permittivity ϵ_i are embedded in the environment with permittivity again being ϵ_e. Let all the ellipsoids be aligned according to Figure 4.5. Then the effective permittivity of the mixture is anisotropic, in other words it has different permittivity components in the different principal directions. The Maxwell Garnett formula for the x-component of the effective permittivity of this mixture is

$$\epsilon_{\text{eff},x} = \epsilon_e + f\epsilon_e \frac{\epsilon_i - \epsilon_e}{\epsilon_e + (1-f)N_x(\epsilon_i - \epsilon_e)} \tag{4.25}$$

---

[3]Note that the dyadic consists of a sum of vector pairs.

and for $\epsilon_{\text{eff},y}$ and $\epsilon_{\text{eff},z}$, replace N_x by N_y and N_z, respectively.

Again, using dyadic notation, the effective permittivity can be written

$$\bar{\bar{\epsilon}}_{\text{eff}} = \sum_{j=x,y,z} \mathbf{u}_j \mathbf{u}_j \, \epsilon_{\text{eff},j} \tag{4.26}$$

Many natural and man-made materials possess this type of anisotropic structure where the materials that compose the mixture are isotropic to begin with but the geometry of the composition is not symmetric, leading to anisotropy in the large scale. For example, many organic substances contain directed sheaths and lamellae and therefore it is to be expected that the macroscopic characteristics of the medium are highly dependent on the field direction. The case where the phases themselves are anisotropic is discussed in Chapter 5.

The relation (4.25) can be given in the form

$$\frac{\epsilon_{\text{eff},x}/\epsilon_e - 1}{\epsilon_{\text{eff},x}/\epsilon_e + u} = f \frac{\epsilon_i/\epsilon_e - 1}{\epsilon_i/\epsilon_e + u} \tag{4.27}$$

with $u = (1 - N_x)/N_x$ being a coefficient that depends on the shape of the ellipsoid. This *Formzahl* [9] is $u = 2$ for spheres and can range from zero to positive infinity for ellipsoids in the various directions.

4.2.4   Random orientation

Another common case for a mixture with ellipsoidal inclusions is that the ellipsoids in the mixture are randomly oriented. This happens for example in a fluid background material when there is no external force orienting the inclusions. Then obviously the previous anisotropy of the ordered mixture will be washed away. Macroscopically, there is no longer any preferred direction. The mixture is isotropic and the effective permittivity ϵ_{eff} is a scalar. One-third of each polarisability component gives equal shares to the macroscopic polarisation density, and the expression for the permittivity reads

$$\epsilon_{\text{eff}} = \epsilon_e + \epsilon_e \frac{\dfrac{f}{3} \displaystyle\sum_{j=x,y,z} \dfrac{\epsilon_i - \epsilon_e}{\epsilon_e + N_j(\epsilon_i - \epsilon_e)}}{1 - \dfrac{f}{3} \displaystyle\sum_{j=x,y,z} \dfrac{N_j(\epsilon_i - \epsilon_e)}{\epsilon_e + N_j(\epsilon_i - \epsilon_e)}} \tag{4.28}$$

For example, the case of randomly oriented needles gives

$$\epsilon_{\text{eff}} = \epsilon_e + f(\epsilon_i - \epsilon_e) \frac{\epsilon_i + 5\epsilon_e}{(3 - 2f)\epsilon_i + (3 + 2f)\epsilon_e} \tag{4.29}$$

and for randomly oriented discs,

$$\epsilon_{\text{eff}} = \epsilon_e + f(\epsilon_i - \epsilon_e) \frac{2\epsilon_i + \epsilon_e}{(3 - f)\epsilon_i + f\epsilon_e} \tag{4.30}$$

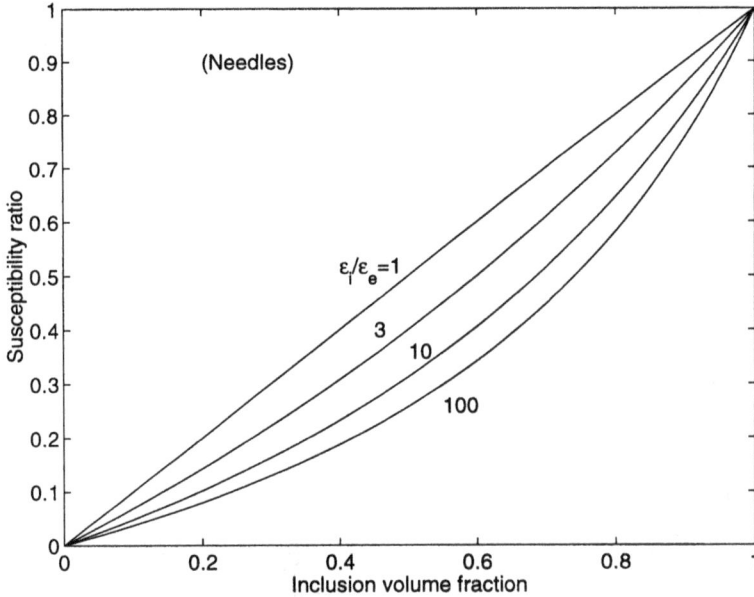

Figure 4.6: *The susceptibility ratio $(\epsilon_{\text{eff}} - \epsilon_e)/(\epsilon_i - \epsilon_e)$ for the Maxwell Garnett prediction of the effective permittivity of a mixture with randomly oriented needle-shaped inclusions of permittivity ϵ_i in a background medium of permittivity ϵ_e.*

To illustrate the effect of the shape of the inclusions upon the effective permittivity, the following Figures 4.6 and 4.7 display ϵ_{eff} for mixtures where the inclusions are randomly oriented needles and discs. The normalised susceptibility

$$(\epsilon_{\text{eff}} - \epsilon_e)/(\epsilon_i - \epsilon_e)$$

is shown. The nonlinearity of the $\epsilon_{\text{eff}}(f)$-curve is again seen to be dependent on the contrast between the inclusion and the environment.

A comparison with Figure 3.5 testifies that the effective permittivity of mixtures is weakly dependent on the shape of inclusions if the dielectric contrast ϵ_i/ϵ_e is small, close to one. But when the contrast increases, the macroscopic permittivity is greatly affected by the inclusion form: spheres give the lowest permittivity, needles a larger permittivity, and discs will provide the largest effect.

If the inclusions have another shape than needles or discs the effective permittivity falls between the extreme limits of discs and spheres [10].

Another interesting comparison of the effective permittivity of different types of mixtures can be seen in Figure 4.8. There the dielectric contrast of the phases is kept constant $(\epsilon_i/\epsilon_e = 50)$. The effect of the inclusion shape can be seen, and again is such that for a given volume fraction, inclusions in the form of needles create a larger effective permittivity than if all inclusions would be in the form of spheres,

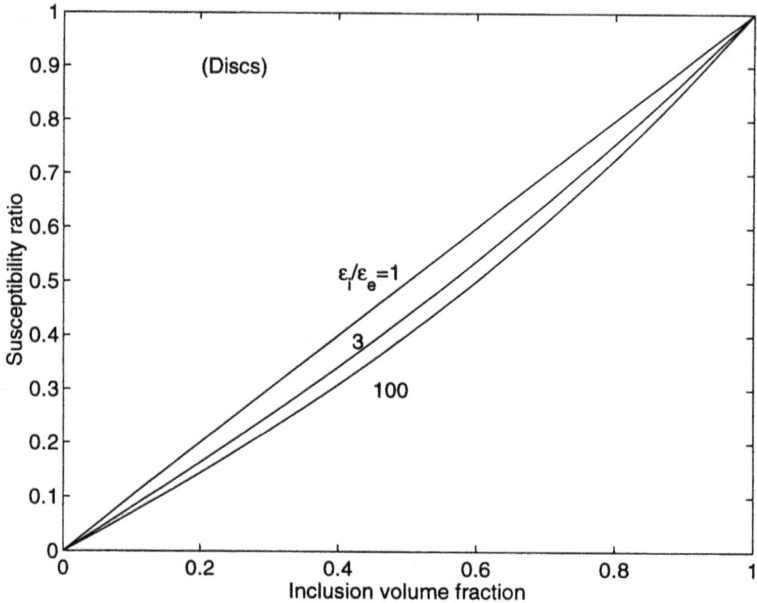

Figure 4.7: *The same as in Figure 4.6, for a mixture where the inclusions are randomly oriented discs.*

and for disc-shaped inclusions this effect is still stronger. The extremum property of the spherical shape is a consequence of the variational principle: the dielectric energy is a stationary functional of the electric field and every deviation from the spherical shape tends to increase its average polarisability [11].

4.2.5   Orientation distribution

If the inclusions are neither aligned nor randomly oriented but rather follow an orientation distribution, the averaging does not necessarily produce isotropy for the final effective permittivity. Then the average polarisation density has to be integrated by weighing the dipole moments with the orientation distribution function. Let us assume for simplicity that all ellipsoids are equal. Let the direction dependences be contained in the function $n(\Omega)$ which generally depends on three angle parameters.[4]

$$< \mathbf{P} > = \int_{4\pi} d\Omega\, n(\Omega) \mathbf{p}_{\mathrm{mix}} \qquad (4.31)$$

Here, the integration has to be performed over the three angular parameters.

---

[4]The number of independent angles on which the orientation distribution function depends is indeed three: two angles fix the correct direction of one of the ellipsoid axes, and around this direction the ellipsoid is rotated by the third angle in order to align the remaining two orthogonal ellipsoid axes.

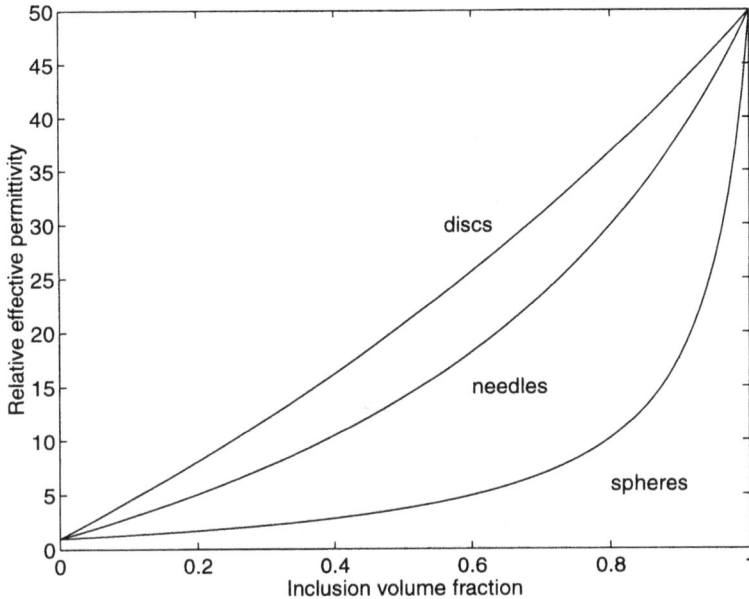

Figure 4.8: *The effective permittivity of a mixture where the contrast between the inclusions and the environment is $\epsilon_i/\epsilon_e = 50$. The three curves (from lowest to highest) correspond to three shapes of inclusions: spheres, needles, and discs. Inclusions are randomly oriented in all cases.*

The final result for the effective permittivity dyadic contains these integrals in the numerator and denominator [12], corresponding to the two sums in Equation (4.28).

4.3   Inhomogeneous inclusions

Until now, all inclusions in the mixture have been assumed themselves to be homogeneous dielectrics. This is a restriction for the applicability range of the resulting mixing models, which, fortunately, can be relaxed to a certain extent.

The calculation of the local field that excites an inclusion was performed in Section 3.3. It was seen to depend only on the shape of the external boundary of the scatterer because this local field was taken to be a cavity-type field; the matter within the inclusion was removed. Therefore the internal structure of the inclusion has no effect on the local field. Subsequently, the problem with nonhomogeneous scatterers is solely dependent on the success of calculating the polarisability of such inclusions.

There are, indeed, certain inhomogeneous structures for which analytical solution can be found in the electrostatic problem.

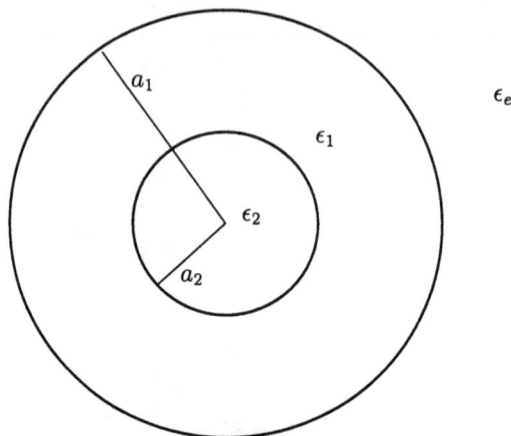

Figure 4.9: *A sphere with a spherical core and a spherical layer.*

4.3.1   Polarisability of a layered sphere

A layered sphere is one example of a partially nonhomogeneous structure for which the exact polarisability can be solved. Let us focus in this section on the characteristics of layered and other radially inhomogeneous spherical structures.

For a homogeneous sphere of permittivity ϵ_i the polarisability was found in Equation (3.10) to be

$$\alpha = 3\epsilon_e V \frac{\epsilon_i - \epsilon_e}{\epsilon_i + 2\epsilon_e} \tag{4.32}$$

in a homogeneous environment of permittivity ϵ_e, and $V = 4\pi a^3/3$ is the volume of this sphere.

Now, let us ask for the polarisability of a partially homogeneous sphere that consists of a spherical core on which there lies a spherical shell of another material, as is shown in Figure 4.9.

The solution for the polarisability of this kind of a sphere is also known in the electromagnetics literature (see, for example, [13]). Of course, more parameters appear in the formula:

$$\alpha = 3\epsilon_e V \frac{(\epsilon_1 - \epsilon_e)(\epsilon_2 + 2\epsilon_1) + \dfrac{a_2^3}{a_1^3}(2\epsilon_1 + \epsilon_e)(\epsilon_2 - \epsilon_1)}{(\epsilon_1 + 2\epsilon_e)(\epsilon_2 + 2\epsilon_1) + 2\dfrac{a_2^3}{a_1^3}(\epsilon_1 - \epsilon_e)(\epsilon_2 - \epsilon_1)} \tag{4.33}$$

where the subscripts 1 and 2 refer to the layer and core, respectively. Obviously (4.33) takes the form of (4.32) for the various possible special cases:

1. $a_2/a_1 \to 0$ leads to a homogeneous sphere with permittivity ϵ_1

2. $a_2/a_1 \to 1$ means a homogeneous sphere with permittivity ϵ_2

3. $\epsilon_2 \to \epsilon_1$ gives a homogeneous sphere with permittivity $\epsilon_1 = \epsilon_2$

4. $\epsilon_1 \to \epsilon_e$ is equivalent to a homogeneous sphere of permittivity ϵ_2, the volume of which is a fraction a_2^3/a_1^3 of the original volume V.

In fact, regardless of the number of layers of a partially homogeneous sphere, the polarisability can be solved analytically [14,15]. For the sphere shown in Figure 4.10, the polarisability is dependent on all the permittivities and radius ratios, and can be written as

$$
\alpha = 3\epsilon_e V \frac{(\epsilon_1 - \epsilon_e) + (2\epsilon_1 + \epsilon_e)\dfrac{(\epsilon_2 - \epsilon_1)a_2^3/a_1^3 + (2\epsilon_2 + \epsilon_1)\dfrac{(\epsilon_3 - \epsilon_2)a_3^3/a_1^3 + \cdots}{(\epsilon_3 + 2\epsilon_2) + \cdots}}{(\epsilon_2 + 2\epsilon_1) + 2(\epsilon_2 - \epsilon_1)\dfrac{(\epsilon_3 - \epsilon_2)a_3^3/a_2^3 + \cdots}{(\epsilon_3 + 2\epsilon_2) + \cdots}}}{(\epsilon_1 + 2\epsilon_e) + 2(\epsilon_1 - \epsilon_e)\dfrac{(\epsilon_2 - \epsilon_1)a_2^3/a_1^3 + (2\epsilon_2 + \epsilon_1)\dfrac{(\epsilon_3 - \epsilon_2)a_3^3/a_1^3 + \cdots}{(\epsilon_3 + 2\epsilon_2) + \cdots}}{(\epsilon_2 + 2\epsilon_1) + 2(\epsilon_2 - \epsilon_1)\dfrac{(\epsilon_3 - \epsilon_2)a_3^3/a_2^3 + \cdots}{(\epsilon_3 + 2\epsilon_2) + \cdots}}}
$$

(4.34)

The numerators and denominators end when finally the core with permittivity ϵ_N is reached, the last term carrying a_N^3/a_1^3 in the numerator and a_N^3/a_{N-1}^3 in the denominator.

Equipped with the dielectric polarisability expression for layered spheres, the mixture problem can be attacked. Suppose that inclusions of the type in Figure 4.10 occupy a volume fraction f in the environment with permittivity ϵ_e. Then, using Equation (3.24), the effective permittivity for our present $N+1$-component mixture obeys the generalised Maxwell Garnett formula

$$
\frac{\epsilon_{\text{eff}} - \epsilon_e}{\epsilon_{\text{eff}} + 2\epsilon_e} = f \frac{(\epsilon_1 - \epsilon_e) + (2\epsilon_1 + \epsilon_e)\dfrac{(\epsilon_2 - \epsilon_1)a_2^3/a_1^3 + (2\epsilon_2 + \epsilon_1)\dfrac{(\epsilon_3 - \epsilon_2)a_3^3/a_1^3 + \cdots}{(\epsilon_3 + 2\epsilon_2) + \cdots}}{(\epsilon_2 + 2\epsilon_1) + 2(\epsilon_2 - \epsilon_1)\dfrac{(\epsilon_3 - \epsilon_2)a_3^3/a_2^3 + \cdots}{(\epsilon_3 + 2\epsilon_2) + \cdots}}}{(\epsilon_1 + 2\epsilon_e) + 2(\epsilon_1 - \epsilon_e)\dfrac{(\epsilon_2 - \epsilon_1)a_2^3/a_1^3 + (2\epsilon_2 + \epsilon_1)\dfrac{(\epsilon_3 - \epsilon_2)a_3^3/a_1^3 + \cdots}{(\epsilon_3 + 2\epsilon_2) + \cdots}}{(\epsilon_2 + 2\epsilon_1) + 2(\epsilon_2 - \epsilon_1)\dfrac{(\epsilon_3 - \epsilon_2)a_3^3/a_2^3 + \cdots}{(\epsilon_3 + 2\epsilon_2) + \cdots}}}
$$

(4.35)

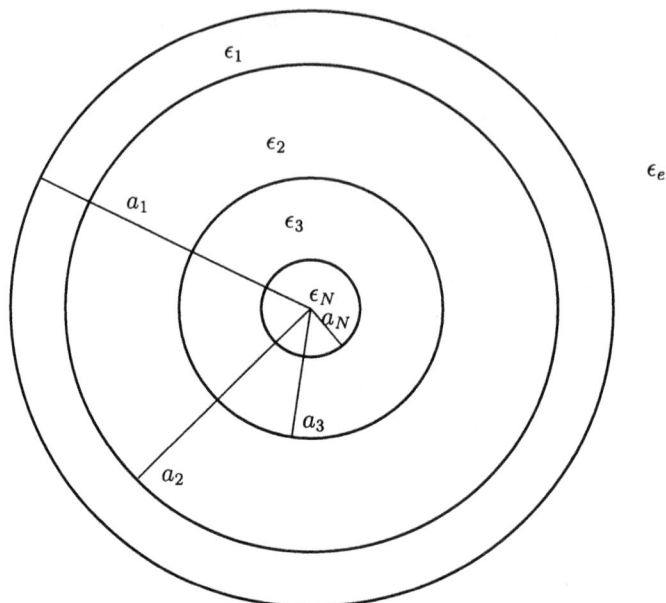

Figure 4.10: *A multilayer sphere with N different component materials.*

The appearance of this expression expands strongly with the number of layers but has, after all, a finite length.

Illustration of the structure effect

How does the substructure of the inclusions affect the effective permittivity for a mixture? Let us study that by comparing two three-phase mixtures, both having the same volume concentration of the component materials but different geometries.

The first type of mixture is one where core-plus-single-layer inclusions are embedded in the background of ϵ_e, which means that the effective permittivity is, using Equation (4.35) for $N = 2$,

$$\frac{\epsilon_{\text{eff}} - \epsilon_e}{\epsilon_{\text{eff}} + 2\epsilon_e} = f \frac{(\epsilon_1 - \epsilon_e)(\epsilon_2 + 2\epsilon_1) + \dfrac{a_2^3}{a_1^3}(\epsilon_2 - \epsilon_1)(\epsilon_e + 2\epsilon_1)}{(\epsilon_1 + 2\epsilon_e)(\epsilon_2 + 2\epsilon_1) + 2\dfrac{a_2^3}{a_1^3}(\epsilon_2 - \epsilon_1)(\epsilon_1 - \epsilon_e)} \tag{4.36}$$

and in fact, this can be rewritten using only the volume fractions f_1, f_2 of the inclusion components of the mixture ϵ_1, ϵ_2, by writing $f = f_1 + f_2$ and $a_2^3/a_1^3 = f_2/(f_1 + f_2)$.

The second mixture is one where the two inclusion components are independent of each other. They appear as separate droplets in the environment. Then we can apply Equation (4.3) for $N = 2$ to get

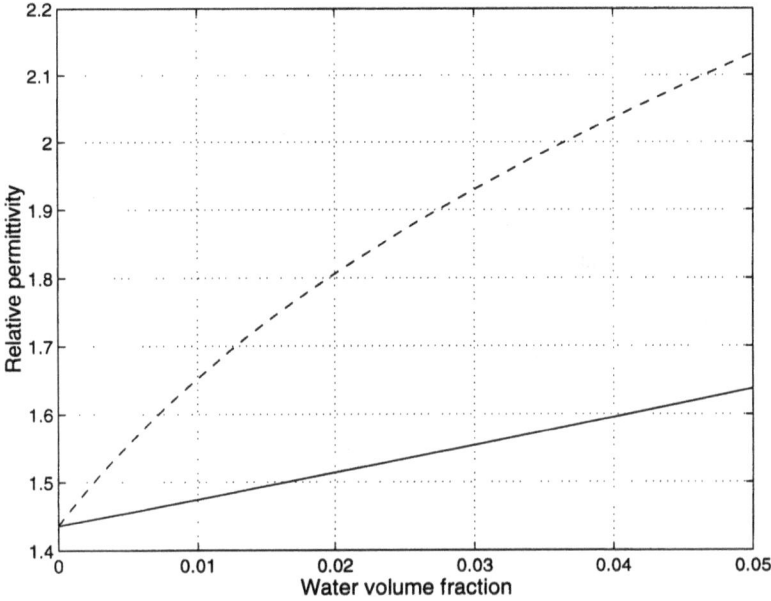

Figure 4.11: *The relative effective permittivity $\epsilon_{\text{eff}}/\epsilon_e$ of two types of three-component mixtures. The volume fractions of all components are the same for both mixtures. Solid line: inclusions form separate spherical ice and water droplets. Dashed line: inclusions are ice spheres covered by a water layer. The ice volume fraction is always 0.3; the liquid water volume fraction changes.*

$$\frac{\epsilon_{\text{eff}} - \epsilon_e}{\epsilon_{\text{eff}} + 2\epsilon_e} = f_1 \frac{\epsilon_1 - \epsilon_e}{\epsilon_1 + 2\epsilon_e} + f_2 \frac{\epsilon_2 - \epsilon_e}{\epsilon_2 + 2\epsilon_e} \qquad (4.37)$$

Figure 4.11 illustrates the difference between the effective properties of these two mixtures. The mixtures would perhaps serve as crude models for wet snow as a three-component composite of air ($\epsilon_e = \epsilon_0$) as background, and the inclusions being liquid water ($\epsilon_1 = 88\epsilon_e$) and ice $\epsilon_2 = 3.2\epsilon_e$. The curves are given as a function of the volume fraction of liquid water in the mixture. The volume fraction of ice remains a constant 0.3. The clear message of the figure is that the mixture where water covers ice particles has higher permittivity than the mixture where the same amounts of water and ice form separate uncorrelated inclusions.

4.3.2 Continuously inhomogeneous inclusions

A limit case of a multilayer sphere is continuously inhomogeneous inclusion. Such an inclusion can also be treated analytically as far as its polarisability is concerned. Then the inclusion structure is described by the permittivity profile, in other words the permittivity as a function of the distance from the core:

$$\epsilon(\mathbf{r}) = \epsilon(r) \tag{4.38}$$

and for $r > a$, the permittivity is that of the environment, $\epsilon(r) = \epsilon_e$. Here a is the radius of the sphere.

If this type of sphere is centred at the origin and exposed to an external uniform electric field $\mathbf{E}_e$ the perturbational field will be that of a dipole $\mathbf{E}_d(\mathbf{r})$, as was before in the case of homogeneous sphere. The scattered field is connected with the scattered potential $\phi_d(\mathbf{r})$ as its negative gradient: $\mathbf{E}_d(\mathbf{r}) = -\nabla\phi_d(\mathbf{r})$.

The total field $\mathbf{E}$ is the sum of these two fields. The equation for the unknown scattered potential can be written from the divergencelessness property of the flux density:

$$\nabla \cdot [\epsilon(r)\mathbf{E}] = \nabla \cdot [\epsilon(r)(\mathbf{E}_e - \nabla\phi_d)] = 0 \tag{4.39}$$

Because $\mathbf{E}_e$ is a constant vector, $\nabla \cdot \mathbf{E}_e = 0$, we can write

$$\nabla \cdot [\epsilon(r)\mathbf{E}_e] = (\nabla\epsilon) \cdot \mathbf{E}_e = \frac{d\epsilon(r)}{dr}\mathbf{u}_r \cdot \mathbf{E}_e \tag{4.40}$$

Now, obviously, the scattered, dipole potential has cosine dependence on θ, and only its radial dependence $f_d(r)$ remains unknown: $\phi_d(\mathbf{r}) = f_d(r)\cos\theta$. Peeling off the θ dependence of all terms in Equation (4.39), we can write

$$\frac{d}{dr}\left[\epsilon(r)\, r^2 \frac{df_d(r)}{dr}\right] - 2\epsilon(r)f_d(r) = r^2 E_e \frac{d\epsilon(r)}{dr} \tag{4.41}$$

where E_e is the amplitude of the external field $\mathbf{E}_e = E_e\mathbf{u}_z$.

This ordinary differential equation has to be solved in order to find the polarisability of the sphere, which is proportional to the amplitude of the function $f_d(r)$. Equation (4.41) simplifies in the external region ($r > a$) into the form

$$\frac{d}{dr}\left[r^2 \frac{df_d(r)}{dr}\right] = 2f_d(r) \tag{4.42}$$

the solution of which, that is localised around the inclusion, is

$$f_d(r) = Dr^{-2} \tag{4.43}$$

where D is connected with the polarisability of the sphere:

$$\alpha = \frac{4\pi\epsilon_e}{E_e}D \tag{4.44}$$

The polarisability comes as the solution of the field external to the scatterer.

Inside the sphere ($r < a$), the function $f_d(r)$ depends on the permittivity profile. Equation (4.41) shows that $f_d(0) = 0$, and the two boundary conditions for this second-order equation are that f_d should be continuous at $r = a$, and also its derivative should be continuous for the case of continuous permittivity across the sphere boundary.[5]

Dielectric sphere with linear permittivity profile

As an example, let us consider a dielectric sphere with linear permittivity profile

$$\epsilon(r) = \epsilon_e \left(2 - \frac{r}{a} \right) \qquad \text{for} \qquad r \leq a \qquad (4.45)$$

The permittivity reaches $2\epsilon_e$ at the core of the sphere, and at the surface, the permittivity has reduced to the environment value and there is no discontinuity at the boundary.

When the differential equation (4.41) is solved for this function, the amplitude of the solution is seen to be $f_d(a) \approx 0.0732\, aE_e$ [15]. This means that the normalised polarisability of the inhomogeneous sphere is

$$\frac{\alpha}{4\pi a^3 \epsilon_e} \approx 0.0732 \qquad (4.46)$$

An interesting observation can be made by comparing the result (4.46) with the polarisability of a homogeneous sphere with the permittivity $2\epsilon_e$. According to (3.10), the polarisability is $\alpha/(4\pi a^3 \epsilon_e) = 0.25$, which is much higher. Very natural is this difference because the homogeneous sphere has a larger "dielectric mass," too.

Let us define the dielectric mass m_ϵ as the volume integral of the susceptibility:

$$m_\epsilon = \int_V [\epsilon(\mathbf{r}) - \epsilon_e]\, dV \qquad (4.47)$$

and we are able to compare in a fair manner the two polarisabilities. If the two spheres (the one with the linear permittivity profile and the homogeneous one) were to have the same volume-averaged susceptibility, the susceptibility $\epsilon - \epsilon_e$ of the homogeneous sphere should be one-fourth of the susceptibility of the core of the inclusion with the linear profile. This means that the homogeneous sphere has a constant permittivity of $\epsilon = 1.25\epsilon_e$. With this permittivity the polarisability is $\alpha/(4\pi a^3 \epsilon_e) = 1/13 \approx 0.0769$.

The polarisability of the homogeneous sphere is higher even after the renormalisation of the relative permittivity from 2 to 1.25. This means that it is possible to decrease the polarisability of a homogeneous inclusion by redistributing the "polarisable mass" more into the core and less into the surface. Consequently, the effective permittivity of a mixture with such inclusions is higher than that of a mixture with homogeneous inclusions.

---

[5]If a discontinuity exists in $\epsilon(r)$ across $r = a$, the remaining condition comes from the continuity of the flux density $\mathbf{D}$.

In [15], mixtures with inclusions having various permittivity profiles are analysed, including parabolic and Gaussian permittivity functions.[6]

4.3.3   Nonhomogeneous ellipsoids

The foregoing analysis with which the polarisabilities of inhomogeneous spheres could be determined can be extended to ellipsoidal geometries. This means that closed-form solutions for layered and even continuously inhomogeneous ellipsoids can be written down, and so also the effective permittivity of mixtures that contain such ellipsoids as inclusions.

However, the inhomogeneity of the ellipsoids cannot be arbitrary. The requirement to manage with the problem is that the Laplace equation $\nabla^2 \phi(\mathbf{r}) = 0$ be separable within the various regions of the problem. This means that the geometry of the multilayer ellipsoid has to be such that all the ellipsoidal boundaries between the layers have to be confocal. Confocal surfaces are constant-co-ordinate surfaces within the ellipsoidal co-ordinate system [1, 16, 17].

The three co-ordinates of the ellipsoidal co-ordinate system (ξ, η, ζ) are connected to the Cartesian co-ordinate system as follows:

$$\frac{x^2}{a_x^2 + u} + \frac{y^2}{a_y^2 + u} + \frac{z^2}{a_z^2 + u} = 1 \tag{4.48}$$

where a_x, a_y, a_z are again the semiaxes of the inclusion ellipsoid. For a given point in space, the three real roots for u of this equation give the three ellipsoidal co-ordinates. The co-ordinate ξ is that root which lies in the range $\xi \geq -d^2$ where d is the smallest of the ellipsoid semiaxes a_x, a_y, a_z. Hence constant-ξ surfaces are ellipsoids, all confocal to the ellipsoid

$$\frac{x^2}{a_x^2} + \frac{y^2}{a_y^2} + \frac{z^2}{a_z^2} = 1 \tag{4.49}$$

which corresponds to $\xi = 0$. The confocality requirement for a multilayer ellipsoid with discontinuity boundaries defined by ellipsoids with semiaxes $a_{x,i}, a_{y,i}, a_{z,i}$ means that

$$a_{x,i}^2 - a_{x,j}^2 = a_{y,i}^2 - a_{y,j}^2 = a_{z,i}^2 - a_{z,j}^2 \tag{4.50}$$

for all pairs i, j. Note that confocal ellipsoids do *not* have the same axis ratios. Instead, the larger ellipsoids are "fatter" than the inner ones.

An extension of the analysis of the continuously inhomogeneous sphere which was sketched above can be generalised to continuously inhomogeneous ellipsoids [18], provided that the permittivity function of the ellipsoid is only dependent on this ξ co-ordinate: $\epsilon(\mathbf{r}) = \epsilon(\xi)$.

---

[6]In [15], also a variational principle is given for the polarisability with which treatment of the differential equation (4.41) can be avioded.

4.4 Lossy materials

The analysis in the previous sections treated the media as plain and pure dielectrics. No charge flow took place when fields were incident on the materials. This is of course an idealised case which has to be opened for practical materials that always display some transport effects.

Conduction of charge is tantamount to electric current. And because for alternating fields, the conduction current and displacement current are out-of-phase from one another, the conductivity term can be identified as the imaginary part of the permittivity in the way as it was shown in Chapter 2:

$$\epsilon_{compl} = \epsilon - \frac{j\sigma}{\omega} \tag{4.51}$$

where ϵ_{compl}, the complex permittivity, contains ϵ, which is the ordinary permittivity of the material, and σ, its conductivity. The angular frequency is denoted by ω. The imaginary part of the complex permittivity now reflects the imperfect character of materials with nonzero conductivity. A complex permittivity leads to absorption of the energy of propagating wave and is a measure for dielectric losses in the material. As a reminder from Section 2.2, we follow here the tradition in electrical engineering [19, 20] of separating the real, imaginary, and complex permittivities on the notational level in the following manner:

$$\epsilon_{compl} = \epsilon' - j\epsilon'' \tag{4.52}$$

where both ϵ' and ϵ'' are real.

Now if a mixture consists of components that are described by a complex permittivity, the mixing rule renders the effective permittivity complex, too. A bold and straightforward application of the Maxwell Garnett mixing rule gives for the complex effective permittivity

$$
\begin{aligned}
\epsilon_{eff} &= \epsilon'_{eff} - j\epsilon''_{eff} \\
&= \epsilon'_e - j\epsilon''_e + 3f(\epsilon'_e - j\epsilon''_e)\frac{\epsilon'_i - \epsilon'_e - j(\epsilon''_i - \epsilon''_e)}{(1-f)\epsilon'_i + (2+f)\epsilon'_e - j[(1-f)\epsilon''_i + (2+f)\epsilon''_e]}
\end{aligned}
\tag{4.53}
$$

where the inclusions are assumed to be spherical.

As numerical examples, Figures 4.12–4.13 show the effective permittivities for a two-phase mixture where the background is lossless (ϵ_e is real) but the inclusion phase has an imaginary part. The result is compared with the corresponding lossless case. The comparison shows (Figure 4.12) clearly how a small imaginary part does not change the behaviour of the real part of the effective permittivity appreciably. A monotonic increase can be observed for both the real and imaginary parts. However, for a very lossy inclusion material (Figure 4.13), the dependence of the effective

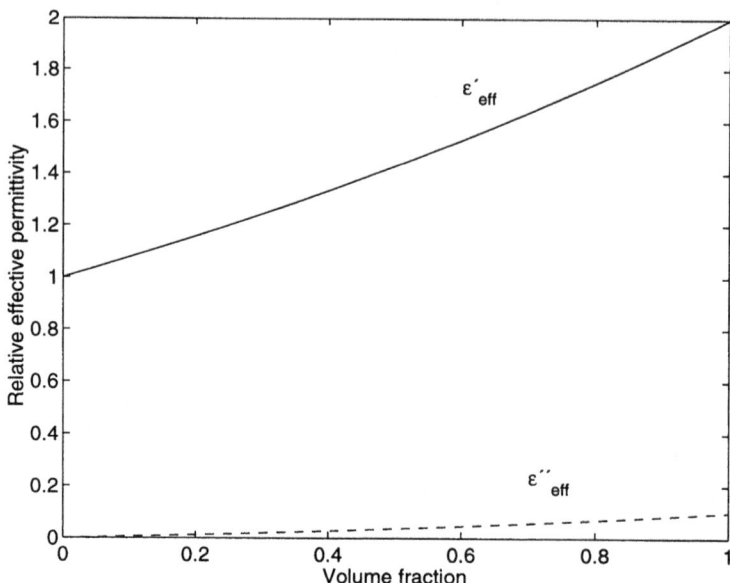

Figure 4.12: *The effective permittivity of a mixture with lossy spherical inclusions. The permittivity of the inclusions relative to background is $\epsilon_i/\epsilon_e = 2 - j0.1$. The ϵ'_{eff} curve is indistinguishable from the corresponding lossless curve ($\epsilon_i/\epsilon_e = 2$).*

permittivity dramatically changes. Although the imaginary part of ϵ_{eff} increases in a regular manner from zero to the imaginary part of the inclusion permittivity, the real part displays a peak. The mixture, for a certain mixing ratio, may possess a larger permittivity value ϵ'_{eff} than either of the components!

It is worth noting that the Maxwell Garnett mixing rule can correctly predict—at least qualitatively—the properties of a mixture where the inclusion phase is conducting. Consider a dielectric mixture with lossless background permittivity ϵ_e, and inclusions with the real part of the permittivity ϵ'_i and conductivity σ_i. If the volume fraction of the inclusion phase is small, the effective conductivity $\sigma_{\text{eff}} = \omega\epsilon''_{\text{eff}}$, calculated from (4.53), is

$$\sigma_{\text{eff}} = \frac{9\epsilon_e^2 f \sigma_i}{(\epsilon'_i + 2\epsilon_e)^2 + \sigma_i^2/\omega^2} \tag{4.54}$$

The result shows that the effective DC conductivity of the mixture vanishes ($\sigma_{\text{eff}} \to 0$ as $\omega \to 0$). This is intuitively acceptable, because conducting particles that do not touch each other in a nonconducting matrix do not make the mixture conducting.[7]

---

[7] The same conclusion can be drawn with a mixing analysis that is based on the homogenisation of the conductivity problem instead of the permittivity problem. The conductivity approach is a more natural approach in geophysical studies of soil and bedrock [21, Section 2.8]. The analogue of the electrical conductivity problem is naturally the heat conduction problem [22] although conduction is only one mechanism of heat transfer beside radiation and convection, etc.

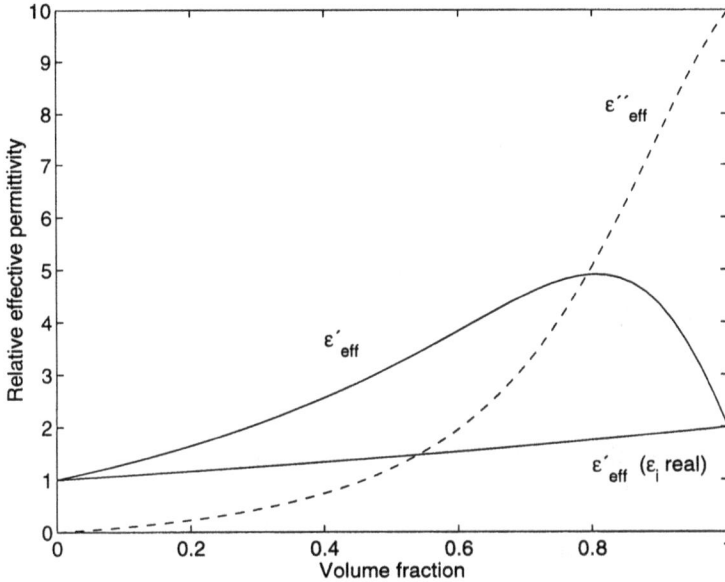

Figure 4.13: *The effective permittivity of a mixture with lossy inclusions. The permittivity of the inclusions relative to background is $\epsilon_i/\epsilon_e = 2 - j10$. The corresponding lossless case (inclusion permittivity $\epsilon_i/\epsilon_e = 2$) is completely different in character.*

The mixing of conducting regions in an insulating background leads to interfacial polarisation accumulation at low frequencies—already mentioned in Section 2.1—which is an important dispersion mechanism for many materials in the kHz region. In many applications, the phenomenon is known as the Maxwell–Wagner effect.

A warning for the use of the mixing rule for complex permittivities is in place for alternating fields. The derivation of the Maxwell Garnett mixing rule was based on the algebraic dependence of the internal field on the parameters of the problem (cf. Equation (3.12)). However, it is known that for time-dependent fields, losses entail exponential attenuation of the field amplitudes which can be considerable if the extent of the lossy medium is large compared with the penetration depth. Hence the requirement of allowed use of the Maxwell Garnett rule for time-dependent fields is that the inclusion size must not be larger than the skin depth of the wave in the lossy medium $\sqrt{2/(\omega\mu_i\sigma_i)}$ with μ_i being the magnetic permeability of the inclusion material.[8]

---

[8]This is a different requirement than the quasistatic one that demands that inclusions be small compared to the wavelength in the effective material. The quasistatic requirement has to be satisfied, too, although the wavelength in the material is different from the background-related wavelength, at least for dense mixtures.

Problems

4.1 Study how the shape of the inclusions affects the effective permittivity of a mixture with inclusion-to-background contrast $\epsilon_i/\epsilon_e = 50$. Draw $\epsilon_{\text{eff}}/\epsilon_e$ as a function of the volume fraction $0 \leq f \leq 1$ for the following five structures. Inclusions are

  (a) spheres

  (b) aligned discs (plot both permittivity components, axial and transversal)

  (c) randomly oriented discs

  (d) aligned needles (plot both permittivity components)

  (e) randomly oriented needles.

4.2 Consider a mixture where spheroids (ellipsoids of revolution), in randomly oriented distribution, occupy a volume fraction f in a background material. Plot the effective permittivity of the mixture as a function of the volume fraction. Assume that the permittivity of the spheroids is 20 times that of the background: $\epsilon_i = 20\epsilon_e$.

Draw two figures: one for prolate spheroids, and another for oblate spheroids. Take many axis ratios including spheres (axis ratio $a_y/a_x = a_z/a_x = 1$) and needles ($a_y/a_x = a_z/a_x = 0$) in the prolate case, and from spheres (axis ratio $a_z/a_x = a_z/a_y = 1$) to discs ($a_z/a_x = a_z/a_y = 0$) in the oblate case.

4.3 Repeat the previous exercise for the Swiss-cheese mixture where the environment is dielectrically heavier than the spheroids: $\epsilon_i = \epsilon_e/20$.

4.4 The depolarisation factors of a general ellipsoid can be calculated from the integral (4.5) as a function of the axis ratios. Study the case when the ellipsoid degenerates to a sphere, in other words, set $a_x = a_y = a_z = a$ in this formula and check that the result equals the depolarisation factor for a sphere, $1/3$.

4.5 Using the integral (4.5), derive the depolarisation factors (4.7)–(4.8) for prolate spheroids, and (4.11)–(4.12) for oblate ones.

4.6 Let us study in this problem the possibility of invisible inclusions. If an inclusion is homogeneous, Equations (3.10) and (4.21) show that they always have a nonzero polarisability if only the permittivity of the inclusion differs from that of the background. However, for inhomogeneous inclusions the situation is different. The polarisability may vanish for some combinations of the permittivities and size ratios of the layers.

Study a spherical inclusion which consists of a spherical core and a single spherical layer. Take two cases for the radius ratios: $a_2/a_1 = 0.5$ and $a_2/a_1 = 0.75$. Plot the possible permittivities for the core and the shell (ϵ_1 and ϵ_2)

such that the "invisibility" requirement is satisfied, in other words such that the polarisability of the composite sphere vanishes.

4.7 Consider a dielectrically inhomogeneous sphere of radius a. Let the permittivity of the sphere be the following parabolic function of the radius:

$$\epsilon(\mathbf{r}) = \left[2 - (r/a)^2\right] \epsilon_e$$

Calculate the polarisability of the sphere.

4.8 Consider a suspension of small inclusions of gold in a transparent matrix material. Plot the real and imaginary parts of the permittivity of the suspension over the whole mixture range (the volume fraction $f = 0 \cdots 1$). Assume that the background medium is lossless and diaphanous ($\epsilon_e \approx \epsilon_0$) and that the dielectric character of gold inclusions are totally determined by its conductivity behaviour. The conductivity of gold is 4.1×10^7 A/Vm. Take the frequency of the field (which is assumed to vary time-harmonically) to be 1 MHz.

4.9 To filter out one polarisation component[9] of an electromagnetic wave one can use a polarising filter. A linearly polarising filter, sometimes also called polaroid, is such that transmits a propagating wave when the electric field points in a certain direction, but attenuates the orthogonal component of the electric field.

A plane wave obeys the distance dependence $\exp(-jkz)$ along the z-direction where $k = \omega\sqrt{\mu\epsilon}$, μ and ϵ being the permeability and permittivity of the material. If the medium is lossy, the material parameters can be complex, which leads to exponential attenuation of the wave.

Let us study in the present problem how a polaroid could be designed by mixing lossy inclusions in a lossless matrix. Let the inclusions all be aligned needles (very prolate ellipsoids, but all small compared to the wavelength). Then the effective permittivity and conductivity are different for the two orthogonal field directions. Calculate how an effective filter can be designed with this principle if the matrix is a lossless plastic with relative permittivity $\epsilon_e = 1.2\epsilon_0$ and the needle-shaped cavities are filled with an ionic liquid that has permittivity $\epsilon_i = (10 - j500)\epsilon_0$. Plot (in decibels) the transmittance of the polarisation orthogonal to the needles and the attenuation for the needle-aligned polarisation, both as functions of the volume fraction of the inclusion liquid. Perform the calculations at the frequency of 10 GHz and for a plate thickness of 5 mm.

---

[9]The term *polarisation* has two distinct meanings in electromagnetics, as was emphasised earlier. In optics and propagation of electromagnetic waves in general, polarisation means—differently from the main text in the present book—the vector direction of the electic field of the wave. One may talk about horizontal, vertical, circular, and elliptical polarisations. It is this meaning of polarisation that is discussed in the present exercise.

References

[1] LANDAU, L.D., and LIFSHITZ, E.M.: 'Electrodynamics of continuous media', Second Edition (Pergamon Press, Oxford, 1984), Section 4

[2] KELLOGG, O.D.: 'Foundations of potential theory' (Dover Publications, New York, 1953), Chapter VII

[3] ABRAMOWITZ, M., and STEGUN, I.A. (Ed.): 'Handbook of mathematical functions' (Dover Publications, New York, 1972)

[4] OSBORN, J.A.: 'Demagnetizing factors of the general ellipsoid', *The Physical Review*, 1945, **67**, (11-12), pp. 351-357

[5] STONER, E.C.: 'The demagnetizing factors for ellipsoids', *Philosophical Magazine*, 1945, Ser. 7, **36**, (263), pp. 803-821

[6] FRICKE, H.: 'The Maxwell–Wagner dispersion in a suspension of ellipsoids', *Journal of Physical Chemistry*, 1953, **57**, pp. 934-937

[7] WEIGLHOFER, W.S.: 'Electromagnetic depolarization dyadics and elliptic integrals', *Journal of Physics A: Math. Gen.*, 28 August 1998, **31**, (34), pp. 7191-7196

[8] LINDELL, I.V.: 'Methods for electromagnetic field analysis' (IEEE Press and Oxford University Press, 1995)

[9] WIENER, O.: 'Zur Theorie der Refractionskonstanten', *Berichte über die Verhandlungen der königlich-sächsischen Gesellschaft der Wissenschaften zu Leipzig*, 1910, Math.-Phys. Klasse, **62**, pp. 256-277

[10] SIHVOLA, A., NYFORS, E., and TIURI, M.: 'Mixing formulae and experimental results for the dielectric constant of snow', *Journal of Glaciology*, 1985, **31**, (108), pp. 163-170

[11] JONES, D.S.: 'Low frequency electromagnetic radiation', *J. Inst. Maths Applications*, 1979, **23**, pp. 421-447

[12] SIHVOLA, A., and KONG, J.A.: 'Effective permittivity of dielectric mixtures', *IEEE Transactions on Geoscience and Remote Sensing*, 1988, **26**, (4), pp. 420-429. Correction, *ibid.*, 1989, **27**, (1), pp. 101-102

[13] STRATTON, J.A.: 'Electromagnetic theory' (McGraw-Hill, New York, 1941)

[14] SIHVOLA, A., and LINDELL, I.V.: 'Transmission line analogy for calculating the effective permittivity of mixtures with spherical multilayer scatterers', *Journal of Electromagnetic Waves and Applications,*, 1988, **2**, (8), pp. 741-756

[15] SIHVOLA, A., and LINDELL, I.V.: 'Polarizability and effective permittivity of layered and continuously inhomogeneous dielectric spheres', *Journal of Electromagnetic Waves and Applications*, 1989, **3**, (1), pp. 37-60

[16] JONES, D.S.: 'The theory of electromagnetism' (Pergamon Press, Oxford, 1964)

[17] MOON, P., and SPENCER, D.E.: 'Field theory handbook' (Springer Verlag, 1971)

[18] SIHVOLA, A., and LINDELL, I.V.: 'Polarizability and effective permittivity of layered and continuously inhomogeneous dielectric ellipsoids', *Journal of Electromagnetic Waves and Applications*, 1990, **4**, (1), pp. 1-26

[19] CHENG, D.K.: 'Field and wave electromagnetics' (Addison Wesley, Reading, Mass., 1989)

[20] COLLIN, R.E.: 'Foundations for microwave engineering' (McGraw–Hill, New York, 1966)

[21] WAIT, J.R.: 'Electromagnetic wave theory' (Harper & Row, New York, 1985)

[22] BURGER, H.C.: 'Das Leitvermögen verdünnter mischkristallfreier Legierungen', *Physikalische Zeitschrift*, 1919, **20**, (4), pp. 73-75

Chapter 5

Anisotropic mixtures

After the treatment in the previous chapter where the geometry of the mixture was allowed to vary, let us now change our focus on complications in matter and medium. In particular, the assumption of isotropy will be relaxed in this chapter. In other words, the material response of the media can depend on the direction of the electric field that acts on it. This phenomenon is called dielectric anisotropy.

5.1 Anisotropy in dielectric materials

The simple relation in homogeneous matter between the electric field $\mathbf{E}$ and displacement (flux density) $\mathbf{D}$

$$\mathbf{D} = \epsilon\mathbf{E} \tag{5.1}$$

is valid when the molecular response, on the average, in the medium is polarised in the same direction as the primary field. This is not often the case in natural materials, strictly taken. For example, many natural media and human-made composites have a fibrous or lamellar structure which breaks the directional symmetry. When the field and displacement point in different directions, as shown in Figure 5.1, the relationship between them can no longer be a scalar multiplication. Rather, it has to be described by a matrix, dyadic, or a second-rank tensor.

Of these representations, let us choose the middle one, and denote the anisotropic relation by a dyadic permittivity $\overline{\overline{\epsilon}}$

$$\mathbf{D} = \overline{\overline{\epsilon}} \cdot \mathbf{E} \tag{5.2}$$

which dyadic is notationally distinguished with the double overbar. It has a matrix representation in any chosen co-ordinate frame. In general, the anisotropic permittivity contains nine independent material parameters. Because the isotropic

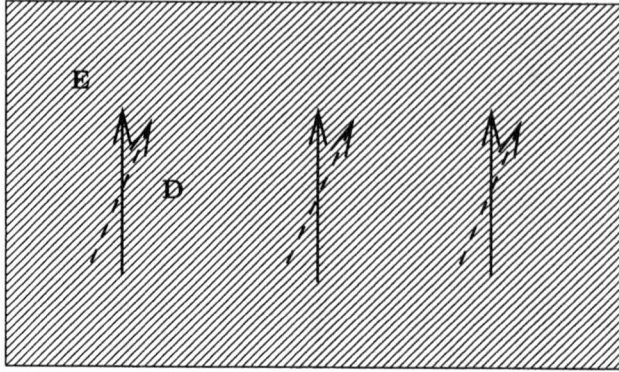

Figure 5.1: *In anisotropic medium, the field and the displacement point in different directions.*

permittivity corresponds to a simple multiplication, in other words has only one degree of freedom, the step into higher dimensions is indeed great when we allow the materials to be anisotropic.[1]

However, in many practical cases of anisotropic materials, the number of independent parameters is much less than 9. A rectangular lattice which forms the basic structure in many crystals determines the three principal directions of its material response. Likewise, a mixture with aligned ellipsoids which was treated in Section 4.2.3 would have three different permittivity parameters along the three orthogonal axes. These are examples of biaxial materials.

In more detail, the crystallographic description of solid state matter distinguishes seven classes [2]. Of these, the ones with least symmetry are biaxial: orthorhombic, monoclinic, and triclinic systems, which have three different permittivity components. More symmetry have the tetragonal, hexagonal, and rhombohedral (trigonal) systems, which are uniaxial. This means that the permittivity component along the axis of the crystal is different from the transversal permittivity. And finally, the cubic crystallographic system has the same permittivities in all three directions which means that its response to a uniform field is isotropic.[2] In solid state physics the tensorial properties of crystals are paid special attention and are formulated into a theorem, the Neumann principle: no asymmetry may be exhibited by any property of a crystal which is not possessed by the crystal itself. In other words, the symme-

---

[1]The choice of the term is somewhat unfortunate, as famously noted by Lord Kelvin [1]: "An anisotropic solid is not an isotropic solid." To prevent misunderstandings that the jangle between the prefix and article may cause, Kelvin suggested the term *aeolotropic* instead, with little success.

[2]In wave propagation, however, one needs to distinguish between the cubic symmetry of solid state from a total spherical symmetry of a fluid. A linear birefringence effect has been observed in cubic crystals [3], and therefore cubic crystals should be strictly termed anisotropic [4,5] although the dyadic relation according to Equation (5.2) is equivalent for cubics and true isotropic media. The difference in material properties of cubics and isotropic media shows itself in the character of their property tensors of rank higher than two.

tries of a particular crystal determine the potential possibility for crystals to have properties like birefringence or optical activity [2, 6].

The crystal's response to electric excitation is symmetric. This means that there exists a co-ordinate system where the matrix corresponding to relation (5.2) is diagonal. But materials exist for which the permittivity is nonsymmetric. An example of such a medium is magnetoplasma where the antisymmetric part arises due to an external magnetic field which has an effect on the movement of the free electrons in the medium.

In real materials losses are present. As for isotropic materials, the losses are taken into account by allowing the permittivity relation to contain complex parameters. The condition that no dissipation in the medium exists can be formally stated [7, p. 51] by $\overline{\overline{\epsilon}} = \overline{\overline{\epsilon}}^\dagger$ where the dagger denotes a hermitian transpose: a complex conjugate of the transposed permittivity dyadic. Therefore for a medium with symmetric permittivity operator, the parameters in a lossless case have to be real. But for the antisymmetric part, the permittivity component has to be purely imaginary if we wish to avoid losses.

Sometimes dielectrically anisotropic materials display anisotropy in their other, non-electrical, physical responses, too. Due to the fact that quite often the electrical anisotropy is a consequence of the geometrical microstructure of the medium, the very same structure is reflected in the reaction of the material to other excitations, like mechanical deformations and thermal gradients, for example. An interesting type of response, from the point of view of electromagnetic wave propagation, is the magnetic behaviour of a medium. Although all media have magnetic properties, being composed of atoms with electronic structure, the macroscopic magnetic response in most materials is weak. But there are media which exhibit strong magnetic response to magnetic excitation. Examples are ferromagnetic and ferrimagnetic materials, shortly introduced in Section 2.3. In these, the response may also be anisotropic which means that the magnetic constitutive parameter, called permeability $\overline{\overline{\mu}}$, cannot be a scalar multiplier of the magnetic field. Also in ferrites, the magnetic permeability may contain an antisymmetric part which is again due to an external magnetic bias field. For lossless magnetic materials, the permeability has to be hermitian dyadic.

Assumptions on the physical behaviour lead to restrictions on the material parameters. In addition to the losslessness condition that restricts the real and imaginary parts of permittivity and permeability of materials, another important special case is formed by *reciprocal* media. Limitation to reciprocity means that the permittivity and permeability dyadics of the material are symmetric: $\overline{\overline{\epsilon}} = \overline{\overline{\epsilon}}^T$, $\overline{\overline{\mu}} = \overline{\overline{\mu}}^T$.

Anisotropy of materials makes it necessary to take a position on many new questions in the homogenisation analysis that were not relevant in the case of isotropic mixtures. In fact, one could say that a sample of anisotropic medium changes its character when it is rotated. This means that a polycrystalline material, although it is formed by many crystals of the same medium which all have the same permittivity matrix, is a mixture of grains of many different materials. This is because the material matrices of the grains are different in the global co-ordinate system even if they are related to each other by an obvious rotational transformation. The present chapter discusses in depth the homogenisation questions involving anisotropy. For that purpose, we need to make a somewhat formal sidestep into dyadic algebra.

5.2   Elementary dyadic analysis

When physical phenomena are being analysed with three-dimensional vectors, a very valuable toolbox is provided in dyadic algebra. A dyadic is a function that operates on one vector. The result of this operation is another vector. The origin of dyadic algebra can be traced back to the 1880s when J.W. Gibbs introduced it in his studies and lectures [8], drawing on some of the quaternion work by W.R. Hamilton.

By the success of the relativity theory, tensor calculus became a major analysis tool in theoretical physics. Dyadic algebra did not manage to keep its supporters in the physics circles; in fact, quite condescending remarks can be found in the literature. "The dyadic is a somewhat clumsy device for extending ordinary vector analysis to cover tensors of second rank" [9, p.136]. However, by use of certain easily memorisable rules of algebra, dyadics are in fact very applicable in electromagnetic calculations. This claim hopefully becomes justified in the course of the present chapter.

The Gibbsian operations on dyadics have been supplemented by dyadic identities that make it possible to perform such operations in compact form that would be otherwise impossible without expanding the formulas in component expressions. The identities have been brought forth by I.V. Lindell in his writings starting from the 1960s [10,11]. It seems that the dyadic algebra is gaining more adherents within the present-day electromagneticists [12]. It may be noted also that in Russian and other former-Soviet-Union electromagnetics literature, dyadic algebra has survived in isolation, although the scientific evolution has led there to a notation that slightly contrasts with that which is widely accepted in the Western tradition. The school established by F.I. Fedorov in the Soviet Union in the 1950s has been actively using dyadics for electromagnetics (see, for example [13,14]).

In the following, a mini-review of dyadic algebra and its operations is given. A more complete exposition of dyadics can be found in [15, Chap. 2]. Appendix A gives a list of some important dyadic identities and rules.

5.2.1   Notation and definitions

The distinction between different types of quantities has been noted by different symbols in the present text. Scalars, like radius a and polarisability α, are in italic typeface. Vectors, like the electric field $\mathbf{E}$ and displacement $\mathbf{D}$ are denoted by boldface characters. Operations between these quantities have been tacitly assumed to be obvious: the multiplication by a scalar is noted by a juxtaposition of two scalars (αa), or a scalar and a vector $(\alpha \mathbf{E} = \mathbf{E}\alpha)$.

In vector algebra, the two well-known products between two vectors $\mathbf{a}$ and $\mathbf{b}$ are the scalar and vector products: the product denoted by $\mathbf{a} \cdot \mathbf{b}$ is a scalar and $\mathbf{a} \times \mathbf{b}$ is a vector. Written in the Cartesian co-ordinate system with unit vectors $\mathbf{u}_x, \mathbf{u}_y, \mathbf{u}_z$ these read

$$\mathbf{a} \cdot \mathbf{b} = (a_x \mathbf{u}_x + a_y \mathbf{u}_y + a_z \mathbf{u}_z) \cdot (b_x \mathbf{u}_x + b_y \mathbf{u}_y + b_z \mathbf{u}_z) = a_x b_x + a_y b_y + a_z b_z \quad (5.3)$$

and

$$\mathbf{a} \times \mathbf{b} = (a_y b_z - a_z b_y)\mathbf{u}_x + (a_z b_x - a_x b_z)\mathbf{u}_y + (a_x b_y - a_y b_x)\mathbf{u}_z \quad (5.4)$$

A familiar geometric interpretation for these products with the help of angles and vector lengths exists for real vectors; for complex vectors, it is perhaps better to split all terms in real and imaginary parts and treat those separately.[3]

A third product between two vectors is the following:

$$
\begin{aligned}
\mathbf{ab} &= (a_x \mathbf{u}_x + a_y \mathbf{u}_y + a_z \mathbf{u}_z)(b_x \mathbf{u}_x + b_y \mathbf{u}_y + b_z \mathbf{u}_z) \quad (5.5) \\
&= a_x b_x \mathbf{u}_x \mathbf{u}_x + a_x b_y \mathbf{u}_x \mathbf{u}_y + a_x b_z \mathbf{u}_x \mathbf{u}_z \\
&\quad a_y b_x \mathbf{u}_y \mathbf{u}_x + a_y b_y \mathbf{u}_y \mathbf{u}_y + a_y b_z \mathbf{u}_y \mathbf{u}_z \\
&\quad a_z b_x \mathbf{u}_z \mathbf{u}_x + a_z b_y \mathbf{u}_z \mathbf{u}_y + a_z b_z \mathbf{u}_z \mathbf{u}_z
\end{aligned}
$$

This is a dyadic product, which combines two vectors without any operation inbetween. Here, the simple vector pair $\mathbf{ab}$ is called a *dyad*. A general *dyadic* is a polynomial of dyads:[4]

$$\overline{\overline{A}} = \sum_{i=1}^{n} \mathbf{a}_i \mathbf{b}_i \quad (5.6)$$

As defined in Equation (5.6), the dyadic character of the quantity $\overline{\overline{A}}$ is emphasised by a double overbar.[5]

From the expansion (5.5) one can guess that there exists a connection between dyadics and matrices. In three-dimensional vector space, a dyadic and a matrix both have nine components. The advantage of dyadics is, however, that the dyadic algebra is independent on any co-ordinate system whereas the numbers inside the matrix box vary if, for example, the co-ordinate basis is rotated.

A dyadic, like a matrix, can be interpreted as an operation on a vector, the result of which is another vector:

$$\mathbf{d} = \overline{\overline{A}} \cdot \mathbf{c} = \left(\sum_i \mathbf{a}_i \mathbf{b}_i \right) \cdot \mathbf{c} = \sum_i \mathbf{a}_i (\mathbf{b}_i \cdot \mathbf{c}) \quad (5.7)$$

---

[3] A geometric interpretation for complex vectors as ellipses can be given, too [15, Chapter 1].

[4] The word "dyadic" is used in these sentences both as an adjective and as a noun, but hopefully nonambigiously.

[5] The reader is certainly familiar with notations where vectors are distinguished from scalars by a single overbar. To clarify terminology: the class of dyads forms a subset of all dyadics, in other words a dyad is always a dyadic but a dyadic—in general—cannot be expressed as a dyad.

From this it can be seen that the vectors **c** and **d** are not necessarily parallel. This makes dyadics very applicable to the analysis of anisotropic materials where the type of general relation (5.2) has to be allowed.

Also, successive operations can be grouped together:

$$\overline{\overline{A}}_1 \cdot (\overline{\overline{A}}_2 \cdot \mathbf{c}) = (\overline{\overline{A}}_1 \cdot \overline{\overline{A}}_2) \cdot \mathbf{c} \tag{5.8}$$

which can be done because the dot-product algebra is associative:

$$\overline{\overline{A}} \cdot (\overline{\overline{B}} \cdot \overline{\overline{C}}) = (\overline{\overline{A}} \cdot \overline{\overline{B}}) \cdot \overline{\overline{C}} \tag{5.9}$$

Another property of this algebra is the distributivity:

$$\overline{\overline{A}} \cdot (\overline{\overline{B}} + \overline{\overline{C}}) = \overline{\overline{A}} \cdot \overline{\overline{B}} + \overline{\overline{A}} \cdot \overline{\overline{C}} \tag{5.10}$$

but it is not commutative: in general

$$\overline{\overline{A}} \cdot \overline{\overline{B}} \neq \overline{\overline{B}} \cdot \overline{\overline{A}} \tag{5.11}$$

One can convince oneself of these properties by expanding the dyadics in dyad sums.[6]

A useful dyadic is the unit dyadic $\overline{\overline{I}}$ with the property that it relates any vector with the same vector:

$$\overline{\overline{I}} \cdot \mathbf{a} = \mathbf{a} \cdot \overline{\overline{I}} = \mathbf{a} \tag{5.12}$$

In the Cartesian system, we obviously have: $\overline{\overline{I}} = \mathbf{u}_x \mathbf{u}_x + \mathbf{u}_y \mathbf{u}_y + \mathbf{u}_z \mathbf{u}_z$.

5.2.2   Operations and invariants

Transpose operation and symmetry

The transpose operation switches between the two vectors in dyads and is denoted by a superscript T:

$$\overline{\overline{A}}^T = \left(\sum_i \mathbf{a}_i \mathbf{b}_i \right)^T = \sum_i \mathbf{b}_i \mathbf{a}_i \tag{5.13}$$

This definition allows us to write $\overline{\overline{A}} \cdot \mathbf{c} = \mathbf{c} \cdot \overline{\overline{A}}^T$ and $(\overline{\overline{A}} \cdot \overline{\overline{B}})^T = \overline{\overline{B}}^T \cdot \overline{\overline{A}}^T$

A symmetric dyadic remains intact in the transpose operation, and an antisymmetric dyadic changes sign when transposed. The nine degrees of freedom in the dyadic space are split such that the space spanned by symmetric dyadics is six-dimensional and the space of antisymmetric dyadics has three dimensions. Because a vector also has three components there is a one-to-one correspondence between vectors and antisymmetric dyadics. Every antisymmetric dyadic can be represented in the form

$$\overline{\overline{A}}_{as} = \mathbf{c} \times \overline{\overline{I}} \tag{5.14}$$

---

[6] Although it is true that the operation on vector: $\overline{\overline{A}}_1 \cdot (\overline{\overline{A}}_2 \cdot \mathbf{c}) = (\overline{\overline{A}}_1 \cdot \overline{\overline{A}}_2) \cdot \mathbf{c}$ is valid, and also $(\mathbf{c} \cdot \overline{\overline{A}}_1) \cdot \overline{\overline{A}}_2 = \mathbf{c} \cdot (\overline{\overline{A}}_1 \cdot \overline{\overline{A}}_2)$ holds, the following must be remembered about associativity: in general, $(\overline{\overline{A}}_1 \cdot \mathbf{c}) \cdot \overline{\overline{A}}_2 \neq \overline{\overline{A}}_1 \cdot (\mathbf{c} \cdot \overline{\overline{A}}_2)$.

which is clear when one notes the anticommutativity of the vector product

$$\overline{\overline{A}}_{as} \cdot \mathbf{a} = \mathbf{c} \times \mathbf{a} = -\mathbf{a} \times \mathbf{c} = -\mathbf{a} \cdot \overline{\overline{A}}_{as} = -\overline{\overline{A}}_{as}{}^T \cdot \mathbf{a} \qquad (5.15)$$

A projection of a vector on a plane can be performed by a two-dimensional unit dyadic

$$\overline{\overline{I}}_t = \overline{\overline{I}} - \mathbf{uu} \qquad (5.16)$$

where $\mathbf{u}$ is the unit vector perpendicular to the plane. An operation $\overline{\overline{I}}_t \cdot \mathbf{a}$ eliminates the $\mathbf{u}$-directed part of the vector $\mathbf{a}$.

Double-dot and double-cross products

The double-dot product $\overline{\overline{A}} : \overline{\overline{B}}$ between two dyadics can be written with the following rule

$$(\mathbf{ab}) : (\mathbf{cd}) = (\mathbf{a} \cdot \mathbf{c})(\mathbf{b} \cdot \mathbf{d}) \qquad (5.17)$$

and the result is a scalar obeying the symmetry relations $\overline{\overline{A}} : \overline{\overline{B}} = \overline{\overline{B}} : \overline{\overline{A}} = \overline{\overline{A}}^T : \overline{\overline{B}}^T$.

Correspondingly, the double-cross product $\overline{\overline{A}}{}^\times_\times\overline{\overline{B}}$ between two dyadics is defined with the rule for dyads:

$$(\mathbf{ab})^\times_\times(\mathbf{cd}) = (\mathbf{a} \times \mathbf{c})(\mathbf{b} \times \mathbf{d}) \qquad (5.18)$$

This product gives as its output another dyadic. The double cross is commutative: $\overline{\overline{A}}{}^\times_\times\overline{\overline{B}} = \overline{\overline{B}}{}^\times_\times\overline{\overline{A}}$.

As examples, the double-dot and double-cross products between two unit dyadics obey

$$\overline{\overline{I}} : \overline{\overline{I}} = 3 \qquad (5.19)$$
$$\overline{\overline{I}}{}^\times_\times\overline{\overline{I}} = 2\overline{\overline{I}} \qquad (5.20)$$

which are transparently satisfied; check, for example, in the expanded component form. Furthermore,

$$\overline{\overline{I}}_t : \overline{\overline{I}}_t = 2 \qquad (5.21)$$
$$\overline{\overline{I}}_t{}^\times_\times\overline{\overline{I}}_t = 2\mathbf{uu} \qquad (5.22)$$

where $\mathbf{u}$ is the unit vector normal to the plane of the transversal unit dyadic.

With the double-cross product, the following square of a dyadic can be defined:

$$\overline{\overline{A}}^{(2)} = \frac{1}{2}\overline{\overline{A}}{}^\times_\times\overline{\overline{A}} \qquad (5.23)$$

which is a very important definition. This cross-product square must be distinguished from the ordinary second power of a dyadic, which is

$$\overline{\overline{A}}^2 = \overline{\overline{A}} \cdot \overline{\overline{A}} \qquad (5.24)$$

The rarer mixed double-products are sometimes useful. In obvious manner, we have

$$(\mathbf{ab})^{\times}_{\cdot}(\mathbf{cd}) = (\mathbf{a} \times \mathbf{c})(\mathbf{b} \cdot \mathbf{d}) \tag{5.25}$$

$$(\mathbf{ab})_{\times}^{\cdot}(\mathbf{cd}) = (\mathbf{a} \cdot \mathbf{c})(\mathbf{b} \times \mathbf{d}) \tag{5.26}$$

These operations can be used to extract the antisymmetric part of a dyadic. The vector $\mathbf{c}$ corresponding to the antisymmetric part of a dyadic $\overline{\overline{A}}$ is

$$\mathbf{c} = \frac{1}{2}\overline{\overline{I}}{}^{\times}_{\cdot}\overline{\overline{A}} = \frac{1}{2}\overline{\overline{A}}{}^{\cdot}_{\times}\overline{\overline{I}} \tag{5.27}$$

because $\overline{\overline{S}}{}^{\cdot}_{\times}\overline{\overline{I}} = 0$ for any symmetric dyadic $\overline{\overline{S}}$, and $(\mathbf{c} \times \overline{\overline{I}})^{\cdot}_{\times}\overline{\overline{I}} = 2\mathbf{c}$. This operation is also connected to the replacement of the dyadic products by cross products:

$$\left(\sum_i \mathbf{a}_i\mathbf{b}_i\right){}^{\times}_{\cdot}\overline{\overline{I}} = \sum_i \mathbf{a}_i \times \mathbf{b}_i \tag{5.28}$$

Trace and determinant

An important scalar function of a dyadic is its *trace* (tr). The trace can be calculated by taking the double-dot product with unit dyadic:

$$\text{tr}\,\overline{\overline{A}} = \overline{\overline{A}} : \overline{\overline{I}} \tag{5.29}$$

and in the matrix algebra it corresponds to the sum of diagonal terms. In practical dyadic work, the trace of a dyadic can be calculated easily by taking the sum of the scalar products of the vector pairs:

$$\text{tr}\,\overline{\overline{A}} = \left(\sum_i \mathbf{a}_i\mathbf{b}_i\right) : \overline{\overline{I}} = \sum_i \mathbf{a}_i \cdot \mathbf{b}_i \tag{5.30}$$

Another invariant is the *sum of principal minors* (spm),

$$\text{spm}\,\overline{\overline{A}} = \frac{1}{2}\overline{\overline{A}}{}^{\times}_{\times}\overline{\overline{A}} : \overline{\overline{I}} \tag{5.31}$$

and it can be seen to be equal to the trace of the $\overline{\overline{A}}{}^{(2)}$:

$$\text{spm}\,\overline{\overline{A}} = \text{tr}\,\overline{\overline{A}}{}^{(2)} \tag{5.32}$$

The *determinant* (det) is a very important cubic function of a dyadic:

$$\det\overline{\overline{A}} = \frac{1}{6}\overline{\overline{A}}{}^{\times}_{\times}\overline{\overline{A}} : \overline{\overline{A}} \tag{5.33}$$

Equipped with a definition for the determinant, the inverse of a dyadic can be written:

$$\overline{\overline{A}}{}^{-1} = \frac{\overline{\overline{A}}{}^{(2)T}}{\det\overline{\overline{A}}} = 3\frac{(\overline{\overline{A}}{}^{\times}_{\times}\overline{\overline{A}})^T}{\overline{\overline{A}}{}^{\times}_{\times}\overline{\overline{A}} : \overline{\overline{A}}} \tag{5.34}$$

Note that the inverse of a dyadic exists only for those dyadics that have a nonzero determinant. Such dyadics are called *complete* dyadics.

The inverse of a dot-product of dyadics can be calculated by inverting the order of inverses:

$$(\overline{\overline{A}} \cdot \overline{\overline{B}})^{-1} = \overline{\overline{B}}^{-1} \cdot \overline{\overline{A}}^{-1} \tag{5.35}$$

One often encounters the term the dyadic *adjoint* to a given dyadic, which simply is

$$\mathrm{adj}\,\overline{\overline{A}} = \overline{\overline{A}}^{(2)T} = \frac{1}{2}(\overline{\overline{A}} \overset{\times}{\times} \overline{\overline{A}})^T \tag{5.36}$$

The adjoint of the unit dyadic is itself: $\mathrm{adj}\,\overline{\overline{I}} = \overline{\overline{I}}$. The determinant of the unit dyadic is unity, and its sum of principal minors is three.

5.2.3 Examples

To gain some understanding of dyadic operations, let us see how these can be applied to material description. An often encountered property of materials is their *uniaxial* character. This means that the structural properties are different in one direction from those in the transverse directions. The dyadic permittivity of a medium of this type is symmetric:

$$\overline{\overline{\epsilon}} = \epsilon_t \overline{\overline{I}}_t + \epsilon_z \mathbf{u}_z \mathbf{u}_z \tag{5.37}$$

where the z-axis is the direction of the special axis (sometimes called the *optical* axis). The transversal unit dyadic is $\overline{\overline{I}}_t = \overline{\overline{I}} - \mathbf{u}_z \mathbf{u}_z$. This uniaxial dyadic has the properties

$$
\begin{aligned}
\mathrm{tr}\,\overline{\overline{\epsilon}} &= \epsilon_z + 2\epsilon_t & (5.38)\\
\mathrm{spm}\,\overline{\overline{\epsilon}} &= \epsilon_t(\epsilon_t + 2\epsilon_z) & (5.39)\\
\det\overline{\overline{\epsilon}} &= \epsilon_z \epsilon_t^2 & (5.40)
\end{aligned}
$$

and its inverse is

$$\overline{\overline{\epsilon}}^{-1} = \frac{1}{\epsilon_t}\overline{\overline{I}}_t + \frac{1}{\epsilon_z}\mathbf{u}_z \mathbf{u}_z \tag{5.41}$$

Another interesting special case of a complete dyadic is the *gyrotropic* dyadic:

$$\overline{\overline{\epsilon}} = \epsilon_t \overline{\overline{I}}_t + \epsilon_z \mathbf{u}_z \mathbf{u}_z + \epsilon_g \mathbf{u}_z \times \overline{\overline{I}} \tag{5.42}$$

where an antisymmetric part is included in the uniaxial permittivity (5.37). The vector of antisymmetry aligns with the special direction of the symmetric part. This dyadic has the properties

$$
\begin{aligned}
\mathrm{tr}\,\overline{\overline{\epsilon}} &= \epsilon_z + 2\epsilon_t & (5.43)\\
\mathrm{spm}\,\overline{\overline{\epsilon}} &= \epsilon_t^2 + \epsilon_g^2 + 2\epsilon_t \epsilon_z & (5.44)\\
\det\overline{\overline{\epsilon}} &= (\epsilon_t^2 + \epsilon_g^2)\,\epsilon_z & (5.45)
\end{aligned}
$$

and

$$\overline{\overline{\epsilon}}^{-1} = \frac{\epsilon_t}{\epsilon_t^2 + \epsilon_g^2}\overline{\overline{I}}_t + \frac{1}{\epsilon_z}\mathbf{u}_z\mathbf{u}_z - \frac{\epsilon_g}{\epsilon_t^2 + \epsilon_g^2}\mathbf{u}_z \times \overline{\overline{I}}_t \qquad (5.46)$$

Furthermore, two-dimensional dyadics are useful in many material modelling problems. A two-dimensional dyadic satisfies

$$\mathbf{u} \cdot \overline{\overline{A}} = \overline{\overline{A}} \cdot \mathbf{u} = 0 \qquad (5.47)$$

where **u** is the unit normal to the plane of two dimensions in which the dyadic exists. An example of a two-dimensional dyadic was introduced before—the transversal unit dyadic $\overline{\overline{I}}_t$.

Because two-dimensional dyadics are spanned by vectors limited to a plane they only contain four of the nine degrees of freedom of the complete three-dimensional dyadic. Subsequently, the three-dimensional determinant of a two-dimensional dyadic is zero, and it does not have the inverse according to the definition (5.34).

However, if the treatment is restricted to vector algebra in the two dimensions spanned by the dyadic, an inverse can be written:

$$\overline{\overline{A}}^{-1} = \frac{\overline{\overline{A}}^T \overset{\times}{\times} \mathbf{uu}}{\mathrm{spm}\overline{\overline{A}}} \qquad \text{(two–dimensional inverse)} \qquad (5.48)$$

with the properties $\overline{\overline{A}}^{-1} \cdot \overline{\overline{A}} = \overline{\overline{A}} \cdot \overline{\overline{A}}^{-1} = \overline{\overline{I}}_t = \overline{\overline{I}} - \mathbf{uu}$.

5.3   Polarisability of anisotropic sphere

With dyadic algebra, the vector equations written for dielectric phenomena of isotropic materials can be retained in format even when the treatment is extended to anisotropic inclusions. What has to be changed is that some scalar multiplications have to be replaced by dyadic operators but, on the level of notation, the only change is that a double bar appears above permittivity and polarisability symbols.

5.3.1   Reinterpretation of scalar polarisability

In Chapter 3, the polarisability of the isotropic sphere of permittivity ϵ_i in isotropic environment ϵ_e was shown to be (cf. Equation (3.10))

$$\alpha = 3\epsilon_e V \frac{\epsilon_i - \epsilon_e}{\epsilon_i + 2\epsilon_e} \qquad (5.49)$$

where V is the volume of the sphere. This quantity, being a scalar, means that the dipole moment **p** induced in the sphere was in the same direction as the external field $\mathbf{E}_e$. This state of affairs is no longer the case for anisotropic spheres, as shown in Figure 5.2.

Not perhaps surprisingly, the polarisability for an anisotropic sphere of the same volume and anisotropic permittivity $\overline{\overline{\epsilon}}_e$ reads

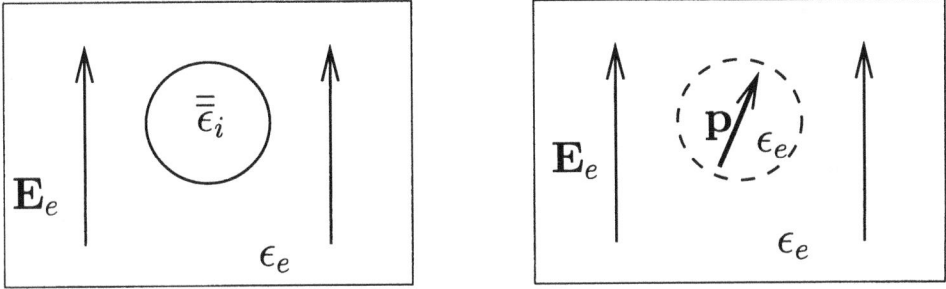

Figure 5.2: *The dipole moment* **p** *induced in an anisotropic sphere is not necessarily aligned with the external field* **E**$_e$.

$$\overline{\overline{\alpha}} = 3\epsilon_e V (\overline{\overline{\epsilon}}_i + 2\epsilon_e \overline{\overline{I}})^{-1} \cdot (\overline{\overline{\epsilon}}_i - \epsilon_e \overline{\overline{I}}) \tag{5.50}$$

where the denominator of (5.49) has been replaced by an inverse of the corresponding dyadic.

What do these double-bars mean in practice? How much does the anisotropy of the sphere complicate calculations? In fact, not too much if the permittivity dyadic $\overline{\overline{\epsilon}}_i$ is real and symmetric. This is the case for many important types of anisotropies. For symmetric permittivities, the dyadic can be written in diagonal form

$$\overline{\overline{\epsilon}}_i = \sum_{j=1}^{3} \epsilon_{i,j} \mathbf{u}_j \mathbf{u}_j \tag{5.51}$$

where $\epsilon_{i,j}$ are the three eigenvalues of the dyadic $\overline{\overline{\epsilon}}_i$, and $\mathbf{u}_j$ the corresponding unit eigenvectors of the dyadic. (A dyadic can be understood as an operator acting on a vector and producing another vector, and hence it has eigenvector characteristics.)

For a real and symmetric dyadic $\overline{\overline{\epsilon}}_i$, the eigenvalues are real and the eigenvectors orthogonal, as in the case of matrices [16].[7] Then the polarisability equation is equivalent to three scalar equations: the eigenvectors for $\overline{\overline{\alpha}}$ are those of $\overline{\overline{\epsilon}}_i$, and the eigenvalues α_j are calculated as in the scalar case, each direction decoupled from the others:

$$\alpha_j = 3\epsilon_e V \frac{\epsilon_{i,j} - \epsilon_e}{\epsilon_{i,j} + 2\epsilon_e} \tag{5.52}$$

and the whole dyadic is simply

$$\overline{\overline{\alpha}} = \sum_{j=1}^{3} \alpha_j \, \mathbf{u}_j \mathbf{u}_j \tag{5.53}$$

---

[7]Strictly taken, the eigenvalues for a real and symmetric definite matrix are orthogonal, for distinct eigenvalues. For degenerate eigenvalues, an orthogonal basis can be constructed.

However, when the dyadic is more general, in other words $\bar{\bar{\epsilon}}_i$ has both symmetric and antisymmetric parts, an inverse makes all the parameters coupled in the remaining expression. The inverse can be calculated using the newly learned rules of dyadic algebra. A condensed form can be written if the permittivity dyadic of the sphere is expressed in a form where the symmetric $\bar{\bar{\epsilon}}_s$ and the antisymmetric parts $\bar{\bar{\epsilon}}_{as}$ are separated:

$$\bar{\bar{\epsilon}}_i = \sum_{j=1}^{3} \epsilon_{i,j} \mathbf{u}_j \mathbf{u}_j + \epsilon_g \mathbf{u}_g \times \bar{\bar{I}} = \bar{\bar{\epsilon}}_s + \bar{\bar{\epsilon}}_{as} \tag{5.54}$$

where the antisymmetric part is denoted by the unit vector $\mathbf{u}_g$ and the gyrotropy amplitude ϵ_g.[8] Using this decomposition, the polarisability dyadic for a generally anisotropic sphere can be written as [17]:

$$\bar{\bar{\alpha}} = 3\epsilon_e V \frac{\bar{\bar{A}} + \bar{\bar{I}} \epsilon_g^2 (\epsilon_c + 2\epsilon_e) - \mathbf{u}_g \mathbf{u}_g 3\epsilon_e \epsilon_g^2 + 3\epsilon_e \epsilon_g [(\bar{\bar{\epsilon}}_s + 2\epsilon_e \bar{\bar{I}}) \cdot \mathbf{u}_g] \times \bar{\bar{I}}}{(\epsilon_{i,1} + 2\epsilon_e)(\epsilon_{i,2} + 2\epsilon_e)(\epsilon_{i,3} + 2\epsilon_e) + \epsilon_g^2 (\epsilon_c + 2\epsilon_e)} \tag{5.55}$$

where ϵ_c is the "average" permittivity in the gyrotropy direction (note that the gyrotropy axis is in general not related to any of the eigenaxes of the symmetric part):

$$\epsilon_c = \bar{\bar{\epsilon}}_s : \mathbf{u}_g \mathbf{u}_g \tag{5.56}$$

and the dyadic $\bar{\bar{A}}$ is defined by

$$\bar{\bar{A}} = \frac{1}{2}(\bar{\bar{\epsilon}}_s - \epsilon_e \bar{\bar{I}}) \cdot (\bar{\bar{\epsilon}}_s + 2\epsilon_e \bar{\bar{I}}) \overset{\times}{\times} (\bar{\bar{\epsilon}}_s + 2\epsilon_e \bar{\bar{I}}) \tag{5.57}$$

$$= \mathbf{u}_1 \mathbf{u}_1 (\epsilon_{i,1} - \epsilon_e)(\epsilon_{i,2} + 2\epsilon_e)(\epsilon_{i,3} + 2\epsilon_e) \tag{5.58}$$

$$+ \mathbf{u}_2 \mathbf{u}_2 (\epsilon_{i,2} - \epsilon_e)(\epsilon_{i,3} + 2\epsilon_e)(\epsilon_{i,1} + 2\epsilon_e) \tag{5.59}$$

$$+ \mathbf{u}_3 \mathbf{u}_3 (\epsilon_{i,3} - \epsilon_e)(\epsilon_{i,1} + 2\epsilon_e)(\epsilon_{i,2} + 2\epsilon_e) \tag{5.60}$$

The earlier case of the polarisability of an anisotropic sphere with symmetric permittivity dyadic can be written evidently from this expression by setting $\epsilon_g = 0$. Another special case, the gyrotropic sphere, with permittivity of the following type

$$\bar{\bar{\epsilon}}_i = \epsilon_i \bar{\bar{I}} + \epsilon_g \mathbf{u}_g \times \bar{\bar{I}} \tag{5.61}$$

possesses polarisability that follows equation

---

[8] A general dyadic in three dimensions has nine independent parameters. Three of these are exhausted for the eigenvalues of the symmetric part, the other three go to define the directions of the orthogonal basis ($\mathbf{u}_1, \mathbf{u}_2, \mathbf{u}_3$), and the three last ones are needed to determine the gyrotropy vector uniquely.

$$\overline{\overline{\alpha}} = 3\epsilon_e V \left[\overline{\overline{I}} \frac{(\epsilon_i - \epsilon_e)(\epsilon_i + 2\epsilon_e) + \epsilon_g^2}{(\epsilon_i + 2\epsilon_e)^2 + \epsilon_g^2} \right.$$

$$- \mathbf{u}_g \mathbf{u}_g \frac{3\epsilon_e}{\epsilon_i + 2\epsilon_e} \frac{\epsilon_g^2}{(\epsilon_i + 2\epsilon_e)^2 + \epsilon_g^2}$$

$$\left. + \mathbf{u}_g \times \overline{\overline{I}} \frac{3\epsilon_e \epsilon_g}{(\epsilon_i + 2\epsilon_e)^2 + \epsilon_g^2} \right] \tag{5.62}$$

showing that the polarisability is also a gyrotropic dyadic, as expected.

5.3.2   Depolarisation and the shape effect

The dyadic polarisability expression (5.50) was written down in the previous section using plain intuitive persuasion. But the field within the scatterer and the corresponding polarisability can also be calculated rigorously and shall be done next. Consider the following geometry where an inclusion is located in unbounded isotropic dielectric of permittivity ϵ_e. In the absence of the scatterer, let there be a uniform electric field $\mathbf{E}_e$ as shown in the left part of Figure 5.3. This is created by a free charge distribution ϱ very far away. This means that the following equation holds:

$$\nabla \cdot (\epsilon_e \mathbf{E}_e) = \varrho \tag{5.63}$$

Then, when the scatterer is present, the total electric field is no longer $\mathbf{E}_e$ but it is perturbed by a "scattered" field $\mathbf{E}_s$. The scattered field is created by a polarisation source $\mathbf{P}$ with which the inclusion can be replaced.[9] The equivalent geometry is shown on the right part of Figure 5.3.

No free charges are created by the introduction of the scatterer; hence the total displacement $\mathbf{D}$ obeys the equation

$$\nabla \cdot \mathbf{D} = \nabla \cdot (\epsilon_e \mathbf{E} + \mathbf{P}) = \nabla \cdot [\epsilon_e(\mathbf{E}_e + \mathbf{E}_s) + \mathbf{P}] = \varrho \tag{5.64}$$

which is equivalent to

$$\nabla \cdot (\epsilon_e \mathbf{E}_s) = -\nabla \cdot \mathbf{P} \tag{5.65}$$

This equation can be interpreted so that the convergence of the polarisation is the source of the scattered field.

According to Coulomb's law, the solution for this equation can be written as an integral over the source (the polarisation) region V (see, for example, [18, Chapter 3]):

---

[9]Note that the polarisation "floats" in the background medium ϵ_e from which it follows that there are no dielectric boundaries in this equivalent problem: The situation is similar to Figure 5.2 where the dipole moment is the volume integral of $\mathbf{P}$.

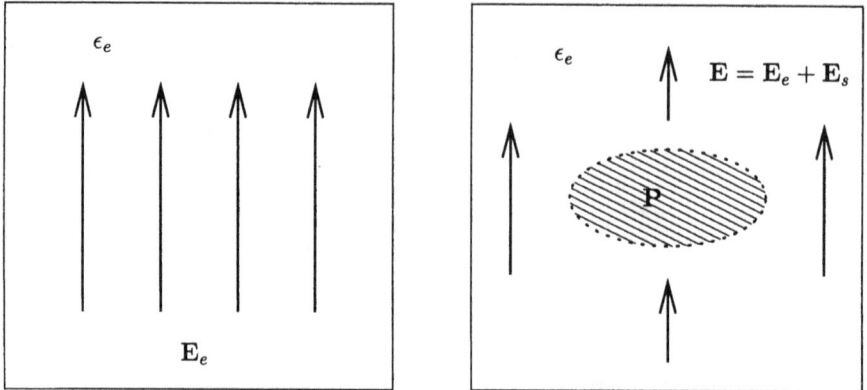

Figure 5.3: *The total field in the presence is treated as a sum of two parts: the external field $\mathbf{E}_e$ which is the field in the absence of the inclusion, and the "scattered field" $\mathbf{E}_s$ which is the field due to the induced polarisation $\mathbf{P}$. Note that in the right-hand figure, there are no material boundaries: the polarisation $\mathbf{P}$ is located "in" the same background permittivity as the environment ϵ_e.*

$$\mathbf{E}_s(\mathbf{r}) = -\int_V \frac{\nabla' \cdot \mathbf{P}(\mathbf{r}')}{4\pi\epsilon_e R^2} \mathbf{u}_R \, dV' \tag{5.66}$$

where the position vector of the field point is $\mathbf{r}$, that of the source point is $\mathbf{r}'$, $R = |\mathbf{r} - \mathbf{r}'|$ is the distance from the source point to the field point, and $\mathbf{u}_R = (\mathbf{r} - \mathbf{r}')/R$ is the unit vector along this direction. Note that the integration is over the source volume, and since the source is the divergence of the polarisation, the arguments have to be primed both in the co-ordinate and in the divergence operation. The only quantity that the nabla operation acts on is the polarisation $\mathbf{P}(\mathbf{r}')$.

Next we make an important assumption. Let us assume—and return later to the justification—that the polarisation density is constant: $\mathbf{P}(\mathbf{r}') = \mathbf{P}$ inside the inclusion (of course, the function $\mathbf{P}(\mathbf{r}') = 0$ outside the inclusion). This means that its divergence vanishes nearly everywhere, both inside and outside the inclusion. Only at the surface is there a sudden drop in its amplitude when we move through the boundary. Then the volume integral shrinks to an integral over the surface S of the inclusion,[10]

$$\mathbf{E}_s(\mathbf{r}) = \int_S \mathbf{P} \cdot \mathbf{n}' \frac{\mathbf{u}_R}{4\pi\epsilon_e R^2} \, dS' = \mathbf{P} \cdot \int_S \frac{\mathbf{n}' \mathbf{u}_R}{4\pi\epsilon_e R^2} \, dS'$$

---

[10]In terms of equations, the divergence of the polarisation contains a delta dunction, which makes the volume integral collapse to a surface integral: $\nabla' \cdot \mathbf{P}(\mathbf{r}') = -\mathbf{n}' \cdot \mathbf{P}\,\delta(\mathbf{r}' - \mathbf{r}_S)$, where $\mathbf{r}_S$ is the position vector defining the surface of the inclusion.

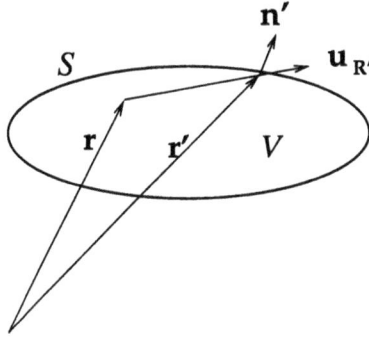

Figure 5.4: *Definition of the variables in the depolarisation dyadic integral.*

$$= -\mathbf{P} \cdot \int_S \frac{\mathbf{n}'\mathbf{u}_{R'}}{4\pi\epsilon_e R^2} \, dS' \tag{5.67}$$

where the last sign change is due to the definition

$$\mathbf{u}_{R'} = \frac{\mathbf{r}' - \mathbf{r}}{R} = -\frac{\mathbf{r} - \mathbf{r}'}{R} = -\mathbf{u}_R \tag{5.68}$$

The unit normal vector $\mathbf{n}'$ points out of the polarisation volume according to Figure 5.4.

If the results are so far correct, the perturbational scattered field can be expressed as

$$\mathbf{E}_s(\mathbf{r}) = -\frac{1}{\epsilon_e}\overline{\overline{L}} \cdot \mathbf{P} \tag{5.69}$$

where the *depolarisation dyadic* is[11]

$$\overline{\overline{L}} = \int_S \frac{\mathbf{n}'\mathbf{u}_{R'}}{4\pi R^2} \, dS' \tag{5.70}$$

The depolarisation dyadic has unit trace: $\mathrm{tr}\overline{\overline{L}} = 1$.

A very important property of the depolarisation dyadic is that for a spherical and for an ellipsoidal volume, $\overline{\overline{L}}$ is independent of the field point $\mathbf{r}$ (see [19, 20]). Therefore, for spheres and ellipsoids, the scattered field is inside the inclusion independent of $\mathbf{r}$ in Equation (5.69): $\mathbf{E}_s(\mathbf{r}) = \mathbf{E}_s$. And therefore the internal field $\mathbf{E}_e + \mathbf{E}_s$ is uniform, and hence the polarisation is also uniform. This observation makes the assumption taken, of uniform $\mathbf{P}$, self-consistent.

---

[11]The depolarisation dyadic is symmetric, which fact allows us to write $\mathbf{P} \cdot \overline{\overline{L}} = \overline{\overline{L}} \cdot \mathbf{P}$.

The internal field of a sphere

In particular, for a sphere with all its symmetry, $\overline{\overline{L}}$ must be a multiple of the unit dyadic, and because the trace of the unit dyadic is 3, the depolarisation dyadic for a sphere is $\overline{\overline{I}}/3$. For a spherical inclusion volume, then, the scattered field is

$$\mathbf{E}_s = -\frac{\mathbf{P}}{3\epsilon_e} \tag{5.71}$$

If we are studying an anisotropic sphere with permittivity dyadic $\overline{\overline{\epsilon}}_i$, the polarisation reads

$$\mathbf{P} = (\overline{\overline{\epsilon}}_i - \epsilon_e \overline{\overline{I}}) \cdot \mathbf{E}_i \tag{5.72}$$

Let us emphasise the fact that in the derivation above of $\mathbf{E}_s$, nothing was assumed of the relation $\mathbf{P} = \mathbf{P}(\mathbf{E}_s)$. The polarisation relation can be isotropic, anisotropic, or even nonlinear; (5.69) remains valid.[12] And now in the anisotropic case, because the internal field is a sum of the external field and the scattered field, we have

$$\mathbf{E}_i = \mathbf{E}_e - \frac{1}{3\epsilon_e} (\overline{\overline{\epsilon}}_i - \epsilon_e \overline{\overline{I}}) \cdot \mathbf{E}_i \tag{5.73}$$

or

$$\mathbf{E}_i = 3\epsilon_e (\overline{\overline{\epsilon}}_i + 2\epsilon_e \overline{\overline{I}})^{-1} \cdot \mathbf{E}_e \tag{5.74}$$

Then, the dipole moment induced in an anisotropic sphere becomes as the volume integral of the polarisation

$$\mathbf{p} = 3V\epsilon_e(\overline{\overline{\epsilon}}_i - \epsilon_e \overline{\overline{I}}) \cdot (\overline{\overline{\epsilon}}_i + 2\epsilon_e \overline{\overline{I}})^{-1} \cdot \mathbf{E}_e \tag{5.75}$$

from which the polarisability can be written, and seen to coincide with the relation (5.50).[13]

The internal field of an ellipsoid

As has already been noted before, the depolarisation dyadic for an ellipsoidal geometry contains the three depolarisation factors as eigenvalues, and we have

$$\overline{\overline{L}} = N_x \mathbf{u}_x \mathbf{u}_x + N_y \mathbf{u}_y \mathbf{u}_y + N_z \mathbf{u}_z \mathbf{u}_z \tag{5.76}$$

again independently on the field position within the ellipsoidal volume. The depolarisation factors as functions of the axis ratios of the ellipsoid were given in (4.5) and their summing to unity politely agrees with the property of unit trace. Then, the internal field of an anisotropic ellipsoid can be written

---

[12]See also the relatively unknown work [21] from 1963 emphasising this point.

[13]One may feel uneasy with the different order of the dyadic multiplication in (5.50) and (5.75). This is relieved by the fact that for any scalars a and b, dyadics of the form $a\overline{\overline{\epsilon}}_i + b\overline{\overline{I}}$ commute with each other since they have the same eigenvectors (those of $\overline{\overline{\epsilon}}_i$).

$$\mathbf{E}_i = \mathbf{E}_e - \frac{1}{\epsilon_e}\overline{\overline{L}} \cdot (\overline{\overline{\epsilon}}_i - \epsilon_e \overline{\overline{I}}) \cdot \mathbf{E}_i \qquad (5.77)$$

from which the polarisability dyadic of an anisotropic ellipsoid can be written as

$$\overline{\overline{\alpha}} = V(\overline{\overline{\epsilon}}_i - \epsilon_e \overline{\overline{I}}) \cdot \left[\epsilon_e \overline{\overline{I}} + \overline{\overline{L}} \cdot (\overline{\overline{\epsilon}}_i - \epsilon_e \overline{\overline{I}})\right]^{-1} \epsilon_e \qquad (5.78)$$

5.4 Mixtures with anisotropic inclusions

With the dielectric responses of anisotropic inclusions known, the mixture with such types of inclusions can be analysed. This mixture is depicted in Figure 5.5.

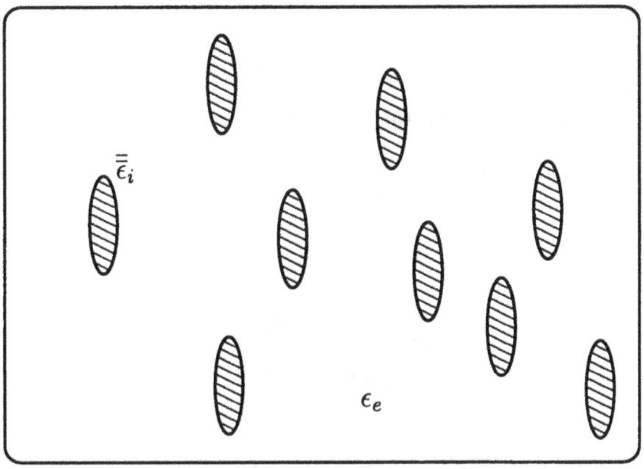

Figure 5.5: *A mixture with aligned anisotropic ellipsoids (permittivity dyadic $\overline{\overline{\epsilon}}_i$) in isotropic environment (permittivity scalar ϵ_e).*

Now let us assume an ordered mixture. This means, first, that all ellipsoids are aligned as shown in the figure, but secondly, that also the material anisotropies of all ellipsoids are assumed to be aligned. The anisotropy can, however, be independent from the geometry of the ellipsoids; in other words, the eigendirections of $\overline{\overline{\epsilon}}_i$ need not be the same as the axis directions of the ellipsoids.

As in the isotropic-inclusion case, the average polarisation $< \mathbf{P} >$ can again be calculated from the dipole moments $\mathbf{p}_{\text{mix}}$ which are now excited by the local field. The local field does not depend on the material or polarisation inside the scatterer, and hence it is the same for isotropic and anisotropic inclusions, or inclusions with even more complicated electromagnetic response. For an ellipsoid, again, it is

$$\mathbf{E}_L = < \mathbf{E} > + \frac{1}{\epsilon_e}\overline{\overline{L}} \cdot < \mathbf{P} > \qquad (5.79)$$

where $\overline{\overline{L}}$ contains the depolarisation factors of the ellipsoid. Therefore the connection between the average polarisation and the average field can be written as a function of the polarisability dyadic of the inclusions. This dyadic connection reads

$$< \mathbf{P} > = \left(\overline{\overline{I}} - \frac{n\overline{\overline{\alpha}}}{\epsilon_e} \cdot \overline{\overline{L}} \right)^{-1} \cdot n\overline{\overline{\alpha}} \cdot < \mathbf{E} > \qquad (5.80)$$

and finally—combining (5.80) with (5.78) and using the dyadic inverse (5.35)—we have the effective permittivity dyadic for this aligned mixture:

$$\overline{\overline{\epsilon}}_{\text{eff}} = \epsilon_e \overline{\overline{I}} + f\epsilon_e \left[\epsilon_e \overline{\overline{I}} + (1 - f)\overline{\overline{L}} \cdot (\overline{\overline{\epsilon}}_i - \epsilon_e \overline{\overline{I}}) \right]^{-1} \cdot (\overline{\overline{\epsilon}}_i - \epsilon_e \overline{\overline{I}}) \qquad (5.81)$$

where $f = nV$ is, once again, the volume fraction of the inclusion phase. It must be remembered that here all ellipsoids have the same permittivity dyadic $\overline{\overline{\epsilon}}_i$ in the global co-ordinate system but this dyadic does not commute in general with $\overline{\overline{L}}$, which dyadic is dictated by the ellipsoid axis directions.

If the geometry of the anisotropic ellipsoids of Figure 5.5 degenerates into spheres, we have $\overline{\overline{L}} = \overline{\overline{I}}/3$, and the effective permittivity reads

$$\overline{\overline{\epsilon}}_{\text{eff}} = \epsilon_e \overline{\overline{I}} + 3\epsilon_e f \left[\overline{\overline{\epsilon}}_i + 2\epsilon_e \overline{\overline{I}} - f(\overline{\overline{\epsilon}}_i - \epsilon_e \overline{\overline{I}}) \right]^{-1} \cdot (\overline{\overline{\epsilon}}_i - \epsilon_e \overline{\overline{I}}) \qquad (5.82)$$

Note the formal similarity of this equation with the isotropic Maxwell Garnett result (3.27).

Then, if the ellipsoids are not in aligned orientation or their anisotropies have an orientation distribution, those have to be averaged by weighting with the distribution functions when the average polarisation $< \mathbf{P} >$ is calculated. Consequently, the final effective permittivity dyadic contains the corresponding integrals that contribute accordingly to the dyadics.

5.5   Mixtures with anisotropic background medium

After having dwelt on the problem of how anisotropic inclusions can be homogenised in an isotropic background, let us pose the next question of how the situation changes if the environment is anisotropic, too. Can one simply replace $\epsilon_e \overline{\overline{I}}$ by $\overline{\overline{\epsilon}}_e$ in all the previous formulas? Although a tempting trick to do, and done, too, several times in the literature, that cannot be justified. Why? In the following, a closer look at the anisotropic-in-anisotropic case is appropriate.

5.5.1   Affine transformation

The Maxwell Garnett homogenisation on which all the previous analysis was based hinged on the equivalence between the inclusion as a dielectric inhomogeneity and a point dipole "radiating" a static field in uniform space. The dipole field in isotropic space could be recognised and separated from the total field solution of the inclusion in uniform external field. However, the solution in anisotropic environment is not the same, and to see how it is different from the isotropic case, a careful analysis has to be performed.

The strategy to solve the dipolar field in a uniform anisotropic medium with permittivity $\bar{\bar{\epsilon}}_e$ is to try to rewrite the field equations in a form from which the Laplace equation $\nabla^2 \phi(\mathbf{r}) = 0$ could be seen. In the anisotropic source-free case, the potential $\phi(\mathbf{r})$ satisfies instead

$$\nabla \cdot [\bar{\bar{\epsilon}}_e \cdot \nabla \phi(\mathbf{r})] = 0 \tag{5.83}$$

This suggests an affine transformation [15, 22], where a new nabla operator ∇_a is defined:

$$\nabla_a = \bar{\bar{\epsilon}}_r^{1/2} \cdot \nabla \tag{5.84}$$

where $\bar{\bar{\epsilon}}_r = \bar{\bar{\epsilon}}_e / \epsilon_0$ is the dimensionless relative permittivity dyadic of the environment, and the power of a half is the square root of the dyadic. Let us assume for simplicity that $\bar{\bar{\epsilon}}_e$ is symmetric. Then the square root is simply

$$\bar{\bar{\epsilon}}_r^{1/2} = \sum_{j=1}^{3} \sqrt{\epsilon_{r,j}} \, \mathbf{u}_j \mathbf{u}_j \tag{5.85}$$

with $\mathbf{u}_j$ being unit vectors along the three orthogonal eigendirections, and $\epsilon_{r,j}$ the corresponding eigenvalues. Then Equation (5.83) can be written

$$\nabla_a^2 \phi(\mathbf{r}_a) = 0 \tag{5.86}$$

which is the familiar Laplace equation. So, in the affinely transformed co-ordinate system $\mathbf{r}_a = \bar{\bar{\epsilon}}_r^{-1/2} \cdot \mathbf{r}$, the solution is the usual isotropic one. But this means that the geometrical interfaces also change. Fortunately, ellipsoidal boundaries remain ellipsoidal. One has to note, however, that the axial ratios change. For example, a sphere becomes squeezed in different orthogonal directions according to the permittivity components of $\bar{\bar{\epsilon}}_e$. So, a sphere in a positively uniaxial environment[14] becomes an oblate spheroid in the affine system, and the depolarisation factors have to be calculated for this affinely transformed shape.

5.5.2   Internal field and polarisability

Consider an ellipsoid for which the depolarisation factors N_n, N_y, N_z are defined as (4.5). Collecting those in a single dyadic, the depolarisation dyadic for an ordinary ellipsoid reads

---

[14]Positive uniaxiality means that in the permittivity dyadic $\bar{\bar{\epsilon}} = \epsilon_z \mathbf{u}_z \mathbf{u}_z + \epsilon_t \bar{\bar{I}}_t$, the axial component is larger than the transversal: $(\epsilon_z > \epsilon_t)$.

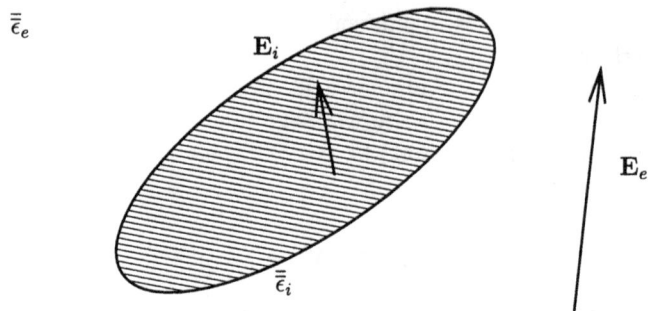

Figure 5.6: *Anisotropic ellipsoid (with permittivity dyadic $\bar{\bar{\epsilon}}_i$) in anisotropic environment ($\bar{\bar{\epsilon}}_e$).* $\mathbf{E}_i$ *refers to the internal electric field.*

$$\bar{\bar{L}} = \sum_{i=x,y,z} N_i \mathbf{v}_i \mathbf{v}_i = \frac{\det \bar{\bar{A}}}{2} \int_0^\infty ds \, \frac{\left(\bar{\bar{A}}^2 + s\bar{\bar{I}}\right)^{-1}}{\sqrt{\det \left(\bar{\bar{A}}^2 + s\bar{\bar{I}}\right)}} \tag{5.87}$$

where the symmetric and positive-definite dyadic

$$\bar{\bar{A}} = \sum_{i=x,y,z} a_i \mathbf{v}_i \mathbf{v}_i \tag{5.88}$$

with

$$\det \bar{\bar{A}} = a_x a_y a_z \tag{5.89}$$

defines the ellipsoid as $\mathbf{r} \cdot \bar{\bar{A}}^{-2} \cdot \mathbf{r} \leq 1$. The semiaxes of the ellipsoid are, as before, a_x, a_y, a_z.

If this ellipsoid, which is now anisotropic with permittivity $\bar{\bar{\epsilon}}_i$, is located in another anisotropic material $\bar{\bar{\epsilon}}_e = \bar{\bar{\epsilon}}_r \epsilon_0$, and exposed to a uniform external field $\mathbf{E}_e$, the internal field $\mathbf{E}_i$ has to be calculated from an affinely transformed geometry. Here the permittivity of free space is denoted by ϵ_0. Figure 5.6 shows the geometry.

The internal field in the ellipsoid can be shown to be [23]

$$\mathbf{E}_i = \left[\bar{\bar{\epsilon}}_e + \bar{\bar{L}}' \cdot (\bar{\bar{\epsilon}}_i - \bar{\bar{\epsilon}}_e)\right]^{-1} \cdot \bar{\bar{\epsilon}}_e \cdot \mathbf{E}_e \tag{5.90}$$

where now one has to remember that the depolarisation factors have to be calculated from the transformed depolarisation dyadic $\bar{\bar{L}}'$, which is that of the real geometry of the ellipsoid after it has been transformed affinely by the anisotropy of the environment. The transformed depolarisation dyadic can be calculated from

$$\bar{\bar{L}}' = \frac{\det \bar{\bar{A}}}{2} \int_0^\infty ds \, \bar{\bar{\epsilon}}_r \cdot \frac{\left(\bar{\bar{A}}^2 + s\bar{\bar{\epsilon}}_r\right)^{-1}}{\sqrt{\det \left(\bar{\bar{A}}^2 + s\bar{\bar{\epsilon}}_r\right)}} \tag{5.91}$$

Note that the difference between the depolarisation dyadic for the "ordinary" ellipsoid (5.87) and this transformed one (5.91) is that $s\overline{\overline{I}}$ is replaced by $s\overline{\overline{\epsilon}}_r$. Then, finally, the polarisability of the ellipsoid with volume $V = 4\pi a_x a_y a_z/3$ reads

$$\overline{\overline{\alpha}} = V\left(\overline{\overline{\epsilon}}_i - \overline{\overline{\epsilon}}_e\right) \cdot \left[\overline{\overline{\epsilon}}_e + \overline{\overline{L'}} \cdot \left(\overline{\overline{\epsilon}}_i - \overline{\overline{\epsilon}}_e\right)\right]^{-1} \cdot \overline{\overline{\epsilon}}_e \qquad (5.92)$$

5.5.3   Homogenisation

The generalisation of the Maxwell Garnett rule (5.81) to the anisotropic-in-anisotropic mixtures follows. Because the local field needed in the excitation in the composite mixture reads

$$\mathbf{E}_L = <\mathbf{E}> + \overline{\overline{\epsilon}}_e^{-1} \cdot \overline{\overline{L}} \cdot <\mathbf{P}> \qquad (5.93)$$

the definition for the effective permittivity dyadic

$$<\mathbf{D}> = \overline{\overline{\epsilon}}_{\text{eff}} \cdot <\mathbf{E}> = \overline{\overline{\epsilon}}_e \cdot <\mathbf{E}> + <\mathbf{P}> \qquad (5.94)$$

leaves us with

$$\overline{\overline{\epsilon}}_{\text{eff}} = \overline{\overline{\epsilon}}_e + f\left(\overline{\overline{\epsilon}}_i - \overline{\overline{\epsilon}}_e\right) \cdot \left[\overline{\overline{\epsilon}}_e + (1-f)\overline{\overline{L'}} \cdot \left(\overline{\overline{\epsilon}}_i - \overline{\overline{\epsilon}}_e\right)\right]^{-1} \cdot \overline{\overline{\epsilon}}_e \qquad (5.95)$$

Here, the lack of orientation integrals indicates that the result applies to a mixture where all ellipsoids have the same orientation. The medium structure in relation to the ellipsoid geometry, too, is the same in all ellipsoids, although the eigendirections of $\overline{\overline{L'}}$, $\overline{\overline{\epsilon}}_e$, and $\overline{\overline{\epsilon}}_i$ need not be related in any way.

As an example that illustrates the affine transformation and its effect on the effective properties of anisotropic mixtures, consider a "Swiss-cheese" mixture where spherical voids are cut into an environment that is uniaxially anisotropic (Figure 5.7). Let the environment permittivity be $\overline{\overline{\epsilon}}_e = (5\mathbf{u}_z \mathbf{u}_z + \overline{\overline{I}}_t)\epsilon_0$, which is positively uniaxial, with the z-direction as the optical axis. The voids have $\epsilon_i = \epsilon_0$. Now the affine transformation "squeezes" the sphere in to an oblate ellipsoid with axial ratio $a_x/a_z = \sqrt{5}$, which means that the depolarisation factor is $N_z \approx 0.558$.

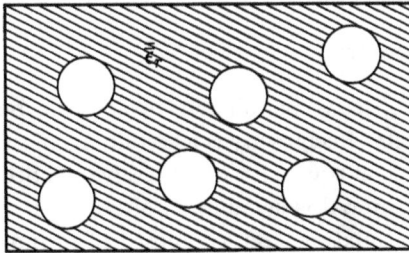

Figure 5.7: *A Swiss-cheese model: spherical voids in anisotropic environment.*

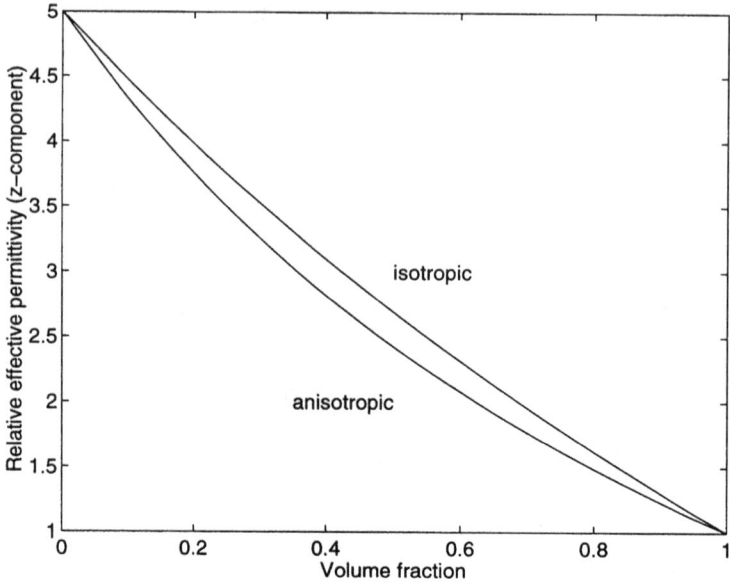

Figure 5.8: *The relative effective permittivity of the Swiss-cheese in Figure 5.7 in the direction of the optical axis as a function of the volume fraction of the voids (lower curve). The upper curve shows the corresponding mixture for the case of isotropic background ($\epsilon_e = 5\epsilon_0$).*

The effective permittivity in the transverse direction is $\epsilon_{\text{eff},x} = \epsilon_{\text{eff},y} = \epsilon_0$ independently of the volume fraction. However, more interesting is the value of the effective permittivity in the direction of the optical axis: as a function of the volume fraction f the effective permittivity component decreases from the environment value $5\epsilon_0$ to the void value ϵ_0. This is illustrated in Figure 5.8 [24]. For comparison, the scalar effective permittivity of the corresponding isotropic mixture ($\overline{\overline{\epsilon}}_e = 5\epsilon_0\overline{\overline{I}}$) is also plotted. The result shows that the effect of the affine transformation is to decrease the effective permittivity component in the z-direction.

Symmetry and reciprocity

The expression for the effective permittivity dyadic can be studied also from the point of view of reciprocity. Reciprocity is a certain manifestation of symmetry with respect to the interchange of transmitter and receiver [7, Sec. 5.5]. The permittivity dyadic of a reciprocal material is symmetric, although it can be anisotropic. It is natural to accept that a mixture, too, composed of reciprocal materials must display reciprocal electromagnetic behaviour; in other words, the effective permittivity dyadic should be symmetric:

$$\bar{\bar{\epsilon}}_e = \bar{\bar{\epsilon}}_e^T \text{ and } \bar{\bar{\epsilon}}_i = \bar{\bar{\epsilon}}_i^T \implies \bar{\bar{\epsilon}}_{\text{eff}} = \bar{\bar{\epsilon}}_{\text{eff}}^T \tag{5.96}$$

From the expression (5.95) for the effective permittivity this symmetry property is not obvious because $\bar{\bar{\epsilon}}_i$ does not commute with $\bar{\bar{\epsilon}}_e$ or $\bar{\bar{L}}'$, even if it is symmetric.[15] However, the property (5.96) can be more easily seen to hold if the effective permittivity dyadic (5.95) is rewritten in the following form:

$$\bar{\bar{\epsilon}}_{\text{eff}} = \bar{\bar{\epsilon}}_e + f \left[\left(\bar{\bar{\epsilon}}_i - \bar{\bar{\epsilon}}_e \right)^{-1} + (1 - f) \bar{\bar{\epsilon}}_e^{-1} \cdot \bar{\bar{L}}' \right]^{-1} \tag{5.97}$$

This form of the Maxwell Garnett equation may be more practical than (5.95) in some applications. The reciprocity of the mixture for reciprocal components is obvious after using the fact that the inverse and transpose operations on a dyadic commute.

Problems

5.1 A dyad, which is only a vector pair, is not general enough to be equal to an arbitrary dyadic in three-dimensional space. A polynomial of dyads is needed:

$$\bar{\bar{A}} = \sum_{i=1}^{n} \mathbf{a}_i \mathbf{b}_i$$

But how many terms are needed? In other words, what is the minimum n in this sum to express a general dyadic?

5.2 Show that a dyad (a dyadic with only one vector pair) can never be antisymmetric but always has a symmetric part.

5.3 A real and symmetric dyadic can be expressed in terms of its orthogonal eigenvectors and eigenvalues:

$$\bar{\bar{A}} = a_x \mathbf{u}_x \mathbf{u}_x + a_y \mathbf{u}_y \mathbf{u}_y + a_z \mathbf{u}_z \mathbf{u}_z$$

Show that for this dyadic, the invariants are

$$
\begin{aligned}
\operatorname{tr} \bar{\bar{A}} &= a_x + a_y + a_z \\
\operatorname{spm} \bar{\bar{A}} &= a_x a_y + a_y a_z + a_z a_x \\
\det \bar{\bar{A}} &= a_x a_y a_z
\end{aligned}
$$

5.4 Show that

$$(\mathbf{a} \times \bar{\bar{I}}) : (\mathbf{b} \times \bar{\bar{I}}) = 2 \mathbf{a} \cdot \mathbf{b}$$

---

[15]The eigenvector directions of $\bar{\bar{\epsilon}}_i$ may be different from those of $\bar{\bar{\epsilon}}_e$ and $\bar{\bar{L}}'$.

5.5 Show that

$$\det(\overline{\overline{A}} + \overline{\overline{B}}) = \det\overline{\overline{A}} + \det\overline{\overline{B}} + \text{tr}[(\text{adj}\overline{\overline{A}}) \cdot \overline{\overline{B}}] + \text{tr}[\overline{\overline{A}} \cdot (\text{adj}\overline{\overline{B}})]$$

5.6 Show that

$$\begin{aligned}
\text{adj}(\overline{\overline{A}} + \overline{\overline{B}}) &= \text{adj}\overline{\overline{A}} + \text{adj}\overline{\overline{B}} + \overline{\overline{B}}^{-1} \cdot \left[\overline{\overline{I}}\,\text{tr}(\overline{\overline{A}} \cdot \text{adj}\overline{\overline{B}}) - \overline{\overline{A}} \cdot \text{adj}\overline{\overline{B}}\right] \\
&= \text{adj}\overline{\overline{A}} + \text{adj}\overline{\overline{B}} + \overline{\overline{A}}^{-1} \cdot \left[\overline{\overline{I}}\,\text{tr}(\overline{\overline{B}} \cdot \text{adj}\overline{\overline{A}}) - \overline{\overline{B}} \cdot \text{adj}\overline{\overline{A}}\right]
\end{aligned}$$

5.7 Show that

$$\text{tr}[(\text{adj}\overline{\overline{A}}) \cdot \overline{\overline{B}}] = \frac{1}{2}\overline{\overline{A}} \overset{\times}{\times} \overline{\overline{A}} : \overline{\overline{B}}$$

5.8 Consider the gyrotropic dyadic

$$\overline{\overline{\epsilon}} = \epsilon_t \overline{\overline{I}}_t + \epsilon_z \mathbf{u}_z \mathbf{u}_z + \epsilon_g \mathbf{u}_z \times \overline{\overline{I}}$$

where $\overline{\overline{I}}_t = \overline{\overline{I}} - \mathbf{u}_z \mathbf{u}_z$ is the transversal unit dyadic in the xy-plane. Calculate

$$\overline{\overline{\epsilon}}^{(2)}$$

5.9 Using the dyadic expression for the polarisability of an anisotropic sphere (5.50) as a function of the permittivity dyadic, derive the expression (5.62) for a gyrotropic sphere for which we have

$$\overline{\overline{\epsilon}}_i = \epsilon_i \overline{\overline{I}} + \epsilon_g \mathbf{u}_g \times \overline{\overline{I}}$$

5.10 A dyadic

$$\begin{aligned}
\overline{\overline{C}} &= a(\mathbf{u}_x \mathbf{u}_x + \mathbf{u}_y \mathbf{u}_y + \mathbf{u}_z \mathbf{u}_z) \\
&+ b(\mathbf{u}_x \mathbf{u}_y + \mathbf{u}_y \mathbf{u}_z + \mathbf{u}_z \mathbf{u}_x + \mathbf{u}_y \mathbf{u}_x + \mathbf{u}_z \mathbf{u}_y + \mathbf{u}_x \mathbf{u}_z)
\end{aligned}$$

is clearly symmetric. But it is in fact uniaxial, of the form

$$\overline{\overline{C}} = C_d \mathbf{u}_d \mathbf{u}_d + C_t \left(\overline{\overline{I}} - \mathbf{u}_d \mathbf{u}_d\right)$$

Calculate the eigenvector $\mathbf{u}_d$ and the eigenvalues C_d and C_t.

5.11 The connection between the local field $\mathbf{E}_L$ acting on an inclusion and the average field $\mathbf{E}$ is dependent on the shape of the assumed cavity:

$$\mathbf{E}_L = \mathbf{E} + \frac{1}{\epsilon_e}\overline{\overline{L}} \cdot \mathbf{P}$$

where **P** is the average polarisation density and $\overline{\overline{L}}$ is the depolarisation dyadic, which is $\overline{\overline{I}}/3$ for a spherical volume. As we now know, the general form for the depolarisation dyadic is [15, 19]

$$\overline{\overline{L}} = \oint \frac{\mathbf{n'}\mathbf{u}_{R'}}{4\pi R'^2} dS'$$

where the integration is over the surface of the cavity, $\mathbf{n'}$ is the outward unit normal vector at the surface, on the integration point, and $R' = |\mathbf{r'} - \mathbf{r}|$, where $\mathbf{r'}$ is the position vector of the integration point (at the surface, again), and $\mathbf{r}$ is the position vector of the field point inside the cavity (i.e. the point where the field is calculated). The unit vector $\mathbf{u}_{R'} = (\mathbf{r'} - \mathbf{r})/R$ is a unit vector pointing from the field point to the source point.

(a) Show that $\overline{\overline{L}} = \overline{\overline{I}}/3$ for a spherical cavity of radius a if the field point is at the centre of the sphere.

(b) Prove that $\overline{\overline{L}} = \overline{\overline{I}}/3$ even if $\mathbf{r}$ is any other point inside the sphere.

5.12 Calculate the depolarisation dyadic $\overline{\overline{L}}$ for a cube-shaped cavity. Put the cube (with edge a) into the centre of the Cartesian co-ordinate system and use the integral given in the previous problem. Now it may happen that the value for the dyadic may depend on the position of the field vector $\mathbf{r}$, unlike in the case of the sphere. But intuition would suggest that if the field point is at the centre of the cube ($\mathbf{r} = 0$), the depolarisation dyadic is again $\overline{\overline{L}} = \overline{\overline{I}}/3$.

Determine $\overline{\overline{L}}$ as the function of $\mathbf{r}$, both along the axis $\mathbf{r} = z\mathbf{u}_z$, $0 \leq z \leq a/2$, along the semidiagonal $\mathbf{r} = \rho(\mathbf{u}_x + \mathbf{u}_y)$, $0 \leq \rho \leq a/2$, and along the diagonal $\mathbf{r} = r(\mathbf{u}_x + \mathbf{u}_y + \mathbf{u}_z)$, $0 \leq r \leq a/2$. Although the calculations can be done analytically for any field point within the cavity, you perhaps might start by doing (at least part of) the integrations numerically.

Hint: Reference [25].

5.13 Prove that the depolarisation dyadic

$$\overline{\overline{L}} = \int_S \frac{\mathbf{n'}\mathbf{u}_{R'}}{4\pi\epsilon_e R^2} dS'$$

is symmetric. Hint: the antisymmetric part of a dyadic corresponds to the "vector of the dyadic" that emerges by replacing the dyadic products by cross-products.

5.14 Show that the trace of the depolarisation dyadic (5.70) is unity.

5.15 The analysis in Subsection 5.3.2 with which the internal field of the inclusion was calculated was based on the integral of a delta function which reduced the volume integration to a two-dimensional integral over the surface. Another method can also be tried, if one is not confident with delta functions or distributions: to apply the divergence theorem. First, partial differentiation

$$\nabla' \cdot (\mathbf{fg}) = (\nabla' \cdot \mathbf{f})\mathbf{g} + \mathbf{f} \cdot \nabla'\mathbf{g}$$

can be applied to Equation (5.66), and then, using Gauss's law, the divergence term can be written as a surface integral [23]. Can you reproduce the result (5.69)?

After having calculated the result, remind yourself of the assumptions with Gauss's law. The integrand contains a polarisation function that is abruptly cut off beyond the inclusion, and its derivatives become infinite. Furthermore, Coulomb's law, also under integral, is very singular, and one may wonder whether the assumptions of Gauss' law have been appropriately respected. Try to save your derivation by taking account of these singularities in an acceptable way.

5.16 Show that the representation (5.87) for the depolarisation factors of an ellipsoid is compatible with the earlier definition (4.5).

5.17 Consider the question how the affine transformation modifies the shape of an ellipsoid. Section 5.5 described how the affine transformation, while rendering the environment isotropic, squeezed and stretched the metric so that the spheres became ellipsoids and ellipsoids turned into other ellipsoids. Take the environment to be uniaxial with $\epsilon_{e,x} = 2\epsilon_0$ and $\epsilon_{e,y} = \epsilon_{e,z} = \epsilon_0$. Consider a prolate ellipsoid with axis ratio 3/1 which is located in the environment such that the axis of revolution makes a 45° degree angle with the x-axis. Determine the depolarisation factors (components of the transformed depolarisation dyadic $\overline{\overline{L}}'$ of (5.91)). What are the axis ratios of the transformed ellipsoid? Compare the calculated axis ratios and depolarisation factors with those of the original prolate spheroid.

5.18 If the anisotropic environment is lossy, the affine transformation defines a complex boundary for the transformed ellipsoid. Obviously the depolarisation dyadic also becomes complex [26]. Calculate the depolarisation factors (eigenvalues of the $\overline{\overline{L}}'$ dyadic) for a sphere in a lossy anisotropic background the permittivity of which is

$$\overline{\overline{\epsilon}}_e/\epsilon_0 = (2 - j3)\mathbf{u}_x\mathbf{u}_x + (5 - j2)\mathbf{u}_y\mathbf{u}_y + (10 - j20)\mathbf{u}_z\mathbf{u}_z$$

References

[1] KELVIN, Lord: 'Baltimore Lectures (Lecture XI)' (London, 1904)

[2] BIRSS, R.R.: 'Symmetry and magnetism' (North Holland, Amsterdam, 1966)

[3] LORENTZ, H.A.: 'Double refraction by regular crystals', *Proceedings K. Akad. Wet. Amsterdam*, 1922, **24**, pp. 333-339

[4] CASIMIR, H.B.G.: 'Note on macroscopic theory of optical rotation and double refraction in cubic crystals', *Philips Research Reports*, 1966, **21**, (6), pp. 417-422

[5] GRAHAM, E.B., and RAAB, R.E.: 'Light propagation in cubic and other anisotropic crystals', *Proc. Royal Society of London, A*, 1990, **430**, pp. 593-614

[6] POST, E.J.: 'Formal structure of electromagnetics' (North-Holland, Amsterdam, 1962)

[7] KONG, J.A.: 'Electromagnetic wave theory' (Wiley, New York, 1986)

[8] GIBBS, J.W., and WILSON, E.B.: 'Vector analysis', Second Edition (Scribner, New York, 1909)

[9] ARFKEN, G.: 'Mathematical methods for physicists', Second Edition (Academic Press, New York, 1970)

[10] LINDELL, I.V.: 'A collection of dyadic identities in three-dimensional vector calculus', *Helsinki University of Technology, Radio Laboratory Report*, 1968, **S57**

[11] LINDELL, I.V.: 'Coordinate independent dyadic formulation of wave normal and ray surfaces of general anisotropic media', *Journal of Mathematical Physics*, 1973, **14**, (1), pp. 65-67

[12] CHEN, H.C.: 'Theory of electromagnetic waves. A coordinate-free approach' (McGraw–Hill, New York, 1983)

[13] FEDOROV, F.I.: 'Optics of anisotropic media' (Academy of Sciences of the BSSR, Minsk, 1958) (in Russian)

[14] FEDOROV, F.I.: 'Theory of gyrotropy' (Nauka i Tekhnika, Minsk, 1976) (in Russian)

[15] LINDELL, I.V.: 'Methods for electromagnetic field analysis' (IEEE Press and Oxford University Press, 1995)

[16] JONES, D.S.: 'Methods in electromagnetic wave propagation' (Clarendon Press, Oxford, 1979)

[17] SIHVOLA, A.H.: 'Dielectric polarizability of a sphere with arbitrary anisotropy', *Optics Letters*, 1994, **19**, (17), pp. 430-432

[18] CHENG, D.K.: 'Field and wave electromagnetics', Second Edition (Addison–Wesley, Reading, Mass., 1989)

[19] YAGHJIAN, A.D.: 'Electric dyadic Green's function in the source region', *Proceedings of the IEEE*, 1980, **68**, (2), pp. 248-263

[20] VAN BLADEL, J.: 'Some remarks on Green's dyadic for infinite space', *IRE Transactions on Antennas and Propagation*, 1961, **9**, pp. 563-566

[21] FRANK, V.: 'On the penetration of a static homogeneous field in an anisotropic medium into an ellipsoidal inclusion consisting of another anisotropic medium', *in* JORDAN, E.C. (Ed.): 'Electromagnetic Theory and Antennas,' Proceedings of a URSI Symposium, Copenhagen, Denmark, June 1962 (Pergamon Press, Oxford, 1963), pp. 615-623

[22] LANDAU, L.D., and LIFSHITZ, E.M.: 'Electrodynamics of continuous media', Second Edition (Pergamon Press, Oxford, 1984), Section 13.

[23] SIHVOLA, A.H., and LINDELL, I.V.: 'Electrostatics of an anisotropic ellipsoid in an anisotropic environment', *AEÜ International Journal of Electronics and Communications*, 1996, **50**, (5), pp. 289-292

[24] SIHVOLA, A.: 'On the dielectric problem of isotropic sphere in anisotropic medium' *Electromagnetics*, **17**, (1), 69-74

[25] AVELIN, J., ARSLAN, A.N., BRÄNNBACK, J., FLYKT, M., ICHELN, C., JUN-TUNEN, J., KÄRKKÄINEN, K., NIEMI, T., NIEMINEN, O., TARES, T., TOMA, C., UUSITUPA, T., and SIHVOLA, A.: 'Electric fields in the source region: the depolarization dyadic for a cubic cavity', *Electrical Engineering – Archiv für Elektrotechnik*, November 1998, **81**, (4), pp. 199-202

[26] WEIGLHOFER, W.S.: 'Electromagnetic depolarization dyadics and elliptic integrals', *Journal of Physics A: Math. Gen.*, 28 August 1998, **31**, (34), pp. 7191-7196

Chapter 6

Chiral and bi-anisotropic mixtures

In the previous chapter quite complicated materials were discussed. In general, anisotropic media form a very wide class of materials, and, as was seen, their analysis required dyadic-algebraic machinery. This is not enough, however. If we look outwards from the familar regime of isotropic materials, anisotropy is not the only direction toward the more general. Materials exist which become polarised not only by the electric field's various components but also by other excitations. These other types of excitations can be thermal, mechanical, or effects that are even more distant from electricity.

The present chapter discusses a very special type of materials: such materials which become electrically excited by magnetic field, and vice versa. These materials are called bi-anisotropic media. A special case of bi-anisotropic media is the class of bi-isotropic media, such materials which are magnetoelectric but not sensitive to field direction. In general, bi-anisotropy could be understood to mean that the electric polarisation is caused by any other type of excitation, in addition to the ordinary electrically caused polarisation component, but commonly adopted use of the term "bi-anisotropy" restricts the cross-polarisation to magnetoelectric effects.

6.1   Bi-anisotropic materials

The radical message of Maxwell equations, that govern the behaviour of electric and magnetic fields, is that when the fields are dynamic, they are coupled. There are no separate electric and magnetic time-variable fields; rather, we have to speak of electromagnetic fields. However, although this coupling between the electric and magnetic fields is nowadays well accepted, one does not so often encounter textbook discussions of the interaction of electricity and magnetism on the level of material response. The constitutive relations usually only relate the electric polarisation to electric excitation and, similarly in magnetic materials, a magnetic field may create magnetic polarisation.[1]

Nevertheless, the *magnetoelectric* effect can be observed in many classes of solid-state materials, the classical example being the antiferromagnetic chromium oxide, Cr_2O_3 [1–3]. Another example of magnetoelectric coupling is the reciprocal chiral effect which is caused by handedness in the geometry of the material. A possibility of nonreciprocal coupling of the electric and magnetic dipoles in matter led B.D.H. Tellegen [4] to suggest a new element, *gyrator*, for circuit theory. For a description of the history of magnetoelectric and chiral materials, see [5, Chapter 1].

For our analysis of linear magnetoelectric phenomena, reciprocal and nonreciprocal, isotropic and anisotropic, we need general bi-anisotropic constitutive relations.

6.1.1   Bi-anisotropic constitutive relations

The constitutive relations for bi-anisotropic materials include, in addition to the electric field **E** and displacement (flux density) **D**, also the magnetic field strength **H** and displacement **B**. The relations look like

$$\mathbf{D} = \bar{\bar{\epsilon}} \cdot \mathbf{E} + \sqrt{\mu_0 \epsilon_0}\, \bar{\bar{\xi}} \cdot \mathbf{H} \tag{6.1}$$

$$\mathbf{B} = \sqrt{\mu_0 \epsilon_0}\, \bar{\bar{\zeta}} \cdot \mathbf{E} + \bar{\bar{\mu}} \cdot \mathbf{H} \tag{6.2}$$

where the four material parameter dyadics are: two co-polarisation dyadics, permittivity $\bar{\bar{\epsilon}}$ and permeability $\bar{\bar{\mu}}$, and two cross-polarisation (magnetoelectric) dyadics $\bar{\bar{\xi}}$ and $\bar{\bar{\zeta}}$. The free-space permittivity ϵ_0 and permeability μ_0 are added to extract the dimensions from the cross-dyadics. If the medium does not have any preferred direction, it is *bi-isotropic*, and all four dyadics are multiples of the unit dyadic. The most general bi-anisotropic material requires 36 parameters in its full constitutive electromagnetic description. Table 6.1 shows bi-anisotropic materials and the subclasses.[2]

In the constitutive relations, the electric and magnetic polarisation effects are embedded in the four material dyadics and it is only through them that the medium response affects the electromagnetic wave interaction. The limitation to dyadics, which is taken in the present chapter, means that higher-order multipole effects are

---

[1] A peculiar unsymmetry remains in the use of terms in electromagnetism. The emancipation of magnetism has brought to our minds the notion of total symmetry between electricity and magnetism and their interaction in full Maxwell equations. This is the reason for talking about "electromagnetic" fields. Why do we not talk about "magnetoelectric" fields? We just do not. Instead, "magnetoelectric" is reserved for certain odd materials which are not commonly described in ordinary electrical engineering texts.

[2] Quite often in the physics literature, it is recognised that the primary fields responsible for the polarisation in matter are **E** and **B**. Indeed, as these fields appear in the Lorentz force, **E** and **B** could be more naturally considered as elementary fields, rather than **E** and **H**. This latter **E**, **H** choice is quite commonly taken within the electrical engineering tradition. The material dyadics in the two systems are connected to each other by simple transformations.

Table 6.1: *Magnetoelectric materials and the number of free material parameters in their full characterisation*

	Direction independence	Direction dependence
No magnetoelectric coupling	$\underline{2}$ (ϵ, μ) (Isotropic)	$\underline{18}$ $(\overline{\overline{\epsilon}}, \overline{\overline{\mu}})$ (Anisotropic)
Magnetoelectric coupling	$\underline{4}$ $(\epsilon, \mu, \xi, \zeta)$ (Bi-isotropic)	$\underline{36}$ $(\overline{\overline{\epsilon}}, \overline{\overline{\mu}}, \overline{\overline{\xi}}, \overline{\overline{\zeta}})$ (Bi-anisotropic)

not contained in this phenomenological description of matter response. This restriction should be kept in mind when interpreting the results. See [6] for a discussion on the balance between the various electric and magnetic multipole contributions.

6.1.2   Dissipation and reciprocity

Physical restrictions set conditions to the material dyadics. If no dissipation is allowed, the medium is lossless. Applied to bi-anisotropic media, this condition means that $\overline{\overline{\epsilon}} = \overline{\overline{\epsilon}}^{\dagger}$, $\overline{\overline{\mu}} = \overline{\overline{\mu}}^{\dagger}$, and $\overline{\overline{\xi}} = \overline{\overline{\zeta}}^{\dagger}$, where the Hermitian operator $\dagger$ denotes a complex conjugate of the transpose.

On the other hand, if bi-anisotropic material is to be reciprocal,[3] the conditions are [7, Section 5.5]: $\overline{\overline{\epsilon}} = \overline{\overline{\epsilon}}^{T}$, $\overline{\overline{\mu}} = \overline{\overline{\mu}}^{T}$, and $\overline{\overline{\xi}} = -\overline{\overline{\zeta}}^{T}$, where T denotes transpose of the dyadic. In other words, permittivity and permeability have to be symmetric dyadics, and the magnetoelectric cross-dyadics have to be one negative transpose of the other. Therefore reciprocal bi-anisotropic materials need 21 (complex) parameters for their characterisation. Nonreciprocal effects are contained in the antisymmetric parts of $\overline{\overline{\epsilon}}$ and $\overline{\overline{\mu}}$, and in the magnetoelectric dyadics, both symmetric and antisymmetric parts can be nonreciprocal. The following representation separates the reciprocal and nonreciprocal parts of the material dyadics:

$$\mathbf{D} \;=\; \overline{\overline{\epsilon}} \cdot \mathbf{E} + \left(\overline{\overline{\chi}}^{T} - j\overline{\overline{\kappa}}^{T}\right) \sqrt{\mu_0 \epsilon_0} \cdot \mathbf{H} \qquad (6.3)$$

---

[3]Reciprocity means an invariance of a system when the transmitter and receiver are interchanged [7, Section 5.5]. Note, however, that this is the common understanding of reciprocity by electrical engineers. Often physicists define a medium to be reciprocal if it is nonmagnetic. This means that their nonreciprocity is a broader concept than what is meant by nonreciprocity in the present text which follows the tradition of the electrical engineering community.

$$\mathbf{B} = \left(\overline{\overline{\chi}} + j\overline{\overline{\kappa}}\right)\sqrt{\mu_0\epsilon_0} \cdot \mathbf{E} + \overline{\overline{\mu}} \cdot \mathbf{H} \tag{6.4}$$

Here, the dyadic $\overline{\overline{\kappa}}$, which is termed the chirality dyadic, is responsible for the reciprocal magnetoelectric phenomena, and $\overline{\overline{\chi}}$ is the nonreciprocal cross-polarisation dyadic. Here the appearance of the imaginary unit j suggests time-harmonic dependence of the fields with the convention $\exp(j\omega t)$. With the definitions of (6.3)–(6.4), the dyadics $\overline{\overline{\kappa}}$ and $\overline{\overline{\chi}}$ are real for lossless materials.

Examples of materials with $\overline{\overline{\kappa}} \neq 0$ are media containing handed (so-called chiral) elements. Omega-medium is a composite of planar Ω-shaped elements that are aligned in uniaxial arrangement [8]. On the other hand, a moving medium is an example of a nonreciprocal magnetoelectric medium [9]. Table 6.2 shows a possible classification of bi-anisotropic materials, according to the principles of symmetry and reciprocity.[4]

Table 6.2: *Bi-anisotropic material classification according to the symmetry properties of the parameter dyadics with material examples*

	$\overline{\overline{\epsilon}}$	$\overline{\overline{\mu}}$	$\overline{\overline{\kappa}}$	$\overline{\overline{\chi}}$
Symmetric part: (6 parameters)	(Reciprocal) Dielectric crystal	(Reciprocal) Magnetic crystal	(Reciprocal) Chiral medium	(Nonreciprocal) Chromium Oxide
Antisymmetric part: (3 parameters)	(Nonreciprocal) Magneto-plasma	(Nonreciprocal) Biased ferrite	(Reciprocal) Omega medium	(Nonreciprocal) Moving medium

6.1.3   Renormalisation of field quantities

Electric and magnetic fields carry different units in the SI system. In the analysis of bi-anisotropic materials, it is advantageous to redimensionalise the quantities such that the units agree, and the material parameters become dimensionless. This can be done with the following definition:

$$c\eta\mathbf{D} = \overline{\overline{\epsilon}}_r \cdot \mathbf{E} + \left(\overline{\overline{\chi}}^T - j\overline{\overline{\kappa}}^T\right) \cdot \eta\mathbf{H} \tag{6.5}$$

$$c\mathbf{B} = \left(\overline{\overline{\chi}} + j\overline{\overline{\kappa}}\right) \cdot \mathbf{E} + \overline{\overline{\mu}}_r \cdot \eta\mathbf{H} \tag{6.6}$$

---

[4]See also [10] for a more detailed classification of bi-anisotropic materials.

where $c = 1/\sqrt{\mu_0\epsilon_0}$ and $\eta = \sqrt{\mu_0/\epsilon_0}$ are the vacuum constants, defined by the free-space permittivity ϵ_0 and permeability μ_0.

Now, in equidimensional form, the electric and magnetic quantities can be easily combined so that the two coupled dyadic relations between vectors can be written as one vector relation with higher dimension:

$$\mathbf{d} = \mathsf{M} \cdot \mathbf{e} \tag{6.7}$$

where the electromagnetic six-vector field $\mathbf{e}$ and six-vector flux density $\mathbf{d}$ combine the respective electric and magnetic quantities:

$$\mathbf{e} = \begin{pmatrix} \mathbf{E} \\ \eta\mathbf{H} \end{pmatrix} \quad \text{and} \quad \mathbf{d} = \begin{pmatrix} c\eta\mathbf{D} \\ c\mathbf{B} \end{pmatrix} \tag{6.8}$$

and the material six-dyadic is

$$\mathsf{M} = \begin{pmatrix} \overline{\overline{\epsilon}}_r & \overline{\overline{\chi}}^T - j\overline{\overline{\kappa}}^T \\ \overline{\overline{\chi}} + j\overline{\overline{\kappa}} & \overline{\overline{\mu}}_r \end{pmatrix} \tag{6.9}$$

Just like the dyadic notation compactifies the analysis of anisotropic materials, with the six-vector notation and six-dyadics, bi-anisotropic media can be treated efficiently. To allow the reader a possibility to become more familiar with six-vector operations, let us next focus on six-vector algebra. For a broader treatment on six-vectors, consult [11].

6.2   Six-vector algebra

The most important operation with six-dyadics and six-vectors is the dot-product. As for three-dyadics, it is associative and distributive (but noncommutative), and written in component form, it follows the law

$$\mathsf{M}_1 \cdot \mathsf{M}_2 = \begin{pmatrix} \overline{\overline{a}}_1 & \overline{\overline{b}}_1 \\ \overline{\overline{c}}_1 & \overline{\overline{d}}_1 \end{pmatrix} \cdot \begin{pmatrix} \overline{\overline{a}}_2 & \overline{\overline{b}}_2 \\ \overline{\overline{c}}_2 & \overline{\overline{d}}_2 \end{pmatrix} = \begin{pmatrix} \overline{\overline{a}}_1 \cdot \overline{\overline{a}}_2 + \overline{\overline{b}}_1 \cdot \overline{\overline{c}}_2 & \overline{\overline{a}}_1 \cdot \overline{\overline{b}}_2 + \overline{\overline{b}}_1 \cdot \overline{\overline{d}}_2 \\ \overline{\overline{c}}_1 \cdot \overline{\overline{a}}_2 + \overline{\overline{d}}_1 \cdot \overline{\overline{c}}_2 & \overline{\overline{c}}_1 \cdot \overline{\overline{b}}_2 + \overline{\overline{d}}_1 \cdot \overline{\overline{d}}_2 \end{pmatrix} \tag{6.10}$$

The unit six-dyadic I, which satisfies $\mathsf{I} \cdot \mathbf{e} = \mathbf{e} \cdot \mathsf{I} = \mathbf{e}$ for any six-vector $\mathbf{e}$, and $\mathsf{I} \cdot \mathsf{M} = \mathsf{M} \cdot \mathsf{I} = \mathsf{M}$ for any six-dyadic M, is defined as

$$\mathsf{I} = \begin{pmatrix} \overline{\overline{I}} & 0 \\ 0 & \overline{\overline{I}} \end{pmatrix} \tag{6.11}$$

The transpose operation, satisfying $(\mathsf{M}_1 \cdot \mathsf{M}_2)^T = \mathsf{M}_2^T \cdot \mathsf{M}_1^T$, reads

$$\mathsf{M}^T = \begin{pmatrix} \overline{\overline{a}} & \overline{\overline{b}} \\ \overline{\overline{c}} & \overline{\overline{d}} \end{pmatrix}^T = \begin{pmatrix} \overline{\overline{a}}^T & \overline{\overline{c}}^T \\ \overline{\overline{b}}^T & \overline{\overline{d}}^T \end{pmatrix} \tag{6.12}$$

The complex conjugate of a six-dyadic is a six-dyadic where all four three-dyadics are complex-conjugated. For lossless materials with material six-dyadic M, the following is valid: $M^{T*} = M$.

The following operation, denoted by ‡, is called *complementary* operation:

$$M^{\ddagger} = \begin{pmatrix} \overline{\overline{a}} & \overline{\overline{b}} \\ \overline{\overline{c}} & \overline{\overline{d}} \end{pmatrix}^{\ddagger} = \begin{pmatrix} \overline{\overline{a}}^T & -\overline{\overline{c}}^T \\ -\overline{\overline{b}}^T & \overline{\overline{d}}^T \end{pmatrix} \tag{6.13}$$

The complementary operation on a material matrix gives another material matrix. Let us call this material the complementary medium.[5] A reciprocal medium is self-complementary; in other words, its materials six-dyadic M satisfies $M^{\ddagger} = M$.

The determinant of a six-dyadic can be calculated as a function of the determinants of component three-dyadics. If the six-dyadic is diagonal, the determinant is obviously

$$\det M = \det \begin{pmatrix} \overline{\overline{a}} & 0 \\ 0 & \overline{\overline{d}} \end{pmatrix} = \det \overline{\overline{a}} \, \det \overline{\overline{d}} \tag{6.14}$$

but if the six-dyadic is full, more algebra is needed. The result is

$$\begin{aligned} \det M &= \det \begin{pmatrix} \overline{\overline{a}} & \overline{\overline{b}} \\ \overline{\overline{c}} & \overline{\overline{d}} \end{pmatrix} \\ &= \det \overline{\overline{a}} \, \det \left(\overline{\overline{d}} - \overline{\overline{c}} \cdot \overline{\overline{a}}^{-1} \cdot \overline{\overline{b}} \right) = \det \overline{\overline{d}} \, \det \left(\overline{\overline{a}} - \overline{\overline{b}} \cdot \overline{\overline{d}}^{-1} \cdot \overline{\overline{c}} \right) \end{aligned} \tag{6.15}$$

Likewise, a very important operation is the inverse of a six-dyadic, defined as the solution to equations $M^{-1} \cdot M = M \cdot M^{-1} = I$. For a diagonal six-dyadic the inverse again is obvious,

$$M^{-1} = \begin{pmatrix} \overline{\overline{a}} & 0 \\ 0 & \overline{\overline{d}} \end{pmatrix}^{-1} = \begin{pmatrix} \overline{\overline{a}}^{-1} & 0 \\ 0 & \overline{\overline{d}}^{-1} \end{pmatrix} \tag{6.16}$$

where the three-dyadic algebra of Section 5.2 is needed to evaluate the inverses. Finally, for the full dyadic,

$$\begin{aligned} M^{-1} &= \begin{pmatrix} \overline{\overline{a}} & \overline{\overline{b}} \\ \overline{\overline{c}} & \overline{\overline{d}} \end{pmatrix}^{-1} \\ &= \begin{pmatrix} \overline{\overline{a}} - \overline{\overline{b}} \cdot \overline{\overline{d}}^{-1} \cdot \overline{\overline{c}} & 0 \\ 0 & \overline{\overline{d}} - \overline{\overline{c}} \cdot \overline{\overline{a}}^{-1} \cdot \overline{\overline{b}} \end{pmatrix}^{-1} \cdot \begin{pmatrix} \overline{\overline{a}} & -\overline{\overline{b}} \\ -\overline{\overline{c}} & \overline{\overline{d}} \end{pmatrix} \cdot \begin{pmatrix} \overline{\overline{a}} & 0 \\ 0 & \overline{\overline{d}} \end{pmatrix}^{-1} \\ &= \begin{pmatrix} \left(\overline{\overline{a}} - \overline{\overline{b}} \cdot \overline{\overline{d}}^{-1} \cdot \overline{\overline{c}} \right)^{-1} & -\overline{\overline{a}}^{-1} \cdot \overline{\overline{b}} \cdot \left(\overline{\overline{d}} - \overline{\overline{c}} \cdot \overline{\overline{a}}^{-1} \cdot \overline{\overline{b}} \right)^{-1} \\ -\overline{\overline{d}}^{-1} \cdot \overline{\overline{c}} \cdot \left(\overline{\overline{a}} - \overline{\overline{b}} \cdot \overline{\overline{d}}^{-1} \cdot \overline{\overline{c}} \right)^{-1} & \left(\overline{\overline{d}} - \overline{\overline{c}} \cdot \overline{\overline{a}}^{-1} \cdot \overline{\overline{b}} \right)^{-1} \end{pmatrix} \end{aligned} \tag{6.17}$$

---

[5] The complementary medium is sometimes called a Lorentz-adjoint medium [12].

This form of the inverse is suitable for bi-anisotropic material analyses because the inverses of the off-diagonal (magnetoelectric) dyadics $\overline{\overline{b}}$ and $\overline{\overline{c}}$ are avoided. Therefore the limit case of vanishing magnetoelectric coupling behaves regularly.

6.3 Chiral mixtures

Let us first analyse bi-isotropic mixtures which have only four parameters out of the 36 possible in the full bi-anisotropic case.

6.3.1 Chiral and bi-isotropic materials

Figure 6.1 shows an example of a man-made chiral sample. Chirality is a reciprocal magnetoelectric phenomenon caused by a handed structure of the material. A canonical example of a handed element is a helix. It is easy to understand that in a helix the linear movement of a charge is inevitably tied to a circular movement.

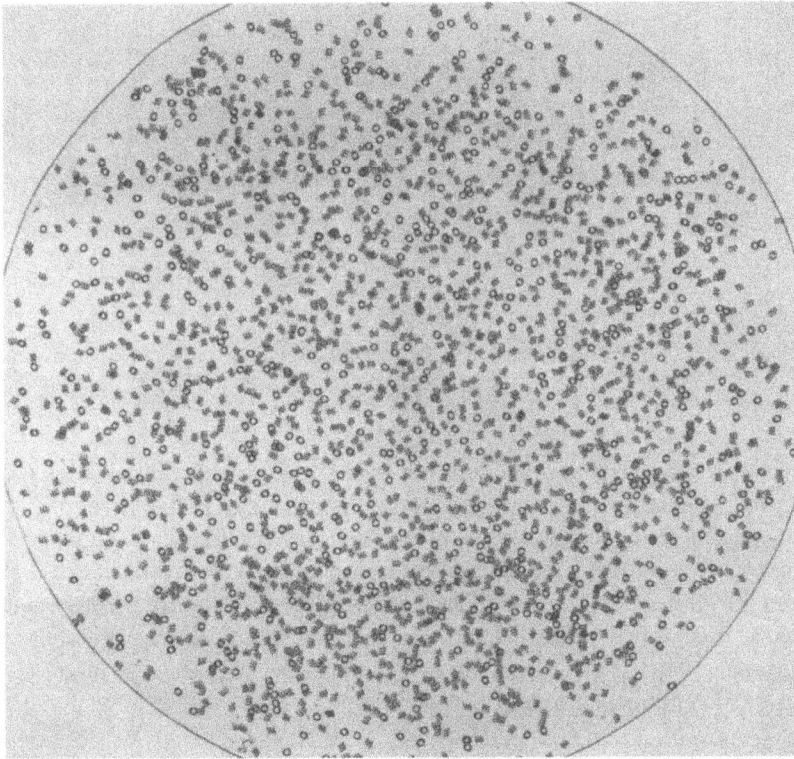

Figure 6.1: *A sample of chiral material manufactured by the company Finnyards Ltd. Material Technology. The diameter is 15 cm.*

Therefore an electric excitation creates a current loop response which is equal to a magnetic dipole. Likewise, a magnetic excitation causes electric polarisation.

Chilar materials, and bi-isotropic materials in general, obey the following constitutive relations:

$$
\begin{pmatrix} c\eta \mathbf{D} \\ c\mathbf{B} \end{pmatrix} = \begin{pmatrix} \epsilon_r & \chi - j\kappa \\ \chi + j\kappa & \mu_r \end{pmatrix} \begin{pmatrix} \mathbf{E} \\ \eta \mathbf{H} \end{pmatrix}
\tag{6.18}
$$

Here, κ is the chirality parameter, and χ is the nonreciprocity parameter. Here, the factor j multiplying the chiral parameter κ is a consequence of the phase difference between the charge density and current.

The effect of chirality on electromagnetic wave propagation is that the plane of polarisation is rotated. This phenomenon—termed *optical activity*—was discovered for optical waves by Biot and Arago in the early part of the 19th century [13, 14]. By the end of the 20th century, chiral materials have experienced a revival as synthetised chiral composites are designed and manufactured for various applications in microwave technology [15]. A detail of the chiral sample in Figure 6.1 is shown in Figure 6.2.

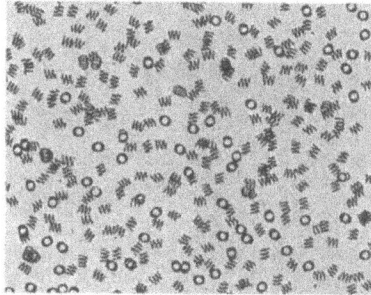

Figure 6.2: *A blow-up of the chiral sample by Finnyards shown in Figure 6.1.*

The chiral effect is reciprocal, in other words, the polarisation rotation unwinds if the wave propagates backwards through the optically active medium.[6] In its stead, a measure for nonreciprocity in bi-isotropic materials is the parameter χ in the relations (6.18). A model for nonreciprocal bi-isotropic medium is such where electric and magnetic dipoles are coupled to each other, so that an electric excitation creates an in-phase magnetic polarisation and vice versa. This model was suggested already by Tellegen in 1948 [4]. Figure 6.3 displays models of such "Tellegen" materials, with nonreciprocal couplings of opposite signs.[7]

---

[6]Note the difference compared to the nonreciprocal *Faraday rotation*, for which the polarisation rotation is not compensated in the returning wave but instead, the rotation angle increases.

[7]The possible existence of nonreciprocal bi-isotropic materials was a source of intensive discussion in the 1990s; see, for example, [16].

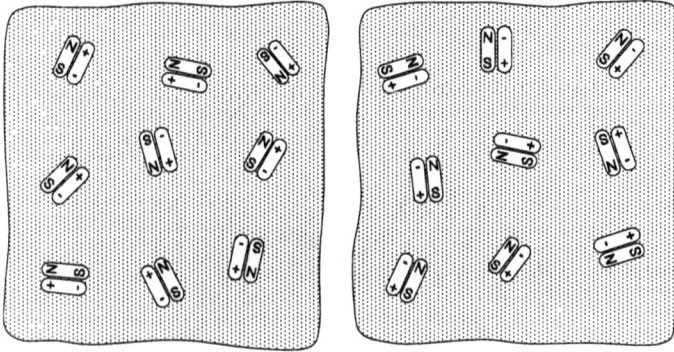

Figure 6.3: *A phenomenological model for the Tellegen material. The two samples are isotropic. The left sample has parallel coupling of the electric and magnetic moments whereas in the right one, the dipoles are antiparallel. The nonreciprocity parameter χ is of the same magnitude but of opposite sign for these samples. Note that the two samples have equal permittivity values, so also the permeability.*

6.3.2   Polarisability of chiral sphere

Consider a bi-isotropic sphere in bi-isotropic background as shown in Figure 6.4. If the sphere is exposed to a uniform electric and magnetic field e, it can be replaced by a six-vector dipole moment (consisting of electric and magnetic dipoles $\mathbf{p}_e$ and $\mathbf{p}_m$)

$$\mathsf{p} = \begin{pmatrix} c\eta\mathbf{p}_e \\ c\mathbf{p}_m \end{pmatrix} = \mathsf{A} \cdot \mathsf{e} \tag{6.19}$$

Note the undimensionalisation of the dipole moment six-vector which renders the polarisability the dimension m^3. Here the polarisability dyadic contains four components

$$\mathsf{A} = \begin{pmatrix} \alpha_{ee}\overline{\overline{I}} & \alpha_{em}\overline{\overline{I}} \\ \alpha_{me}\overline{\overline{I}} & \alpha_{mm}\overline{\overline{I}} \end{pmatrix} = \begin{pmatrix} \alpha_{ee} & \alpha_{em} \\ \alpha_{me} & \alpha_{mm} \end{pmatrix} \overline{\overline{I}} \tag{6.20}$$

Note that all components are multiples of the unit dyadic because of the bi-isotropy of the problem.

Let the environment and the inclusion materials be denoted by M_e and M_i:

$$\mathsf{M}_e = \begin{pmatrix} \epsilon_{r,e} & \chi_e - j\kappa_e \\ \chi_e + j\kappa_e & \mu_{r,e} \end{pmatrix} \overline{\overline{I}} \tag{6.21}$$

$$\mathsf{M}_i = \begin{pmatrix} \epsilon_{r,i} & \chi_i - j\kappa_i \\ \chi_i + j\kappa_i & \mu_{r,i} \end{pmatrix} \overline{\overline{I}} \tag{6.22}$$

The polarisability six-dyadic is

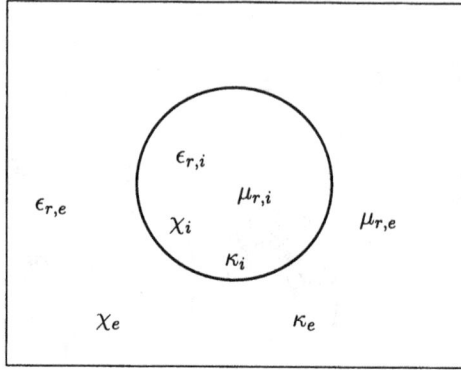

Figure 6.4: *A bi-isotropic sphere in bi-isotropic environment.*

$$\mathbf{A} = 3V(\mathbf{M}_i - \mathbf{M}_e) \cdot (\mathbf{M}_i + 2\mathbf{M}_e)^{-1} \cdot \mathbf{M}_e \tag{6.23}$$

where V is the volume of the sphere.

For example, the polarisabilities of a bi-isotropic sphere in vacuum ($\epsilon_{r,e} = \mu_{r,e} = 1$, $\chi_e = \kappa_e = 0$) can be explicitly written:

$$\alpha_{ee} = 3V\frac{(\epsilon_{r,i} - 1)(\mu_{r,i} + 2) - (\chi_i^2 + \kappa_i^2)}{(\epsilon_{r,i} + 2)(\mu_{r,i} + 2) - (\chi_i^2 + \kappa_i^2)} \tag{6.24}$$

$$\alpha_{em} = 3V\frac{3(\chi_i - j\kappa_i)}{(\epsilon_{r,i} + 2)(\mu_{r,i} + 2) - (\chi_i^2 + \kappa_i^2)} \tag{6.25}$$

$$\alpha_{me} = 3V\frac{3(\chi_i + j\kappa_i)}{(\epsilon_{r,i} + 2)(\mu_{r,i} + 2) - (\chi_i^2 + \kappa_i^2)} \tag{6.26}$$

$$\alpha_{mm} = 3V\frac{(\mu_{r,i} - 1)(\epsilon_{r,i} + 2) - (\chi_i^2 + \kappa_i^2)}{(\epsilon_{r,i} + 2)(\mu_{r,i} + 2) - (\chi_i^2 + \kappa_i^2)} \tag{6.27}$$

For lossless inclusion material, $\epsilon_{r,e}, \mu_{r,e}, \chi_e, \kappa_e$ are real. This gives us real co-polarisabilities α_{ee}, α_{mm}, and conjugate cross-polarisabilities $\alpha_{me} = \alpha_{em}^*$.

6.3.3   Chiral Maxwell Garnett mixing formula

The six-vector notations makes it easy to transform the isotropic and anisotropic results for mixtures to bi-isotropic mixtures. Consider a mixture where bi-isotropic spheres with polarisability $\mathbf{A}$ and material dyadic $\mathbf{M}_i$ occupy randomly a volume fraction f in another bi-isotropic environment $\mathbf{M}_e$. The Maxwell Garnett mixing rule for the effective material matrix then reads

$$\mathbf{M}_{\text{eff}} = \mathbf{M}_e + 3f\mathbf{M}_e \cdot [\mathbf{M}_i + 2\mathbf{M}_e - f(\mathbf{M}_i - \mathbf{M}_e)]^{-1} \cdot (\mathbf{M}_i - \mathbf{M}_e)$$

$$= \ M_e + 3f(M_i - M_e) \cdot [M_i + 2M_e - f(M_i - M_e)]^{-1} \cdot M_e \qquad (6.28)$$

Note that these two expressions are equal although the six-dyadics M_e and M_i do not commute in general.

And given the case that the bi-isotropic inclusions are ellipsoids, all aligned, a closed-form solution exists. The polarisability six-dyadic for the ellipsoid reads

$$A = V(M_i - M_e) \cdot [I + M_e^{-1} \cdot L \cdot (M_i - M_e)]^{-1} \qquad (6.29)$$

where the depolarisation six-dyadic is

$$L = \begin{pmatrix} \overline{\overline{L}} & 0 \\ 0 & \overline{\overline{L}} \end{pmatrix} \qquad (6.30)$$

with $\overline{\overline{L}}$ being the earlier depolarisation three-dyadic of the ellipsoid (5.87).

Now the mixture is effectively bi-anisotropic, the material six-dyadic being

$$M_{\text{eff}} = M_e + f(M_i - M_e) \cdot [M_e + (1-f)L \cdot (M_i - M_e)]^{-1} \cdot M_e \qquad (6.31)$$

6.3.4   Example: a racemic mixture

As an example, let us calculate the effective macroscopic parameters of a special type of symmetric mixture where isotropic chiral inclusions are embedded in an isotropic chiral environment. The material parameters of the components, relative to a vacuum, are $\epsilon_r = 2$, $\mu_r = 1.5$, and $\kappa = 1$ for the inclusions, and $\epsilon_r = 2$, $\mu_r = 1.5$, and $\kappa = -1$ for the background medium. In other words, the permittivity, permeability, and chirality parameters are the same for the inclusions and the environment; the only difference is the sign of κ. This means that the media are mirror images of each other, otherwise they are chemically the same. The inclusion medium is left-handed, the environment right-handed. Note that here both components are assumed to be reciprocal, in other words, $\chi = 0$ for both the inclusions and environment, and hence the mixture is also reciprocal.

The results are shown in Figures 6.5–6.7. Several interesting conclusions can be drawn from these figures.

Figure 6.5 shows that although both components have the same electric permittivity, the mixture permittivity ϵ_{eff} is not the same. It is lower than that of the components, but of course approaches that for the limiting cases $f = 0$ (no inclusions but homogeneous background) and $f = 1$ (no background, everything just inclusions). The strange effect on permittivity is caused by the magnetoelectric coupling. If the chirality of the components vanished, the effective permittivity of the mixture would be constant, that of background and inclusion, independent of the

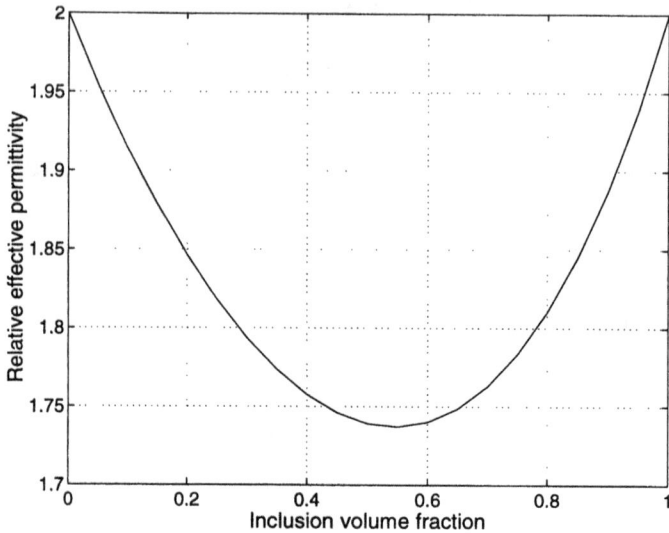

Figure 6.5: *Macroscopic permittivity* ϵ_{eff} *of a chiral-in-chiral isotropic mixture where the background is right-handed material with parameters are* $\epsilon_r = 2, \mu_r = 1.5, \kappa_r = -1$ *and parameters* $\epsilon_r = 2, \mu_r = 1.5, \kappa_r = +1$ *for the left-handed inclusions. The mixing is according to the Maxwell Garnett principle.*

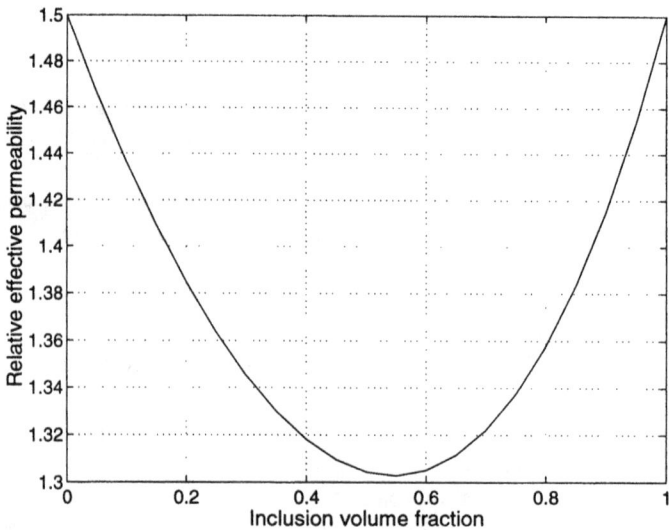

Figure 6.6: *The macroscopic permeability* μ_{eff} *of the same mixture as considered in Figure 6.5.*

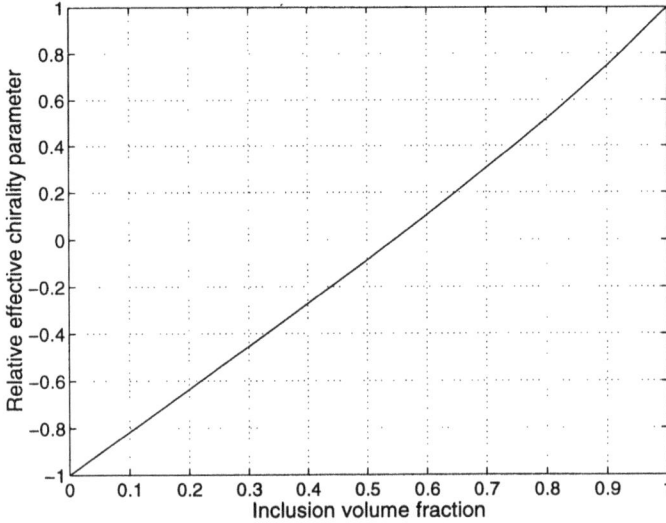

Figure 6.7: *The macroscopic chirality parameter κ_{eff} of the same mixture as considered in Figure 6.5.*

volume fraction. A similar behaviour can be observed for the effective permeability in Figure 6.6.

Interesting information can also be gleaned from Figure 6.7 which displays the effective chirality parameter of this mixture where left-handed inclusions occupy a certain space in a similar but right-handed ambient medium. Of course the curve runs from right-handed (negative chirality parameter) behaviour to left-handed (positive chirality parameter) behaviour as the volume fraction of the inclusions increases. But we can observe a nonsymmetry between the two phases in the mixture. It is not a 50/50 mixture that would be racemic.[8] Rather, slightly over half of the mixture volume has to be filled with the inclusion spheres in order to racemise the composite, so that $\kappa_{\text{eff}} = 0$.

6.4 Bi-anisotropic mixtures

The bi-isotropic media in the previous section had all the material dyadics as multiples of the unit dyadic. But the six-vector analysis is capable of treating more complex material dyadics, too. Let us consider in the following some of the results for mixtures of fully bi-anisotropic materials.

---

[8] A racemic mixture does not exhibit optical activity, in other words its effective chirality parameter vanishes.

6.4.1   Polarisability of a bi-anisotropic sphere

Consider a bi-anisotropic inclusion with material dyadic

$$
\mathsf{M}_i = \begin{pmatrix} \bar{\bar{\epsilon}}_{r,i} & \bar{\bar{\xi}}_i \\ \bar{\bar{\zeta}}_i & \bar{\bar{\mu}}_{r,i} \end{pmatrix}
\tag{6.32}
$$

in a bi-isotropic environment for which the material six-dyadic is

$$
\mathsf{M}_e = \begin{pmatrix} \epsilon_{r,e} & \xi_e \\ \zeta_e & \mu_{r,e} \end{pmatrix} \bar{\bar{I}}
\tag{6.33}
$$

Now the polarisability six-dyadic of the sphere obeys the same relation (6.23). Similarly, the expression (6.29) can be used to calculate the polarisabilities of a bi-anisotropic ellipsoid in bi-isotropic background.

The four three-dyadics, written explicitly, are long expressions. As a special case, the following are the polarisability dyadics for the case when the environment is vacuum ($\mathsf{M} = \mathsf{I}$):

$$
\bar{\bar{\alpha}}_{ee} = 3V \left[\bar{\bar{\epsilon}}_{r,i} - \bar{\bar{I}} - \bar{\bar{\xi}}_i \cdot \left(\bar{\bar{\mu}}_{r,i} + 2\bar{\bar{I}} \right)^{-1} \cdot \bar{\bar{\zeta}}_i \right] \cdot \bar{\bar{D}}_E^{-1}
\tag{6.34}
$$

$$
\bar{\bar{\alpha}}_{em} = 9V \left(\bar{\bar{\epsilon}}_{r,i} + 2\bar{\bar{I}} \right)^{-1} \cdot \bar{\bar{\xi}}_i \cdot \bar{\bar{D}}_M^{-1}
\tag{6.35}
$$

$$
\bar{\bar{\alpha}}_{me} = 9V \left(\bar{\bar{\mu}}_{r,i} + 2\bar{\bar{I}} \right)^{-1} \cdot \bar{\bar{\zeta}}_i \cdot \bar{\bar{D}}_E^{-1}
\tag{6.36}
$$

$$
\bar{\bar{\alpha}}_{mm} = 3V \left[\bar{\bar{\mu}}_{r,i} - \bar{\bar{I}} - \bar{\bar{\zeta}}_i \cdot \left(\bar{\bar{\epsilon}}_{r,i} + 2\bar{\bar{I}} \right)^{-1} \cdot \bar{\bar{\xi}}_i \right] \cdot \bar{\bar{D}}_M^{-1}
\tag{6.37}
$$

where

$$
\bar{\bar{D}}_E = \bar{\bar{\epsilon}}_{r,i} + 2\bar{\bar{I}} - \bar{\bar{\xi}}_i \cdot \left(\bar{\bar{\mu}}_{r,i} + 2\bar{\bar{I}} \right)^{-1} \cdot \bar{\bar{\zeta}}_i
\tag{6.38}
$$

$$
\bar{\bar{D}}_M = \bar{\bar{\mu}}_{r,i} + 2\bar{\bar{I}} - \bar{\bar{\zeta}}_i \cdot \left(\bar{\bar{\epsilon}}_{r,i} + 2\bar{\bar{I}} \right)^{-1} \cdot \bar{\bar{\xi}}_i
\tag{6.39}
$$

6.4.2   Bi-anisotropic mixing rules

A mixture where bi-anisotropic spheres (M_i) are located in bi-isotropic background medium (M_e) can be treated with formula (6.28), and the case of ellipsoids with (6.31). Of course, then the inclusions have to be aligned again both in geometry and the medium anisotropy. If, on the other hand, the ellipsoids are oriented according to a distribution $n(\Omega)$, their contributions have to be averaged. The material parameters read in this case

$$M_{\text{eff}} = M_e + \left(I - M_e^{-1} \cdot \int_\Omega L \cdot nA \, d\Omega \right) \cdot \int_\Omega nA \, d\Omega \qquad (6.40)$$

where $L = L(\Omega)$ is the depolarisation dyadic of the ellipsoids, the variation of which is determined by the orientation distribution. Likewise, the number density n gives the relative amounts of various ellipsoid polarisabilities in different directions. The integral has to be normalised so that

$$\int_\Omega nV \, d\Omega = f \qquad (6.41)$$

The orientation integral is in the most general case three-dimensional because it gives the number of ellipsoids with a given "state," or the orientation in space. The orientation of a general ellipsoid with respect to any global co-ordinate system is uniquely defined by three Eulerian angles.

If the environment is also fully bi-anisotropic, the depolarisation dyadic of an inclusion is not solely determined by the geometry of the ellipsoid. Instead, the anisotropy of the environment has to be taken into account through the affine transformation in a way similar to the method in Section 5.5. In this vein, the reader may find the homogenisation study [17] interesting where, instead of the affine transformation, the effect of the environment anisotropy is accounted for by a careful treatment of the singularity of the Green dyadic.

Problems

6.1 Show, by substituting the constitutive relations in Maxwell equations, that a medium with antisymmetric nonreciprocal magnetoelectric dyadic ($\overline{\overline{\chi}} = \mathbf{v} \times \overline{\overline{I}}$) corresponds to a moving medium. Determine the velocity vector of the movement.

6.2 Show that the six-dyadic inverse (6.17) satisfies $M^{-1} \cdot M = I$ and also $M \cdot M^{-1} = I$.

6.3 Show that the Tellegen parameter χ does not have effect when the constitutive relations of homogeneous bi-isotropic medium are substituted in Maxwell equations. (Delightful hint for the student with inclination for Lewis Carroll: [18].)

6.4 Prove that the two forms for the bi-isotropic Maxwell Garnett mixing rule (6.28) and (6.28) are equal also in the case when M_i and M_e do not commute.

6.5 Consider a mixture where spherical chiral inclusions $(\epsilon_i, \mu_i, \kappa_i)$ occupy a volume fraction f in the chiral environment $(\epsilon_e, \mu_e, \kappa_e)$. Calculate explicitly the Maxwell Garnett prediction for the three effective parameters $\epsilon_{\text{eff}}, \mu_{\text{eff}}$, and κ_{eff}.

6.6 Consider a mixture where chiral left-handed spheres are located in a right-handed chiral environment. Let the inclusion and environment materials be mirror-images of each other, which means that their material parameters are exactly the same except for a sign change in the chirality parameter, as for the case in the example of Section 6.3.4. Calculate the racemisation density, in other words, the volume fraction f for which the mixture is effectively racemic ($\kappa_{\text{eff}} = 0$). Use the Maxwell Garnett mixing principle. Let the relative permittivity and permeability of the phases be $\epsilon_{r,i} = \epsilon_{r,e} = 5$ and $\mu_{r,i} = \mu_{r,e} = 2$. Calculate the following three cases: $\kappa_i = -\kappa_e = 0.1, 1$, and 3.

References

[1] LANDAU, L.D., and LIFSHITZ, E.M.: 'Electrodynamics of continuous media', Second Edition (Pergamon Press, Oxford, 1984), Section 13.

[2] DZYALOSHINSKII, I.E.: 'On the magneto-electrical effects in antiferromagnets', *Soviet Physics JETP*, 1960, **10**, pp. 628-629

[3] ASTROV, D.N.: 'Magnetoelectric effect in chromium oxide', *Soviet Physics JETP*, 1961, **13**, (4), pp. 729-733

[4] TELLEGEN, B.D.H.: 'The gyrator, a new electric network element', *Philips Research Reports*, 1948, **3**, (2), pp. 81-101

[5] LINDELL, I.V., SIHVOLA, A.H., TRETYAKOV, S.A., and VIITANEN, A.J.: 'Electromagnetic waves in chiral and bi-isotropic media' (Artech House, Boston and London, 1994)

[6] RAAB, R.E., and GRAHAM, E.B.: 'Universal constitutive relations for optical effects in transmission and reflection in magnetic crystals', 1997, *Ferroelectrics*, **204**, pp. 157-171

[7] KONG, J.A.: 'Electromagnetic wave theory' (Wiley, New York, 1986)

[8] SOCHAVA, A.A., SIMOVSKI, C.R., and TRETYAKOV, S.A.: 'Chiral effects and eigenwaves in bi-anisotropic Omega structures', *in* PRIOU, A., SIHVOLA, A., TRETYAKOV, S., and VINOGRADOV, A. (Eds.): 'Advances in complex electromagnetic materials', NATO ASI Series, 3. High Technology, **28**, pp. 85-102 (Kluwer, Dordrecht, 1997)

[9] SIHVOLA, A.H., and LINDELL, I.V.: 'Material effects in bi-anisotropic electromagnetics', 1995, *IEICE Trans. Electron. (Japan)*, **E-78-C**, (10), pp. 1383-1390

[10] TRETYAKOV, S.A., SIHVOLA, A.H., SOCHAVA, A.A., and SIMOVSKI, C.R.: 'Magnetoelectric interactions in bi-anisotropic media, *Journal of Electromagnetic Waves and Applications*, 1998, **12**, (4), pp. 481-497. Correction: *ibid.*, 1999, **13**, (2), p. 225

[11] LINDELL, I.V., SIHVOLA, A.H., and SUCHY, K.: 'Six-vector formalism in electromagnetics of bi-anisotropic media', *Journal of Electromagnetic Waves and Applications*, 1995, **9**, (7/8), pp. 887-903

[12] ALTMAN, C., and SUCHY, K.: 'Reciprocity, spatial mapping and time reversal in electromagnetics' (Kluwer, Dordrecht, 1991)

[13] HEGSTROM, R.A., and KONDEPUNDI, D.K.: 'The handedness of the universe', *Scientific American*, January 1990, pp. 108-115

[14] BARRON, L.D.: 'Molecular light scattering and optical activity' (Cambridge University Press, 1982)

[15] LAKHTAKIA, A. (Editor): 'Selected papers on natural optical activity', **15** (SPIE Optical Engineering Press, Bellingham, Washington, 1990)

[16] LAKHTAKIA, A., and WEIGLHOFER, W.S.: 'Are linear, nonreciprocal, bi-isotropic media forbidden?', *IEEE Transactions on Microwave Theory and Techniques*, September 1994, **42**, (9), pp. 1715-1716. SIHVOLA, A.: 'Are nonreciprocal, bi-isotropic media forbidden indeed?', *ibid.*, September 1995, **43**, (9), pp. 2160-2162. LAKHTAKIA, A., and WEIGLHOFER, W.S.: 'Comment', and SIHVOLA, A.: 'Reply': *ibid.*, December 1995, **43**, (12), pp. 2722-2724.

[17] MICHEL, B., LAKHTAKIA, A., and WEIGLHOFER, W.S.: 'Homogenization of linear bianisotropic particular composite media—Numerical studies', *Int. Journal of Applied Electromagnetics and Mechanics*, 1998, **9**, (2), pp. 167-178

[18] LAKHTAKIA, A.: 'The Tellegen medium is "a Boojum, you see" ', *International Journal of Infrared and Millimeter Waves*, 1994, **15**, (10), pp. 1625-1630

Chapter 7

Nonlinear mixtures

The electrical response of materials is sometimes complicated, and in the previous chapters examples of such behaviours were mentioned. And still all the preceding discussion was simple in one respect: the polarisation was a linear response of its cause. Linearity means that if the cause were doubled, the response would be exactly twice the original. In other words, permittivity is independent of the field amplitude.

In the present chapter discussion shall be directed into the regime of nonlinear materials. When this is done, the characterisation of the polarisation of materials becomes much more difficult because of very many new possible connections between the various components of the fields and polarisations. Therefore the discussion of the mixing and homogenisation principles that follows cannot be as exhaustive as it has been for isotropic and anisotropic media. Nevertheless, let us consider some of the salient points in the character of nonlinear materials and the manner in which their properties affect the character of a mixture.

7.1 The characterisation of nonlinearity

7.1.1 Examples of nonlinear mechanisms in matter

Materials exist which are so characteristically nonlinear that the deviation from the linear behaviour is their distinctive property. But it is also true that all material media become nonlinear in their response if they are exposed to a sufficiently strong field. This can be understood by considering the classical picture of polarisable atoms and molecules. There, positive and negative centres of charge are displaced by the force of an external electric field, and the system works like a mechanical spring, or a collection of such springs. The charges are bound together and obviously their displacement cannot be arbitrarily large. Therefore for large field strengths the polarisation does not follow linearly the field amplitude. It may happen that

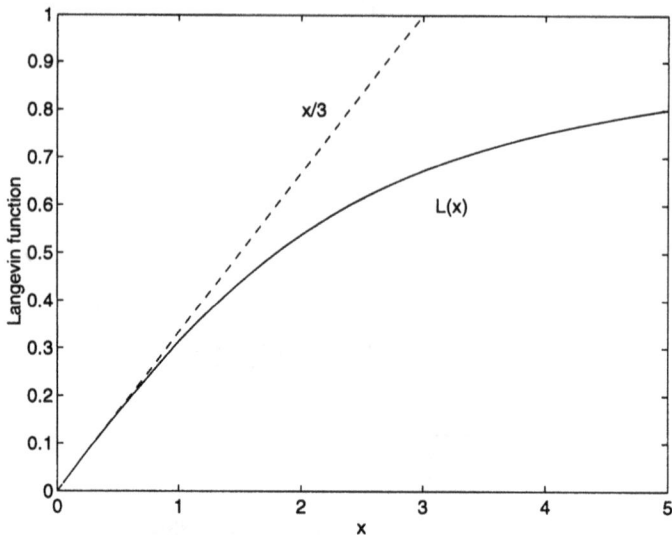

Figure 7.1: *The Langevin function* L(x) *shows how a paramagnetic polarisation deviates from the linear behaviour, shown with the dashed line. The argument* x *of the function is linearly proportional to the field amplitude and inversely proportional to temperature.*

the dielectric strength of the material is exceeded and an irreversible breakdown takes place. Another possible type of nonlinear behaviour is that the polarisation saturates.

For example, the orientational polarisation discussed in Section 2.1 was a result of the permanent dipole moments that tend to orient themselves with the external field. Suppose that a certain field has aligned all dipoles. Then the saturation level is reached and an increase in field amplitude cannot create any additional polarisation. A similar mechanism is the paramagnetic polarisation in many gases and alkali metals which have small permanent magnetic moments that are affected by an external magnetic field. A quantitative model for this paramagnetism is the Langevin theory [1] where the polarisation curve follows the function

$$L(x) = \coth(x) - \frac{1}{x} \tag{7.1}$$

This function is shown in Figure 7.1. For small values of the argument x, the function is linear, $L(x) \approx x/3$, but it saturates for large x: $L(x) \to 1$.

A Langevin model also suits dielectric orientational polarisation well although the field amplitudes where the nonlinearity has to be accounted for are very high for substances like water.

In ferromagnetic and ferrimagnetic substances, very highly nonlinear effects take place. In extreme cases, the magnetic permeability is not a very suitable concept to characterise the relation between the magnetic field and the flux density. The magnetisation depends on the history of the field within the material and the connection between the magnetic field and the magnetisation that can be described using the hysteresis curve. The physics behind the hysteresis is connected to the domain structure of magnetic materials. The field exerts forces on the Bloch walls that separate these magnetic domains (Weiss domains) in such a material, and high fields deform these walls. This leads to nonlinearity and losses. The distinction is often made between soft and hard magnetic materials. Soft magnetic materials have "slim" hysteresis loops whereas hard magnetic materials deviate more from the linear behaviour. Hard materials have large remanent magnetic polarisation, and also the coercive field amplitude is large which is the opposite-directed magnetic field that is needed to cancel the remanent magnetic induction. Metal industries have been able to produce materials with saturation magnetic fluxes of the order of teslas, and with coercive fields of tens of thousands A/m.

The analogue of ferromagnetic materials in the dielectrics are the ferroelectric materials. These highly nonlinear materials also follow hysteretic behaviour, and can be treated with similar analysis principles as ferromagetic media.

Another example of a nonlinear medium is plasma, ionised but macroscopically neutral gas. A dramatic effect on radio wave propagation is caused by the plasma in the ionosphere, first of all by the fact that electromagnetic waves longer than a certain wavelength are reflected from the ionosphere. Another effect of the ionosphere is that radio signals may be transferred to other frequencies due to the nonlinear mixing in plasma. This phenomenon goes under the name "Luxembourg effect."

In many applications where materials are exposed to optical or microwave signals the local field amplitudes within the media are relatively small. The response remains linear as long as the fields are small compared to those fields internal to the atoms and molecules that are responsible for the binding of valence electrons. For example, the optical radiation from the Sun may carry electric fields with magnitudes of the order of a hundred V/m. But the invention of laser in 1958 has made it possible to create very high electromagnetic power intensities, and nowadays materials can be exposed to electric fields with amplitudes that locally attain values of the order of 10^{10} V/m. With such fields, the quantitative description of wave–matter interaction exceeds the limits of the linear regime.

7.1.2 Nonlinear susceptibilities

To be able to treat quantitatively the susceptible properties of nonlinear materials, let us start with the relation between the electric flux density and the electric field at a given point in the material medium:

$$\mathbf{D} = \epsilon_0 \mathbf{E} + \mathbf{P} \tag{7.2}$$

The character of nonlinear materials is given by the relation between the polarisation and the electric field $\mathbf{P}(\mathbf{E})$. For linear materials, the relation is simple:

$$\mathbf{P}^{L} = \epsilon_0 \chi \mathbf{E} \tag{7.3}$$

where χ is the (linear) susceptibility of the medium and the free-space permittivity ϵ_0 is traditionally included in this relation. A possibility to describe a more general relation is to represent the polarisation as a sum of terms proportional to the powers of the electric field:

$$\mathbf{P} = \epsilon_0 \left(\chi^{[1]} \mathbf{E} + \chi^{[2]} \mathbf{E}^2 + \chi^{[3]} \mathbf{E}^3 + \cdots \right) \tag{7.4}$$

where now $\chi^{[1]}$ is the linear susceptibility, and $\chi^{[2]}, \chi^{[3]}, \ldots$ are higher-order (nonlinear) susceptibilities.

A problem in (7.4) compared with linear polarisation is how to define the powers of the electric field vector. One way to handle the nonlinear polarisation is to interpret the susceptibilities as polyadics: $\chi^{[1]}$ as a dyadic, $\chi^{[2]}$ as a triadic, $\chi^{[3]}$ as a tetradic, and so on. Then the operation on the electric field is via a polydot product:

$$\mathbf{P} = \epsilon_0 \left(\chi^{[1]} \cdot \mathbf{E} + \chi^{[2]} : \mathbf{E}\mathbf{E} + \chi^{[3]} \circledS \mathbf{E}\mathbf{E}\mathbf{E} + \cdots \right) \tag{7.5}$$

Here the notation $\circledS$ is introduced for the polydot operation: the number n inside the circle stands for the number of dots in the dot product. This contraction operation lowers the rank of the polyadics on both sides by its number.

Another possibility to describe the susceptibility operations is to treat those as tensors [2, 3]. Then the connection (7.4) can be written as

$$P_i = \epsilon_0 \left(\chi^{[1]}_{ij} E_j + \chi^{[2]}_{ijk} E_j E_k + \chi^{[3]}_{ijkl} E_j E_k E_l + \cdots \right) \tag{7.6}$$

where a repeated index in the product means that the product includes summation over the three spatial components.

If the fields are sinusoidal that are exciting a nonlinear medium it is practical to give the polarisation relations in the frequency domain. In linear media the electric field and polarisation vibrate at the same frequency:

$$\mathbf{P}(\omega) = \epsilon_0 \chi^{[1]}(\omega) \cdot \mathbf{E}(\omega) \tag{7.7}$$

but the situation is more complicated because of the multiplication process inherent in nonlinear phenomena; sum and difference frequencies of the various spectral components of the electric field appear in the response of the material. This is often emphasised by showing explicitly the frequencies in the nonlinear susceptibilities. For the lowest-order nonlinearity,

$$\mathbf{P}(\omega) = \epsilon_0 \chi^{[2]}(\omega, \omega_1, \omega_2) : \mathbf{E}(\omega_1)\mathbf{E}(\omega_2) \tag{7.8}$$

with $\omega = \omega_1 + \omega_2$, and for the next-order nonlinear term,

$$\mathbf{P}(\omega) = \epsilon_0 \chi^{[3]}(\omega, \omega_1, \omega_2, \omega_3) \, ③ \, \mathbf{E}(\omega_1)\mathbf{E}(\omega_2)\mathbf{E}(\omega_3) \tag{7.9}$$

where $\omega = \omega_1 + \omega_2 + \omega_3$. Although the polarisation frequency is a sum of the frequencies of the field terms, also difference frequencies appear in the polarisation. This is because the complex representation of frequency-domain signals includes both positive and negative frequencies. Another matter of convention is that since the sum rule determines the possible frequencies of the output polarisation, quite often the first frequency argument in $\chi^{[2]}$ and $\chi^{[3]}$ is omitted.

7.1.3   Quadratic and cubic nonlinearities

The lowest-order nonlinear susceptibility is determined by the quadratic term $\chi^{[2]}$, which is responsible for several effects. For a single-frequency signal, the nonlinearity gives rise to the frequency doubling of the incident signal, also known as second-harmonic generation. But mixing also leads to the difference frequency which falls into a zero-frequency signal (dc). This means that materials that display quadratic nonlinearity can be used in detectors of amplitude-modulated electromagnetic waves, and in the rectification of light. The carrier wave is eliminated by the mixing process.

If the incident electric field is composed of two sinusoidal components, the second-order mixing creates sum and difference signals of these two frequencies, and parametric amplification of signals is possible. The Pockels effect is a special case of this phenomenon: if one of the frequency components is of much lower frequency than the other, the output is at nearly the same frequency as the high-frequency input component. This could be considered as a linear mechanism where the optical susceptibility is controlled by an independent electric field, which leaves the input and output high-frequency fields standing in a linear relation with each other.

For materials with spatial inversion symmetry in their susceptibility response, $\chi^{[2]}$ vanishes, meaning that for such media, the lowest order nonlinearity is the cubic term.[1] This is characterised by the components of the tetradic $\chi^{[3]}$. Many types of phenomena are possible due to the cubic nonlinearity. Examples of such processes are third-harmonic generation, optical phase conjugation, Raman scattering, and self-focusing of waves [4].

The analogue to the Pockels effect on the level of cubic nonlinearity is the Kerr effect. In that mechanism, a constant (or low-frequency) electric field is mixed with a higher frequency, for example optical, signal. Then, by the sum rule, one of the resulting output components is (nearly) at the same signal frequency. But because this polarisation amplitude is affected by the square of the constant electric field amplitude, the "permittivity" of such a medium can be controlled by the intensity

---

[1] The inversion symmetry is connected to the symmetry of the point group of the material. Therefore, for systems with inversion centre, the second-order susceptibility tensor (and also other even tensors, $\chi^{[4]}, \chi^{[6]}, \ldots$) vanish. However, the inversion symmetry of the susceptibility tensors can be destroyed by, for example, a static electric field.

of this low-frequency or dc electric field.[2] Another name for such a Kerr effect is "static-field-induced birefringence" because of the anisotropic connection of the refractive index to the static field.

7.2   Mixing rules for nonlinear materials

After introducing the basic quantities that can be used to characterise continuous nonlinear dielectric materials, let us take a look at how the nonlinear parameters can be determined for a nonhomogeneous medium.

7.2.1   Polarisability components for a nonlinear sphere

As in the earlier chapters where isotropic, anisotropic, and bi-anisotropic inclusions were analysed as parts of mixtures, here also, in the nonlinear case, it is important to start by studying the dipole moment and polarisability of a spherical inclusion. Figure 7.2 illustrates the situation where a nonlinear sphere is exposed to a uniform electric field $\mathbf{E}_e$. The sphere is located in an isotropic environment of permittivity ϵ_e.

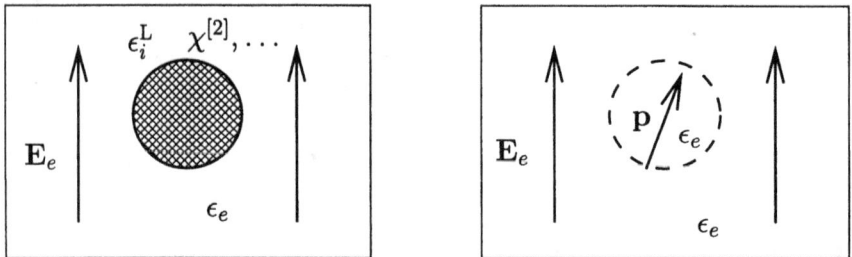

Figure 7.2: *The dipole moment* $\mathbf{p}$ *induced in a nonlinear sphere, characterised by the linear permittivity component* ϵ_i^L *and nonlinear terms* $\chi^{[2]}, \cdots$. *Even in the case of isotropic linear susceptibility, the dipole moment of the sphere* $\mathbf{p}$ *is not necessarily aligned with the external field* $\mathbf{E}_e$, *due to the generally anisotropic nature of the nonlinear terms of the susceptibility.*

The nonlinear material in the sphere is characterised by its response to electric fields according to relation (7.5). Let us assume for simplicity that the linear susceptibility of the material is isotropic, in other words $\chi^{[1]}$ is a scalar. Then the

---

[2]Sometimes the Pockels effect is called the "linear Kerr effect." This may be because the "real" Kerr effect is a quadratic effect, if the low-frequency or dc field is considered. Another possible cause for confusion in the Kerr terminology is the fact that in addition to the classical "static Kerr effect," another mechanism is distinguished: the "optical Kerr effect." The optical Kerr effect represents the change of the refractive index of a nonlinear material which is caused by radiation in the visible region [5].

polarisation inside the sphere, referred to the environment,[3] is

$$\mathbf{P}_i = (\epsilon_i - \epsilon_e)\mathbf{E}_i + \epsilon_0 \left(\chi^{[2]} : \mathbf{E}_i\mathbf{E}_i + \chi^{[3]} \circled{3} \mathbf{E}_i\mathbf{E}_i\mathbf{E}_i + \cdots\right) \tag{7.10}$$

where $\epsilon_i = \epsilon_0(1 + \chi^{[1]})$, and $\mathbf{E}_i$ is the internal field.

The dipole moment of the sphere is obviously also a nonlinear function of the external field:

$$\mathbf{p} = \alpha\mathbf{E}_e + \beta : \mathbf{E}_e\mathbf{E}_e + \gamma \circled{3} \mathbf{E}_e\mathbf{E}_e\mathbf{E}_e + \cdots \tag{7.11}$$

where the contributions by the various powers of the field are given by hyperpolarisabilities.[4] The hyperpolarisabilities are denoted by $\alpha, \beta, \gamma, \ldots$, and are, of course, polyadics. Although α was assumed scalar in the present case, in general it is a dyadic. β is a triadic, γ a tetradic, and so on to increasing order.

The next step is to find the connection between the internal and external fields. In fact, it was already derived in Section 5.3.2 for a general polarisation relation:

$$\mathbf{E}_i = \mathbf{E}_e - \frac{1}{3\epsilon_e}\mathbf{P}_i \tag{7.12}$$

Equations (7.10) and (7.12) are connected by anisotropy and nonlinearity to infinite order, and a solution for the internal field as a function of the external field is not easy.[5] The relation reads

$$\mathbf{E}_i = \frac{3\epsilon_e}{\epsilon_i + 2\epsilon_e}\mathbf{E}_e - \frac{\epsilon_0}{\epsilon_i + 2\epsilon_e}\left(\chi^{[2]} : \mathbf{E}_i\mathbf{E}_i + \chi^{[3]} \circled{3} \mathbf{E}_i\mathbf{E}_i\mathbf{E}_i + \cdots\right) \tag{7.13}$$

Let us now solve this relation with the assumption that in the nonlinear correction on the right-hand side of (7.13), the linear approximation for the internal field is used. We have therefore

$$
\mathbf{E}_i \approx \frac{3\epsilon_e}{\epsilon_i + 2\epsilon_e}\mathbf{E}_e
$$
$$
- \frac{\epsilon_0}{\epsilon_i + 2\epsilon_e}\left[\left(\frac{3\epsilon_e}{\epsilon_i + 2\epsilon_e}\right)^2 \chi^{[2]} : \mathbf{E}_e\mathbf{E}_e + \left(\frac{3\epsilon_e}{\epsilon_i + 2\epsilon_e}\right)^3 \chi^{[3]} \circled{3} \mathbf{E}_e\mathbf{E}_e\mathbf{E}_e + \cdots\right]
$$
$$\tag{7.14}$$

Because the field and polarisation inside the sphere are constant, the following holds: $\mathbf{p} = V\mathbf{P}_i$ with the volume V of the sphere, and a combination of (7.11) and

---

[3]Note that the environment need not be a vacuum, and $\epsilon_e \neq \epsilon_0$ in general.

[4]The term "hyperpolarisability" is most commonly used to descibe the nonlinear response of an atom or molecule to the electric field, and in the tensor form, often the coefficients $1/2$ and $1/6$ are included in the second- and third-order terms [6, 7].

[5]But on the other hand it is important to realise that despite the complicated connection between the field and polarisation, nonlinearity does not affect the uniformity of the internal field. If the sphere is homogeneous, the uniformity of the field guarantees the same for the polarisation, and Equation (7.12) is consistent.

(7.14) with $\mathbf{P}_i = 3\epsilon_e(\mathbf{E}_e - \mathbf{E}_i)$ gives us a perturbative solution for the hyperpolarisabilities. Finally, we have

$$\alpha = 3\epsilon_e \frac{\epsilon_i - \epsilon_e}{\epsilon_i + 2\epsilon_e} V \tag{7.15}$$

$$\beta = \left(\frac{3\epsilon_e}{\epsilon_i + 2\epsilon_e}\right)^3 \epsilon_0 \chi^{[2]} V \tag{7.16}$$

$$\gamma = \left(\frac{3\epsilon_e}{\epsilon_i + 2\epsilon_e}\right)^4 \epsilon_0 \chi^{[3]} V \tag{7.17}$$

In these expressions for the hyperpolarisabilities, only the leading terms are included.[6]

7.2.2   Dilute mixtures

What, then, about a mixture? If a heterogeneous material contains nonlinear phases, we must be able to estimate the effective higher-order susceptibilities of the mixture, in addition to the effective permittivity. To determine such macroscopic quantities, we define the average quantities with the relation

$$
\begin{aligned}
< \mathbf{D} > &= \epsilon_e < \mathbf{E} > + < \mathbf{P} > \\
&= \epsilon_{\text{eff}}^L < \mathbf{E} > + \epsilon_0 \left(\chi_{\text{eff}}^{[2]} : < \mathbf{E} >< \mathbf{E} > \right. \\
&\left. + \chi_{\text{eff}}^{[3]} \textcircled{3} < \mathbf{E} >< \mathbf{E} >< \mathbf{E} > + \cdots \right)
\end{aligned}
\tag{7.18}
$$

where the superscript L has been added to the effective permittivity to stress the fact that ϵ_{eff}^L only stands for the linear part of the polarisation response of the system.

Let the mixture consist of nonlinear spheres in the background with permittivity ϵ_e. In the dilute-mixture limit, where the volume fraction of the inclusions f is low, we can assume that the field that excites an inclusion is the average field $< \mathbf{E} >$, and the effective parameters for the mixture can be given as

$$\epsilon_{\text{eff}}^L = \epsilon_e + n\alpha = \epsilon_e + 3f\epsilon_e \frac{\epsilon_i - \epsilon_e}{\epsilon_i + 2\epsilon_e} \tag{7.19}$$

$$\chi_{\text{eff}}^{[2]} = \frac{n\beta}{\epsilon_0} = f \left(\frac{3\epsilon_e}{\epsilon_i + 2\epsilon_e}\right)^3 \chi^{[2]} \tag{7.20}$$

$$\chi_{\text{eff}}^{[3]} = \frac{n\gamma}{\epsilon_0} = f \left(\frac{3\epsilon_e}{\epsilon_i + 2\epsilon_e}\right)^4 \chi^{[3]} \tag{7.21}$$

where n is the number density of the spheres and $f = nV$.

These relations can be easily generalised for ellipsoidal inclusions and cases where the linear part of the susceptibility $\chi^{[1]}$ is anisotropic.

---

[6]For example, in the next order, the cubic hyperpolarisability (7.17) contains a term which is quadratic to the second-order susceptibility $\chi^{[2]}$.

7.2.3 Towards denser mixtures

If the mixture becomes denser, the distinction has to be made between the average and local fields. From the classical mixing analysis, the Lorentz correction to the average field gave for the local field

$$\mathbf{E}_L = <\mathbf{E}> + \frac{1}{3\epsilon_e} <\mathbf{P}> \tag{7.22}$$

A good approximation for the local field is to take only the linear interaction into account, which gives

$$\mathbf{E}_L = \frac{1}{1 - \dfrac{n\alpha}{3\epsilon_e}} <\mathbf{E}> \tag{7.23}$$

Using this correction, the linear part of the effective permittivity for this dense mixture becomes

$$\epsilon_{\mathrm{eff}}^L = \epsilon_e + \frac{n\alpha}{1 - \dfrac{n\alpha}{3\epsilon_e}} = \epsilon_e + 3f\epsilon_e \frac{\epsilon_i - \epsilon_e}{\epsilon_i + 2\epsilon_e - f(\epsilon_i - \epsilon_e)} \tag{7.24}$$

which is the correct Maxwell Garnett result. The nonlinear effective susceptibilities are

$$\chi_{\mathrm{eff}}^{[2]} = f\left(\frac{3\epsilon_e}{\epsilon_i + 2\epsilon_e - f(\epsilon_i - \epsilon_e)}\right)^3 \chi^{[2]} \tag{7.25}$$

$$\chi_{\mathrm{eff}}^{[3]} = f\left(\frac{3\epsilon_e}{\epsilon_i + 2\epsilon_e - f(\epsilon_i - \epsilon_e)}\right)^4 \chi^{[3]} \tag{7.26}$$

These results have the delightful property that for high-density limit ($f = 1$), the nonlinear susceptibilities of the mixture become those of the inclusions, as the situation indeed should be.

The effective susceptibility behaviours of a mixture are illustrated in Figure 7.3. There the nonlinear susceptibilities of the mixture (relative to the corresponding inclusion susceptibilities) are plotted as functions of the volume fraction of the spherical inclusions, and also of the permittivity ratio ϵ_i/ϵ_e. The figures show that for increasing dielectric contrast, the relative nonlinearity will be weaker. Another observation is that the mixing process attenuates the third-order component more than the second-order susceptibility. This is an evident characteristic of a mixture with spherical inclusions. Due to depolarisation, the field inside the sphere is smaller than the external field, and because here the nonlinear behaviour only comes from within the spheres, the nonlinearity is damped in the homogenisation. For inclusions with nonspherical shapes, this damping effect is smaller, and can be estimated for ellipsoids using the shape factors given in Section 4.2.

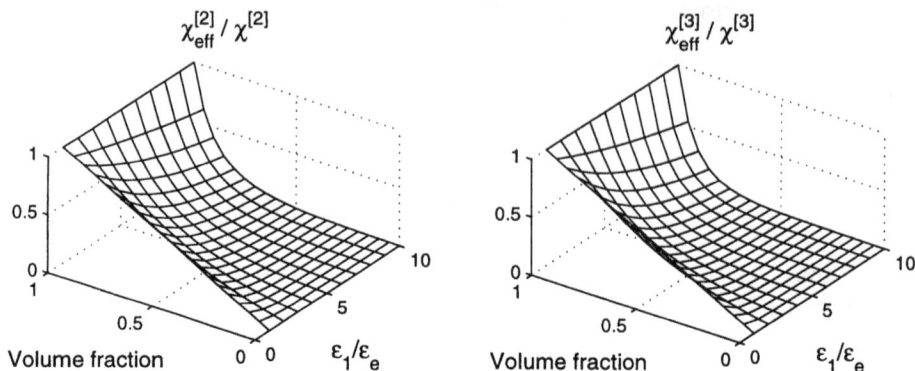

Figure 7.3: *The relative nonlinear susceptibilities $\chi^{[2]}_{\text{eff}}/\chi^{[2]}$ and $\chi^{[3]}_{\text{eff}}/\chi^{[3]}$ of a mixture with nonlinear spheres in linear host medium. The nonlinearity parameters are shown as functions of the volume fraction f of the spheres and the permittivity ratio between the spheres and the background ϵ_i/ϵ_e.*

7.2.4   Nonlinearity as perturbation to permittivity

For weakly nonlinear materials we do not need to take into account the full interaction equations in the homogenisation process. A very effective way in such cases to solve the effective properties of nonlinear mixtures is to treat the nonlinearity as a perturbational correction to the basic mechanism which is the linear permittivity. Then, using the known mixing rules from the basic linear permittivity, the nonlinear case arises as its Taylor expansion.

This approach is illustrated by the treatment of the weakly cubic nonlinearity in [8]. There, the isotropic nonlinear material is defined with the constitutive relation

$$\mathbf{D} = \epsilon^{\text{L}}_i \mathbf{E} + \xi_i |\mathbf{E}|^2 \mathbf{E} \qquad (7.27)$$

where ξ_i is now a scalar, a measure for the norm of the susceptibility tetradic $\chi^{[3]}$. Although misunderstandings can result from the following notation, it is rather common in the literature: the permittivity of the nonlinear medium of the type of (7.27) is often written as the scalar

$$\epsilon_i = \epsilon^{\text{L}}_i + \xi_i |\mathbf{E}|^2 \qquad (7.28)$$

Now, if nonlinear inclusions of this type are embedded in another, linearly responding material with permittivity ϵ_e, the characteristics of the mixture are calculated approximately in the following way. The linear effective permittivity of such a mixture can be calculated according to a given mixing rule:

$$\epsilon^{\text{L}}_{\text{eff}} = F(\epsilon^{\text{L}}_i, \epsilon_e, f) \qquad (7.29)$$

where f is the volume fraction of the inclusions. The function F reflects the chosen mixing model. It could be a Maxwell Garnett model with the given geometry of the mixture, or it could be any of the other mixing rules which will be introduced in Chapter 9.

The idea of [8] is to invoke the nonlinear relation according to the form (7.29). Expanding it as a Taylor expansion about the linear effective permittivity the result is the nonlinear effective permittivity, where the first-order correction gives the field-dependent term,

$$\epsilon_{\text{eff}} = \epsilon_{\text{eff}}^{\text{L}} + \frac{\xi_i}{f}F'|F'| \, | < \mathbf{E} > |^2 \tag{7.30}$$

where $< \mathbf{E} >$ is the external field, and $F' = \partial F/\partial \epsilon_i^{\text{L}}$. This means that the effective cubical nonlinearity parameter of the mixture is

$$\xi_{\text{eff}} = \frac{\xi_i}{f} \frac{\partial \epsilon_{\text{eff}}^{\text{L}}}{\partial \epsilon_i^{\text{L}}} \left| \frac{\partial \epsilon_{\text{eff}}^{\text{L}}}{\partial \epsilon_i^{\text{L}}} \right| \tag{7.31}$$

As a special case, the low-density limit ($f \ll 1$) of (7.30)–(7.31) for spherical inclusions, in the Maxwell Garnett model, reads

$$\epsilon_{\text{eff}}^{\text{L}} = \epsilon_e + 3f\epsilon_e \frac{\epsilon_i^{\text{L}} - \epsilon_e}{\epsilon_i^{\text{L}} + 2\epsilon_e} \tag{7.32}$$

$$\xi_{\text{eff}} = f\xi_i \left(\frac{3\epsilon_e}{\epsilon_i^{\text{L}} + 2\epsilon_e} \right)^2 \left| \frac{3\epsilon_e}{\epsilon_i^{\text{L}} + 2\epsilon_e} \right|^2 \tag{7.33}$$

This result is in agreement with the earlier dilute-limit estimates (7.19) and (7.21) for real isotropic material parameters.

7.3 Characteristics of nonlinear mixtures

In composite materials, nonlinearity may play an important role, like for example in dielectric breakdown in metal-insulator mixtures, or percolation properties. From the homogenisation point of view, it is of course to be expected that the nonlinear properties of a mixture differ from those of its constituents. In some cases the difference in the material character is striking when the bulk and particulate materials are compared.

Nonlinear effects start to dominate when the field magnitudes are large. Therefore it is easy to understand that the difference between linear and nonlinear dielectric mixtures is enhanced if the mixture favours large field variations. In the calculations of the previous section, the spherical geometry tended to lead to "soft" behaviour of the effective properties. However, if the inclusion shapes are nonspherical, and especially if they contain sharp corners, field amplitudes are increased. With this "lightning rod effect" close to inclusion corners the local response goes much easier into the nonlinear regime than near boundaries that have a large radius of curvature. For needle-shaped inclusions, this enhancement effect requires that the background medium be nonlinear because for ellipsoidal shapes only the external fields grow in amplitude.[7]

The nonlinear properties of matter may also be especially enhanced near percolation threshold of metal-insulator composites. Also, inclusions and clusters may experience a dramatic increase in the polarisability close to the so-called surface plasmon resonance. This has been shown for gold and silver nanosphere suspensions in linear dielectric. In that case, the nonlinear susceptibility component of the gold and silver spheres may increase a thousandfold and more [9, 10]. In [11], an analysis on composite optical materials shows that the relative strengths of different nonlinear susceptibilities of a mixture change noticeably along with the volume fractions of the constituents. In other words, not only the strength of the nonlinearity in general increases but the character and type of the nonlinear anisotropy of the materials changes when they are mixed.

Aside from the quantitative changes in nonlinearity caused by mixing, another important phenomenon is that in the mixture such nonlinear properties may be observed that are totally absent in the components. For example, it can be shown from the symmetry properties that the nonlinear susceptibilities of even order, like for example $\chi^{[2]}$, have to vanish for materials with centre of symmetry. Consequently, phenomena like second-harmonic generation are not possible without external influences. However, if a composite of such material and a nonlinear dielectric host is designed in a manner such that the microgeometry breaks the inversion symmetry, the homogenised medium can display effective second-order susceptibility $\chi_{\text{eff}}^{[2]}$. Another example of a qualitative change in the behaviour of the material is presented in [12]. There the effective properties of a mixture are studied with nonlinear spherical grains. The nonlinearity of the inclusions is of cubic type. The resulting mixture, too, is a nonlinear Kerr-type medium but in addition to the cubic nonlinearity $\chi_{\text{eff}}^{[3]}$ it also possesses fifth-order nonlinearity $\chi_{\text{eff}}^{[5]}$. Furthermore, strong resonances in the nonlinear susceptibilities exist that are absent in the bulk behaviour of the grains.

A suggestion for the design of an artifical nonlinear composite material is given in [13]. The material consists of polarisable elements, each loaded with a nonlinear circuit, in the background matrix. A clamping circuit is an example of such a loading: a diode and a resistor are connected in series to an electrically small dipole antenna.

Problems

7.1 Determine the hyperpolarisabilities β and γ of an ellipsoid with susceptibility polyadics $\chi^{[n]}$, $n = 1, 2, 3$. Let the axes of the ellipsoid be a, b, and c.

7.2 Calculate the effective linear permittivity and the effective nonlinear suscep-

---

[7]Assuming that the permittivity of the background medium is smaller than the permittivity of the inclusion.

tibility polyadics $\chi_{\text{eff}}^{[n]}$ for a mixture where ellipsoids of the previous problem are embedded in a linear isotropic host material with permittivity ϵ_e. Assume an ordered mixture: all the ellipsoids are aligned with similar axes pointing in the same direction.

7.3 The calculation of the hyperpolarisabilities in Section 7.2.1 assumed that the linear part of the inclusion permittivity $\epsilon_L = (1 + \chi^{[1]})\epsilon_0$ was isotropic. Derive the hyperpolarisabilities α, β, γ if this is not the case, in other words the first-order susceptibility of the sphere is a general dyadic $\overline{\overline{\chi}}^{[1]}$.

References

[1] ROBERT, P.: 'Electrical and magnetic properties of materials' (Artech House, Norwood, Mass., 1988)

[2] BLOEMBERGEN, N.: 'Nonlinear optics' (World Scientific, Singapore, 1996, 4th edn.)

[3] SAUTER, E.G.: 'Nonlinear optics' (Wiley, New York, 1996)

[4] YARIV, A.: 'Quantum electronics' (Wiley, New York, 1989, 3rd edn.)

[5] KREIBIG, U., and VOLLMER, M.: 'Optical properties of metal clusters', Materials Science Series, **25** (Springer, Berlin, 1995)

[6] BUCKINGHAM, A.D., and ORR, B.J.: 'Molecular hyperpolarisabilities', *Quart. Rev. Chem. Soc.*, 1967, **21**, pp. 195-212

[7] EVANS, M., and KIELICH, S. (Eds.): 'Modern nonlinear optics', Parts 1–3. Advances in Chemical Physics, **LXXXV** (Wiley–Interscience, New York, 1997)

[8] ZENG, X.C., BERGMAN, D.J., HUI, P.M., and STROUD, D.: 'Effective-medium theory for weakly nonlinear composites', *Physical Review B*, **38**, (15), pp. 10970-10973

[9] RICARD, D., ROUSSIGNOL, Ph., and FLYTZANIS, Chr.: 'Surface-mediated enhancement of optical phase conjugation in metal colloids', *Optics Letters*, 1985, **10**, (10), pp. 511-513

[10] NEEVES, A.E., and BIRNBOIM, M.H.: 'Composite structures for the enhancement of nonlinear-optical susceptibility', *J. of the Optical Society of America, B*, 1989, **6**, (4), pp. 787-796

[11] SIPE, J.E., and BOYD, R.W.: 'Nonlinear susceptibility of composite materials in the Maxwell Garnett model', *Physical Review A*, 1992, **46**, (3), pp. 1614-1629

[12] AGARWAL, G.S., and GUPTA, S.D.: '*T*-matrix approach to the nonlinear susceptibilities of heterogeneous media', *Physical Review A*, 1988, **38**, (11), pp. 5678-5687

[13] AUZANNEAU, F., and ZIOLKOWSKI, R.W.: 'Microwave signal rectification using artificial composite materials composed of diode-loaded electrically small dipole antennas', *IEEE Transacations on Microwave Theory and Techniques*, 1998, **46**, (11), pp. 1628-1637

Part II

To transgress the pattern:
Functionalistic and modernist mixing

Chapter 8

Difficulties and uncertainties in classical mixing

The permeating principle in all chapters throughout Part I was Maxwell Garnett philosophy. First, the response of a free scatterer was analysed, and then in the mixing case, the inclusions were assumed to be excited by the local field that was deterministically calculated wth the average polarisation. But mixtures are often completely random in their structure. It may therefore look contradictory to try to analyse properties of such media with the approach taken so far in the present book. Our calculations have now treated idealised geometries and well-defined boundaries and materials, and have led to exact predictions for the effective permittivity of a mixture which, in fact, is not well-defined in the sense we try to treat it.

The following objection can indeed be advanced against the Maxwell Garnett principle of homogenisation: we are claiming to assign a quantitative estimate for a macroscopic property to a particular sample of a material which could be changed with another member of an infinite ensemble. To the degree that we describe this whole ensemble—by merely one structural parameter, the volume fraction of the inclusion phase—all of the members of this ensemble are identical although their real microstructure can be of extreme diversity. And our description does not make any distinction between the individuals within this infinite variety.

The present chapter is an attempt to dispel impressions that conceptual difficulties regarding the randomness are not given attention. Let us try to articulate the weaknesses and problems of the classical approach to mixing rules. Also the question is addressed between which limits the effective properties of heterogeneous media can vary if the case is really so that we cannot predict an exact macroscopic permittivity for a mixture.

8.1   Weak links in the mixing rule derivation

The various forms of Maxwell Garnett mixing formula that were derived in the previous chapters were constructed by making use of solutions of the following subproblems:

(i) What is the polarisability of a single inclusion in homogeneous environment?

(ii) What is the field that excites the scatterer in a mixture?

(iii) What is the average polarisation density in the mixture as a function of the induced dipoles?

These were solved by finding

(a) the exact solution of the Laplace equation for a single inclusion in a homogeneous environment

(b) the local field within a cavity in homogeneously polarised matter

(c) the volume average of the dipole moment vectors.

While few readers would loudly raise their voices against the first and third answers to these three items,[1] the solution to the second question is perhaps least satisfactory even if it is bolstered with qualitative arguments in the style of the previous text. Admittedly, the local field defined in the Lorentzian sense (Equation (3.21)) is an approximation, although an approximation which leads to a flourishing Maxwell Garnett formula that has been applied for decades in many fields of materials science. Solutions to these three basic questions other than those listed above can also result in meaningful mixing formulas.

8.1.1   Interaction between the scatterers

It is difficult not to have one's intuition upset by the very coarse modelling of the situation of an inclusion in the mixture that is illustrated in Figure 8.1. All the randomness of the neighbourhood with its strong inhomogeneity is replaced by uniform polarisation density, in other words the volume average of the dipole moments. In this floating polarisation, a cavity is set up which has the shape of the inclusion. Then we are left with a deterministic and straightforward static problem, a solution to which gives us the familiar relations between the local field $\mathbf{E}_L$ and the average polarisation $< \mathbf{P} >$:

$$\mathbf{E}_L = < \mathbf{E} > + \frac{1}{\epsilon_e} \overline{\overline{L}} \cdot < \mathbf{P} > \tag{8.1}$$

The field relation uses the depolarisation dyadic $\overline{\overline{L}}$ of the volume under consideration which determines the excess field over the average field.

---

[1] The mixing literature, however, contains approaches which do not take even these two answers for granted.

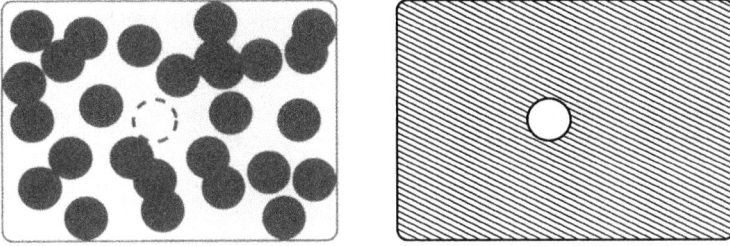

Figure 8.1: *When the local field that excites a single scatterer is calculated, the whole heterogeneity of the medium is replaced by the average polarisation density in which a cavity is carved.*

What else could be done? A rigorous attack to analyse the fields in a random medium requires a complete definition of the many-body problem [1–3]. In this formulation, the problem is to solve for the average electric field $< \mathbf{E} >$ in the case

$$\mathbf{D}(\mathbf{r}) = \epsilon(\mathbf{r})\mathbf{E}(\mathbf{r}) \tag{8.2}$$
$$\nabla \times \mathbf{E}(\mathbf{r}) = 0 \tag{8.3}$$
$$\nabla \cdot \mathbf{D}(\mathbf{r}) = 0 \tag{8.4}$$

at each point within the medium, where the permittivity function is a two-valued function: $\epsilon(\mathbf{r}) = \epsilon_i$ within the inclusions and ϵ_e outside those. Once $\mathbf{E}(\mathbf{r})$ is known, so is the flux density $\mathbf{D}(\mathbf{r})$ at each point, and their averages are calculable. Finally the relation between them is the sought-for effective permittivity:

$$< \mathbf{D} > = \epsilon_{\text{eff}} < \mathbf{E} > \tag{8.5}$$

To solve the field, differentiating operators have to be inverted. This is often done by the so-called T-matrix formulation which can be understood as each scatterer "transferring" a field to the surroundings. The field at each point then, finally, is an infinite sum of multiply scattered fields, and the solution requires a consistently calculated addition of all contributions. Feynman diagrams (see, for example, [4]) are often used in the calculations to help with bookkeeping of the infinities. Rigorous perturbation methods quite often fail, unfortunately, and in the end one needs to resort to approximations to cope with the averaging.

Although one were able to write an exact solution for the mixing problem where spheres or other simple scatterers randomly occupy a given volume fraction of the medium, another problem is caused by the fact that inclusions are not necessarily separated from one another. Especially in dense mixtures the inclusions touch each other and form larger clusters as shown even in the simple mixture in Figure 8.1. Then in fact the whole approach of splitting the problem into a single-scattering part and the averaging of the dipole moments fails because a cluster is already a complex singly scattering object which should be averaged against other, possibly also quite complicated, inclusions.

8.1.2   Quasi-static approximation

Another limitation of the homogenisation approach that we have been following in the present book is its static character. For someone who desires to have a description of a mixture that would be valid through the whole electromagnetic spectrum, the static approach is surely a handicap. The polarisabilities and local fields were calculated from the properties of the Laplace equation. And as is well known, the Laplace equation neglects the coupling effects of time-variation of the electric and magnetic fields. This means that the wave-propagation properties of the fields are excluded from the treatment.

The approach nonetheless can be valuable for use in connection with time-dependent electromagnetic fields, too. This claim can be paraphrased by saying that the assumptions in the treatment were static but also quasi-static. The results, although derived according to static analysis, can be used with very good accuracy for a certain range of dynamic fields. But then one has to be aware of the limits on how quick the variation of the fields can be before the effective permittivities lose their meaning.

For electromagnetic fields, the temporal and spatial dependence are connected. In a homogeneous medium, fields that are slow in time also have small gradients in space. This means that in small regions, the behaviour of a high-frequency field becomes identical to a static field at a given time instant. Of course, at every point the fields undergo a sinusoidal change, but the time-variation only is a separable function of the total solution. Therefore the question of how dynamically the fields must behave before our homogenisation description breaks down, distils down to the question of how large the spatial "wave" variation of the field is in this homogenised medium.

Of course this question is ill-posed. We cannot know the wave variation of the field in the homogenised medium because we could not even solve the static problem in the mixing case but approximately. However, we can draw conclusions from the corresponding deterministic problem. The case of an electrodynamic field inside a spherical scatterer can be calculated by operating with the Green dyadic on the source [5]. If the size of the scatterer decreases sufficiently, the name "scatterer" loses its meaning. At a certain stage, it only becomes an inclusion with a uniform internal field, as has been discussed several times in the previous sections.

When can we forget the wave character and be satisfied with the static solution? A useful answer can be found from [6] where the singularity of the Green dyadic is studied numerically along with the orders of magnitude of its terms. Interpreting the simulations and illustrations of [6] we can estimate that the quasi-static limit can be formulated approximately as

$$\frac{\lambda}{2\pi} > \delta \tag{8.6}$$

where δ is a measure for the size of the inclusion, and λ is the wavelength of the field.

This is a good rule of thumb. But it is not exact and this approximateness is furthered by the fact that in a mixture, no unique wavelength can be defined. It could be that of a wave propagating in homogeneous environment, inclusion, or the effective medium itself. This is a question one needs to address when approaching a given type of mixture.

8.1.3   Correlation length

As has been stated many times, a rough definition for the quasistatic requirement is that the size of the scatterers is considerably smaller than the wavelength of the operating field. Aside from the problem of how to define the wavelength in a heterogeneous medium this definition is problematic also in another sense: in a polydisperse mixture, scatterers have different sizes, and even for a single, non-spherical inclusion one may wonder what is its effective size. Furthermore, in a random medium, the concept of a scatterer is not unique. If the mixture does not contain discrete homogeneous particles but rather a continuously varing permittivity profile we have to give up the tangible concept of inclusion size. In such a case a better-defined concept is the correlation length.

The correlation length is a measure of the inhomogeneity scale of a mixture or a random medium in general. To calculate the correlation length, one needs to evaluate the autocorrelation function of the spatial permittivity function $\epsilon(\mathbf{r})$ of the mixture. Of course, the correlation length can be different for different directions in space. This happens when we have an anisotropic mixture.

It is perhaps important to note that the correlation length may be much smaller than the apparent inclusion size for the case of an ordinary two-component disrete random mixture. From the structural snowpack study [7], one can see that the correlation length of dry snow may be five to ten times smaller than the visually estimated grain size of the ice particles in this two-component mixture of ice and air.

8.2   Limits for the effective permittivity

If the question about the exact solution for the field behaviour, even in the static limit, is unsolvable, we can say more if the problem is posed in another way. Suppose that we need to find bounds for the effective material properties. Then, we may make bolder claims. Stationary quantities, like, for example, energy integrals can be formulated for the mixture in electric field. Variational principles exist for these stationary quantities with which upper and lower limits for the effective permittivity can be calculated. And crucially, by increasing the effort in evaluating the variational functionals, the limits can be tightened.

8.2.1   General bounds

Consider a two-phase mixture, where permittivities of the phases are ϵ_e and ϵ_i. Searching for absolutely loosest bounds, in other words bounds that cannot be exceeded no matter which volume fractions and geometries the phases take, one might intuitively have the courage to say that the effective permittivity of the mixture has to fall in between the two component permittivities:

$$\min\{\epsilon_e, \epsilon_i\} \leq \epsilon_{\text{eff}} \leq \max\{\epsilon_e, \epsilon_i\} \tag{8.7}$$

Ironically enough, one needs to be careful with the absolute character of these natural bounds. As we saw in Section 4.4, for lossy mixtures the real and imaginary parts of the effective permittivity were not limited by the real and imaginary parts of the component phases (see Figure 4.13). Likewise, the example with the mirror-image materials in Section 6.3.4 showed that even in the case of two phases having the same permittivity, the permittivity of the mixture was different (lower) from that of the phases (see Figure 6.5).

But if we concentrate on purely real- and scalar-valued material parameters, the limit (8.7) should hold. Perhaps the best way to convince oneself of this would be to consider the analogous problem of the conductivity of a two-phase mixture. If we have a good conductor, and mix into it pieces of another material that has smaller conductivity, the result cannot be a better conductor. Rather, the effect is to increase the resistivity. Likewise, if we start with a homogeneous sample of an insulating material, and allow it to contain inclusions of a good conductor, the conductivity should increase; at least it cannot decrease.

But in fact we can write stricter bounds than these. We might think which is the most effective way of increasing the conductivity (or permittivity) of a given material by mixing into it inclusions that have a different material parameter. The answer to this problem is to try to form 'easy paths' to the flux or current to flow. This means that the boundaries between the inclusion and the matrix should be parallel to the flow, as shown on the left side of Figure 8.2. Here, the field is constant throughout the mixture but the flow is different in the host and inclusion phase.

Likewise, how do we minimise the increase of the flow when we add more conducting material to the background? Then the interfaces should block the flow as effectively as possible. The best effective medium in this respect is again a stacked-plate medium but the boundaries are perpendicular to the flow according to the right side of Figure 8.2. In this case the flow (or flux in the dielectric case) density is uniform across the mixture, but the field is different within the phases.

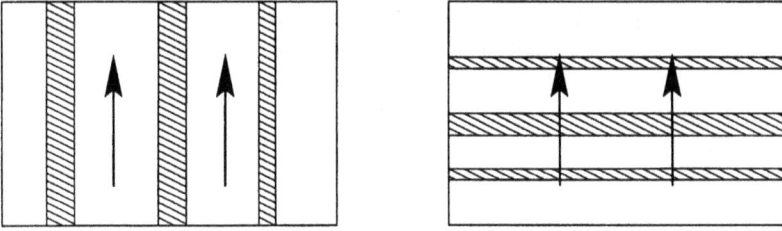

Figure 8.2: *The maximum effective permittivity or conductivity for a given volume fraction of inclusions comes if the inclusions are along the flux according to the left side. Likewise, the minimum effective corresponds to the right-hand side where the flux (flow) is forced to pass through the phase with lower permittivity (conductivity).*

The effective permittivities for these two cases are

$$\epsilon_{\text{eff,max}} = f\epsilon_i + (1 - f)\epsilon_e \tag{8.8}$$

and

$$\epsilon_{\text{eff,min}} = \frac{\epsilon_i \epsilon_e}{f\epsilon_e + (1 - f)\epsilon_i} \tag{8.9}$$

These two cases correspond to capacitors (conductances) that are connected in parallel or series in a circuit. It is worth noting also that these two cases are the effective permittivities from the mixing formulas with aligned ellipsoids (4.25), where the depolarisation factors are 0 and 1, respectively. Note that the bounds retain the minimum and maximum character independently of the type of the mixture, in other words (8.8) is the maximum for both $\epsilon_i > \epsilon_e$ and $\epsilon_i < \epsilon_e$. Also, (8.9) is the minimum for both cases.

These two 'absolute bounds' have been sometimes called *Wiener bounds* [8, 9]. In fact, the Wiener bounds can be continued into complex-valued permittivities, and then the relations (8.8)–(8.9) define a straight line and a circular arc within the complex plane. Within the segment between these limits lie the allowed values of ϵ_{eff} [10].

8.2.2 Hashin–Shtrikman bounds

But bounds stricter to Wiener have been proposed. As is obvious from Figure 8.2, the mixture corresponding to the absolute bounds (8.8)–(8.9) is anisotropic. If we assume that the mixture is macroscopically isotropic, we can indeed refine the bounds.

Although very many authors have discussed the problem of bounds in the literature, quite often the famous work by Hashin and Shtrikman [11] is quoted in connection with the main result concerning the bounds for the effective material parameters of a mixture. The analysis by Hashin and Shtrikman establishes variational theorems which are then used to derive limits for the magnetic permeability of a mixture. Of course, by the duality between the electrostatic and magnetostatic problems, the results are equally valid for the effective permittivity of a mixture.

A functional can be written for the electrostatic energy within the material.[2] This functional is a volume integral of the fields and polarisation densities. The stationary value of the integral can be shown to be the correct energy within the integration volume. This stationary property can be exploited by using various trial field distributions and calculating the value of the functional. Depending on which of the component phases is dielectrically denser, the values of the functional give upper and lower bounds for the energy. And because of the stationary character of the functional, with approximate trial fields a value for the energy can be calculated in which the error is much smaller than in the fields themselves.

This approach leads to the following Hashin–Shtrikman bounds for the effective permittivity of a mixture of two components ϵ_e and ϵ_i, with f being the volume fraction of the phase ϵ_i:[3]

$$\epsilon_{\text{eff},(\text{ext},1)} = \epsilon_e + \frac{f}{\dfrac{1}{\epsilon_i - \epsilon_e} + \dfrac{1-f}{3\epsilon_e}} \tag{8.10}$$

and

$$\epsilon_{\text{eff},(\text{ext},2)} = \epsilon_i + \frac{1-f}{\dfrac{1}{\epsilon_e - \epsilon_i} + \dfrac{f}{3\epsilon_i}} \tag{8.11}$$

Note that these two formulas can be transformed to each other with the complemetary change $\epsilon_i \to \epsilon_e, \epsilon_e \to \epsilon_i, f \to 1 - f$.

Now the question which of the extrema is maximum and which one represents the minimum depends on the relation between the component permittivities. For the case $\epsilon_i > \epsilon_e$, (8.10) is the minimum for the effective permittivity, and (8.11) gives the maximum attainable effective permittivity. For the case $\epsilon_e > \epsilon_i$, the sitation is exactly the opposite.

Writing the Hashin–Shtrikman bounds in the form

$$\epsilon_{\text{eff},(\text{ext},1)} = \epsilon_e + 3f\epsilon_e \frac{\epsilon_i - \epsilon_e}{\epsilon_i + 2\epsilon_e - f(\epsilon_i - \epsilon_e)} \tag{8.12}$$

and

$$\epsilon_{\text{eff},(\text{ext},2)} = \epsilon_i + 3(1-f)\epsilon_i \frac{\epsilon_e - \epsilon_i}{\epsilon_e + 2\epsilon_i - (1-f)(\epsilon_e - \epsilon_i)} \tag{8.13}$$

it can be seen that the bounds are equal to the Maxwell Garnett mixing rules for isotropic spherical inclusions! The limit $\epsilon_{\text{eff},(\text{ext},1)}$ is the ordinary case of spherical inclusions ϵ_i in environment ϵ_e, and $\epsilon_{\text{eff},(\text{ext},2)}$ corresponds to the case spherical inclusions ϵ_e in environment ϵ_i.

---

[2]For variational methods in practical electromagnetics engineering, see [12, 13].

[3]Note that we cannot now talk about 'inclusion' and 'environment,' nor about the shape of the scatterers. We only treat an arbitrary mixture which is statistically homogeneous and isotropic; in fact we talk about all possible such mixtures.

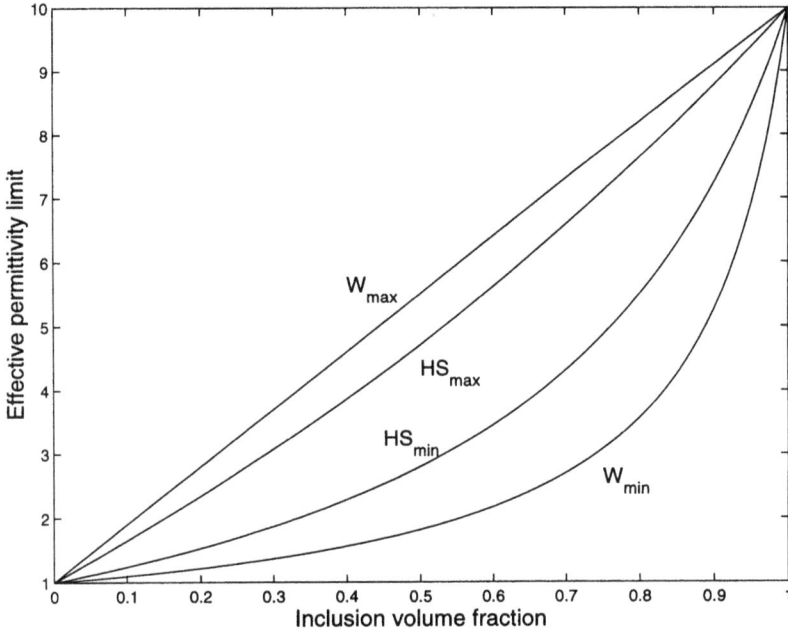

Figure 8.3: *Wiener bounds and Hashin–Shtrikman bounds for the relative effective permittivity of a mixture where the environment has relative permittivity 1 and the inclusion 10. Three-dimensionally isotropic mixture.*

This result gives further confirmation to the property which was noted in earlier chapters that if one wishes to produce the smallest possible polarisability with a given amount of polarisable material, it should be moulded in spherical form. The deviations into ellipsoids will increase the polarisability. From the existence of the complementary bound (8.11) one might now expect that the maximum effect of polarisability comes from geometries where the polarisable material forms vault-like concave shapes.

To compare how strict the Wiener and Hashin–Shtrikman bounds are, Figure 8.3 shows the limits for a mixture where the inclusion–environment contrast is $\epsilon_i/\epsilon_e = 10$. The stricter character of Hashin–Shtrikman bounds is evident.

The analysis leading to the bounds (8.10)–(8.11) was assuming isotropy in the three spatial dimensions. If the isotropy is only required in a plane, in other words in two dimensions, the Hashin–Shtrikman bounds could be expected to read

$$\epsilon_{\text{eff},(\text{ext},1)} = \epsilon_e + \frac{f}{\dfrac{1}{\epsilon_i - \epsilon_e} + \dfrac{1 - f}{2\epsilon_e}} \tag{8.14}$$

and

$$\epsilon_{\text{eff},(\text{ext},2)} = \epsilon_i + \cfrac{1 - f}{\cfrac{1}{\epsilon_e - \epsilon_i} + \cfrac{f}{2\epsilon_i}} \tag{8.15}$$

8.2.3   Higher-order bounds

The theoretical foundations for bounds for the effective parameters of heterogeneous materials have been studied extensively. Classifications for various limits can be done according to the order of the bounds, according to their realisability, optimality, or attainability [14]. Let us not go into details of these various meanings for the characterisation of the limits, but just give an example how the Hashin–Shtrikman bounds are sharpened.[4]

Although many of the studies of higher-order bounds are written for the effective conductivity of the mixture, let us here modify the formulas for the effective permittivity ϵ_{eff}. Assume again a mixture of two components: ϵ_e and ϵ_i, with f_e and f_i being the corresponding volume fractions.[5] Assuming $\epsilon_i > \epsilon_e$, the following are the accepted third-order bounds for the effective permittivity for three-dimensional isotropic mixtures:

$$\epsilon_{\text{eff},\min} = \epsilon_e \frac{(\epsilon_e + 2\epsilon_i)(\epsilon_i + 2f_e\epsilon_e + 2f_i\epsilon_i) - 2f_e\zeta(\epsilon_i - \epsilon_e)^2}{(\epsilon_e + 2\epsilon_i)(2\epsilon_e + f_e\epsilon_i + f_i\epsilon_e) - 2f_e\zeta(\epsilon_i - \epsilon_e)^2} \tag{8.16}$$

and

$$\epsilon_{\text{eff},\max} = f_e\epsilon_e + f_i\epsilon_i - \frac{f_e f_i(\epsilon_i - \epsilon_e)^2}{3\epsilon_e + (f_e + 2\zeta)(\epsilon_i - \epsilon_e)} \tag{8.17}$$

where for symmetric inclusion cells we have

$$\zeta = f_i + \frac{(f_e - f_i)(9G - 1)}{2} \tag{8.18}$$

and G is the Miller parameter depending on the cell shape. It is limited within the interval $[1/9, 1/3]$, and for cubical cells, Helsing has calculated it to be around $G \approx 0.13751643$.

8.2.4   Anisotropic bounds

For the case of polycrystals which are mixtures of anisotropic inclusions, bounds can also be derived for the macroscopic permittivity. Assume the eigenvalues of the (symmetric) permittivity dyadic of a single crystal to be ϵ_x, ϵ_y, and ϵ_z. If this anisotropic material forms a quasi-isotropic polycrystal, its effective permittivity is

---

[4]See, for example, the works by Beran, Miller, Bergman, Milton, and Avellaneda ([15–19], and references therein).

[5]Of course, $f_e + f_i = 1$.

a scalar. This effective permittivity is limited by the classical so-called Voigt and Reuss bounds:

$$3\left(\frac{1}{\epsilon_x} + \frac{1}{\epsilon_y} + \frac{1}{\epsilon_z}\right)^{-1} \leq \epsilon_{\text{eff}} \leq \frac{1}{3}(\epsilon_x + \epsilon_y + \epsilon_z) \tag{8.19}$$

These bounds have found tighter forms [20, 21]. Choose the axes of the crystal in such a way that $\epsilon_x \leq \epsilon_y \leq \epsilon_z$. Then the lower bound is estimated to be

$$\epsilon_{\text{eff,min}} = \epsilon_x \frac{4\epsilon_x^2 + 8\epsilon_x\epsilon_y + 7\epsilon_y\epsilon_z + 8\epsilon_z\epsilon_x}{16\epsilon_x^2 + 5\epsilon_x\epsilon_y + \epsilon_y\epsilon_z + 5\epsilon_z\epsilon_x} \tag{8.20}$$

and the upper bound

$$\epsilon_{\text{eff,max}} = \epsilon_z \frac{4\epsilon_z^2 + 8\epsilon_z\epsilon_x + 7\epsilon_x\epsilon_y + 8\epsilon_y\epsilon_z}{16\epsilon_z^2 + 5\epsilon_z\epsilon_x + \epsilon_x\epsilon_y + 5\epsilon_y\epsilon_z} \tag{8.21}$$

Note that the polycrystal is macroscopically isotropic.

Problems

8.1 Plot the various bounds for the effective permittivity of a mixture as a function of the volume fraction. Use Equations (8.7), (8.8)–(8.9), (8.10)–(8.11), (8.14)–(8.15), and (8.16)–(8.17). Treat the following mixtures:

(a) $\epsilon_i = 2\epsilon_e$

(b) $2\epsilon_i = \epsilon_e$ (not the bounds (8.16)–(8.17))

(c) $\epsilon_i = 20\epsilon_e$

(d) $20\epsilon_i = \epsilon_e$ (not the bounds (8.16)–(8.17))

8.2 Study how the bounds can be illustrated in the complex plane. Assume a two-component mixture where both the phases have complex permittivities. Then the limits for the complex permittivity of a mixture define a region in the complex plane with axes ϵ'_{eff} and ϵ''_{eff}. Plot the allowed regions for the mixture permittivity, if the environment has permittivity $\epsilon_e/\epsilon_0 = 2 - \text{j}1$ and the inclusions $\epsilon_i/\epsilon_0 = 5 - \text{j}4$, determined by the

(a) Wiener bounds

(b) Hashin–Shtrikman bounds.

8.3 Given the upper and lower limits for the effective permittivity for various volume fractions f, $\epsilon_{\text{eff,max}}(f)$ and $\epsilon_{\text{eff,min}}(f)$, a good measure for the looseness or sharpness of these limits is the area between these two curves. Let us define the weakness of a pair of bounds by

$$W = \int_0^1 [\epsilon_{\text{eff,max}}(f) - \epsilon_{\text{eff,min}}(f)] \, df \tag{8.22}$$

Calculate the weaknesses for all the various absolute and Hashin–Shtrikman bounds. If analytical results cannot be achieved, calculate the numerical value for the weakness parameter for the dielectric constrasts given in the previous problem.

8.4 Compare the various bounds in the dilute-mixture limit. Let us assume that the mixture mainly consists of material with permittivity ϵ_e, and there is only a trace fraction f of the other phase, ϵ_i. Then constant and linear terms in f of the Taylor series of the effective permittivity expansions

$$\epsilon_{\text{eff}} \approx \epsilon_e + \frac{\partial \epsilon_{\text{eff}}(f)}{\partial f}\bigg|_{f=0} \cdot f \qquad (8.23)$$

give a reasonable approximation for the effective permittivity, because $f \ll 1$. Calculate the different limits (Equations (8.8)–(8.9), (8.10)–(8.11), and (8.16)–(8.17)) to this approximation. Assume $\epsilon_i > \epsilon_e$.

Using the results, take also the strong-contrast limit. In other words, in your perturbation expansions, let $\epsilon_i \gg \epsilon_e$, and compare the bounds.

References

[1] TSANG, L., KONG, J.A., and SHIN, R.T.: 'Theory of microwave remote sensing' (Wiley, New York, 1985)

[2] ISHIMARU, A.: 'Wave propagation and scattering in random media' (Academic Press, New York, 1982)

[3] KOHLER, W.E., and PAPANICOLAOU, G.C.: 'Some applications of the coherent potential approximation', *in* CHOW,P.L., KOHLER, W.E., and PAPANICOLAOU, G.C. (Eds.): 'Multiple scattering and waves' (North Holland, New York, 1981), pp. 199-223

[4] HALZEN, F., and MARTIN, A.D.: 'Quarks and leptons: and introductory course in modern particle physics' (Wiley, New York, 1984)

[5] YAGHJIAN, A.D.: 'Electric dyadic Green's function in the source region', *Proceedings of the IEEE*, 1980, **68**, (2), pp. 248-263

[6] YAGHJIAN, A.D.: 'Reply', *Proceedings of the IEEE*, 1981, **69**, (2), pp. 283-285

[7] WIESMANN, A., MÄTZLER, C., and WEISE, T.: 'Radiometric and structural measurements of snow samples', *Radio Science*, March–April 1998, **33**, (2), pp. 273-289

[8] BRUGGEMAN, D.A.G.: 'Berechnung verschiedener physikalischer Konstanten von heterogenen Substanzen, I. Dielektrizitätskonstanten und Leitfähigkeiten der Mischkörper aus isotropen Substanzen', *Annalen der Physik*, 1935, Series 5, **24**, pp. 636-664

[9] GRIMVALL, G: 'Thermophysical properties of materials' (North-Holland, Amsterdam, 1986)

[10] ASPNES, D.E.: 'Local-field effects and effective-medium theory: a microscopic perspective', *American Journal of Physics*, 1982, **50**, (8), pp. 704-709

[11] HASHIN, Z., and SHTRIKMAN, S.: 'A variational approach to the theory of the effective magnetic permeability of multiphase materials', *Journal of Applied Physics*, 1962, **33**, (10), pp. 3125-3131

[12] HARRINGTON, R.: 'Time-harmonic electromagnetic fields' (Mc-Graw–Hill, New York, 1961)

[13] COLLIN, R.E.: 'Field theory of guided waves' (IEEE Press, New York, 1991)

[14] HELSING, J.: 'Models for conduction and estimates of conductivity in composite materials', (1991), Doctoral Dissertation, The Royal Institute of Technology, Dept. Theoretical Physics, Sweden

[15] BERAN, M.J.: 'Statistical continuum theories' (Interscience, New York, 1968)

[16] MILLER, M.N.: 'Bounds for effective electrical, thermal, and magnetic properties of heterogeneous materials', *Journal of Mathematical Physics*, 1969, **10**, (11), pp. 1988-2004; 'Bounds for effective bulk modulus of heterogeneous materials', pp. 2005-2013

[17] BERGMAN, D.J.: 'The dielectric constant of a composite material — a problem in classical physics', *Phys. Rep. C*, 1978, **43**, pp. 377-407

[18] MILTON, G.W.: 'Bounds on the complex permittivity of a 2-component composite material', *Journal of Applied Physics*, 1981, **52**, (8), pp. 5286-5293.

[19] HELSING, J.: 'Bounds to the conductivity of some two-component composites', *Journal of Applied Physics*, 1993, **73**, (3), pp. 1240-1245

[20] HASHIN, Z., and SHTRIKMAN, S.: 'Conductivity of polycrystals', *Physical Review*, 1963, **130**, pp. 129-133

[21] HELSING, J.: 'Improved bounds on the conductivity of composites by translation in a variational principle', *Journal of Applied Physics*, 1993, **74**, (8), pp. 5061-5063

Chapter 9

Generalised mixing rules

The reader may have grown exhausted by this time by the endless emphasis on the Maxwell Garnett-type mixtures. Indeed, the discussion has heretofore concentrated heavily on a single approach to the homogenisation. One of the components is treated as environment, and the inclusion phase is considered as a perturbation against this background. And in particular, the local field that excites a single scatterer in the mixture was calculated by replacing all neighbours by a uniform polarisation density in which a hole was carved. This treatment is clearly approximate, as was pointed out in Chapter 8 where the limitations of Maxwell Garnett philosophy were discussed.

But on the other hand, no exact solution exists for the electrostatic problem in a random heterogeneous geometry. Multiple opinions can flourish when nobody knows for certain. It is evident that the Maxwell Garnett model cannot remain at the scene as the only ruling formula. Many rival mixing rules are being used in the modelling of heterogeneous materials. In the present chapter, some of the most common competitor models are presented.

9.1 Bruggeman formula

Starting the list with a successful homogenisation formula, the *Bruggeman formula* is an important mixing rule that is widely used in electromagnetics literature [1]. It is also known by other names: in the remote sensing community it goes under the names *Polder–van Santen formula* [2] and *de Loor formula* [3]. Also the name *Böttcher formula* [4] can be found in textbooks. In other fields of materials science, the corresponding formula often carries the name *effective-medium model* [5].

The essence of the Bruggeman formalism is the absolute equality between the phases in the mixture. There is no more host-versus-guest hierarchy. Instead, the homogenised medium itself is considered as the background against which polarisations are measured. And now polarisation with respect to the effective medium emerges as well from the environment as from the inclusions.

If N isotropic phases, with permittivities ϵ_j, form a mixture, each occupying a volume fraction f_j with $\sum_{j=1}^{N} f_j = 1$, the consistency requirement for the effective permittivity ϵ_{eff} according to the Bruggeman philosophy is

$$\sum_{j=1}^{N} f_j \frac{\epsilon_j - \epsilon_{\text{eff}}}{\epsilon_j + 2\epsilon_{\text{eff}}} = 0 \qquad (9.1)$$

where spherical geometry is assumed, as is obvious from the coefficients in the quotient.

Specialising the mixing rule for an ordinary two-phase mixture, the Bruggeman formula reads

$$(1 - f)\frac{\epsilon_e - \epsilon_{\text{eff}}}{\epsilon_e + 2\epsilon_{\text{eff}}} + f\frac{\epsilon_i - \epsilon_{\text{eff}}}{\epsilon_i + 2\epsilon_{\text{eff}}} = 0 \qquad (9.2)$$

where now (according to a Maxwell-Garnett-minded picture) spherical inclusions (ϵ_i), with volume fraction f, are located in homogeneous environment (ϵ_e). Although this form is not explicit for the effective permittivity ϵ_{eff}, the Bruggeman formula has the appeal in the very property that it treats the inclusions and the environment symmetrically. There is no difference between the two phases. A mixture and its complement (which emerge through the transformation $\epsilon_i \rightarrow \epsilon_e$, $\epsilon_e \rightarrow \epsilon_i$, $f \rightarrow 1-f$) have exactly the same effective permittivity.[1]

The interpretation of Equation (9.2) is that the formula balances both mixing components with respect to the unknown effective medium, using the volume fraction of each component as weight (f for the inclusions and $1 - f$ for the environment). This symmetry property of (9.2) makes the radical distinction between the Maxwell Garnett rule and the Bruggeman formula. The Maxwell Garnett approach is inherently nonsymmetric.

The Bruggeman formula for the case when the inclusions are randomly oriented ellipsoids is

$$\epsilon_{\text{eff}} = \epsilon_e + \frac{f}{3}(\epsilon_i - \epsilon_e) \sum_{j=x,y,z} \frac{\epsilon_{\text{eff}}}{\epsilon_{\text{eff}} + N_j(\epsilon_i - \epsilon_{\text{eff}})} \qquad (9.3)$$

where N_j are the depolarisation factors of the inclusion ellipsoids in the three orthogonal directions. These can be calculated using the methods of Section 4.2.1.[2]

---

[1]In fact, why many authors label this single formula after Bruggeman can be questioned. Bruggeman in the extensive article [1] presented very many different mixing rules, including the present one. Some investigators indeed call the formula (9.2) a 'symmetric Bruggeman' formula, to distinguish it from other Bruggeman formulas that do not share this property.

[2]Perhaps it would be more correct, historically, to call the formula with ellipsoidal inclusions after Polder and van Santen rather than Bruggeman since [1]—although full of suggestions for various mixing principles—did not contain the case (9.3).

In snow studies, and dielectric modelling of geophysical media at large, the Bruggeman/Polder–van Santen formula has enjoyed popularity and been successful. In Chapter 13, we shall return to the application of this formula on real-life materials.

9.2   Coherent potential formula

Another well-known formula which is relevant in the theoretical studies of wave propagation in random media is the so-called *Coherent potential formula* [6, p. 475], [7, 8]. This formula for spherical inclusions can be presented in several forms, of which one of the beautiful ones looks like

$$\epsilon_{\text{eff}} = \epsilon_e + f(\epsilon_i - \epsilon_e)\frac{3\epsilon_{\text{eff}}}{3\epsilon_{\text{eff}} + (1 - f)(\epsilon_i - \epsilon_e)} \tag{9.4}$$

which is again an implicit formula for the effective permittivity ϵ_{eff}. The formula has been generalised for mixtures with ellipsoidal inclusions [9]. If the ellipsoids are randomly oriented, the Coherent potential rule reads

$$\epsilon_{\text{eff}} = \epsilon_e + \frac{f}{3}(\epsilon_i - \epsilon_e) \sum_{j=x,y,z} \frac{(1 + N_j)\epsilon_{\text{eff}} - N_j\epsilon_e}{\epsilon_{\text{eff}} + N_j(\epsilon_i - \epsilon_e)} \tag{9.5}$$

The philosophy behind the approaches that lead to Coherent potential mixing formulas is that one should not treat a single scatterer floating in isolation in the environment when the dipole moment and the local field are calculated. Instead, the Green function which is used to enumerate the field of a given polarisation density is taken to be that of the effective medium, not that of the background. One might attach the 'coherence' of the mixing rule to this approach. The reason for the word 'potential' in the mixing rule name is that in solid state physics[3] and scattering studies of random media, the problem is formulated with the inhomogeneous potential which is a function proportional to the difference in permittivity at a given point in space from the permittivity of a reference medium.

From a comparison of the three mixing principles it is worth noting that for dilute mixtures ($f \ll 1$), all formulas, Maxwell Garnett, Bruggeman, and Coherent potential, predict the same results. Up to the first order in f, the formulas are the same, $\epsilon_{\text{eff}} \approx \epsilon_e + n\alpha$, which is for spherical inclusions

$$\epsilon_{\text{eff}} \approx \epsilon_e + 3f\epsilon_e\frac{\epsilon_i - \epsilon_e}{\epsilon_i + 2\epsilon_e} \tag{9.6}$$

---

[3]In solid state physics studies, the Maxwell Garnett rule goes under the name ATA (average T-matrix approximation), the Bruggeman rule has the label EMA (effective-medium approximation), and Coherent potential may be found under the name GKM (after Gyorffy, Korringa, and Mills). Note, however, that the EMA is sometimes also termed the Coherent potential principle [10].

9.3 Unified mixing rule

The chaotic absence of order that is allowed in the structure of arbitrarily random media is a natural reason for the variety of mixing rules that have survived in the evolution of scientific materials modelling and analyses. But on the other hand, it is fair to wish to see the different homogenising principles arising from a common ground. Indeed, there exists hope for unification, at least to a certain degree. This is discussed next.

9.3.1 Spherical inclusions

The mixing approach presented in [11] collects all the previous aspects of dielectric mixing rules into one family. For the case of isotropic spherical inclusions ϵ_i in the isotropic environment ϵ_e, the formula looks like

$$\frac{\epsilon_{\text{eff}} - \epsilon_e}{\epsilon_{\text{eff}} + 2\epsilon_e + \nu(\epsilon_{\text{eff}} - \epsilon_e)} = f \frac{\epsilon_i - \epsilon_e}{\epsilon_i + 2\epsilon_e + \nu(\epsilon_{\text{eff}} - \epsilon_e)} \tag{9.7}$$

This formula contains a dimensionless parameter ν. For different choices of ν, the previous mixing rules are recovered: $\nu = 0$ gives the Maxwell Garnett rule, $\nu = 2$ gives the Polder–van Santen formula, and $\nu = 3$ gives the Coherent potential approximation.

Figure 9.1 shows the predictions of different models. It can be seen that for low inclusion–background contrast, the predictions are similar. Likewise, for dilute mixtures ($f \ll 1$), the value of ν does not matter much.

That for dilute mixtures the type of the mixing rule is immaterial, can be seen from the perturbation expansion:

$$\begin{aligned}
\epsilon_{\text{eff}} &= \epsilon_e + 3\epsilon_e \frac{\epsilon_i - \epsilon_e}{\epsilon_i + 2\epsilon_e} f + 3\epsilon_e \left(\frac{\epsilon_i - \epsilon_e}{\epsilon_i + 2\epsilon_e} \right)^2 \left(1 + \nu \frac{\epsilon_i - \epsilon_e}{\epsilon_i + 2\epsilon_e} \right) f^2 \\
&\quad + 3\epsilon_e \left(\frac{\epsilon_i - \epsilon_e}{\epsilon_i + 2\epsilon_e} \right)^2 \left(1 + \nu \frac{\epsilon_i - \epsilon_e}{\epsilon_i + 2\epsilon_e} \right) \left(1 + \nu \frac{\epsilon_i - 4\epsilon_e}{\epsilon_i + 2\epsilon_e} \right) f^3 + \cdots
\end{aligned} \tag{9.8}$$

where the linear term in f is independent of the parameter ν.

The unified mixing formula (9.7), assumed a spherical geometry in three dimensions. To modify it to, for example, two-dimensional spheres (the transverse direction of aligned circular cylinders), in the denominators the term $2\epsilon_e$ has to be replaced by ϵ_e.

9.3.2 Ellipsoidal inclusions

The unification of the mixing rules can also be done for mixtures with ellipsoidal inclusions. Consider a mixture where the ellipsoids (permittivity ϵ_i) are randomly oriented, with the three depolarisation factors N_j. Then the Maxwell Garnett

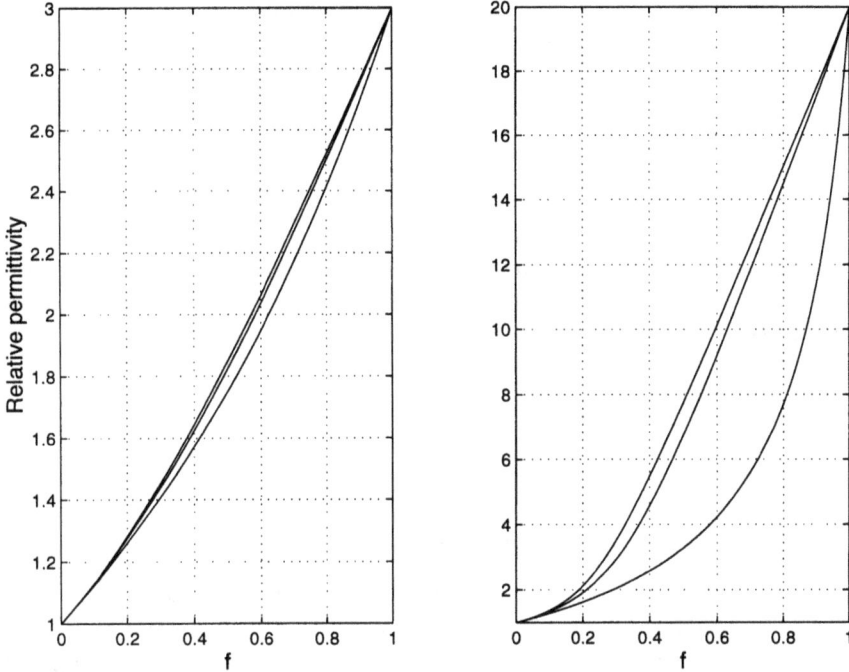

Figure 9.1: *The relative permittivity of a mixture (spherical inclusions) for two dielectric contrasts: $\epsilon_i/\epsilon_e = 3$ (left curves) and 20 (right curves). The lowest curve is Maxwell Garnett prediction ($\nu = 0$), the next is the Bruggeman rule ($\nu = 2$), and the highest is the Coherent potential formula ($\nu = 3$).*

prediction follows the formula (4.28):

$$\epsilon_{\text{eff}} = \epsilon_e + \epsilon_e \frac{\dfrac{f}{3} \sum\limits_{j=x,y,z} \dfrac{\epsilon_i - \epsilon_e}{\epsilon_e + N_j(\epsilon_i - \epsilon_e)}}{1 - \dfrac{f}{3} \sum\limits_{j=x,y,z} \dfrac{N_j(\epsilon_i - \epsilon_e)}{\epsilon_e + N_j(\epsilon_i - \epsilon_e)}} \tag{9.9}$$

A modification was suggested in [9] to the Maxwell Garnett derivation. An apparent permittivity ϵ_a was introduced, which is defined as the permittivity which an inclusion "feels" in its surroundings in the mixture. This means that the depolarisation fields are $-\overline{\overline{L}} \cdot \mathbf{P}/\epsilon_a$ instead of the Maxwell Garnett-like $-\overline{\overline{L}} \cdot \mathbf{P}/\epsilon_e$. Then the apparent permittivity ϵ_a creeps into the mixing formula in the following manner:

$$\epsilon_{\text{eff}} = \epsilon_e + \epsilon_a \frac{\dfrac{f}{3} \sum\limits_{j=x,y,z} \dfrac{\epsilon_i - \epsilon_e}{\epsilon_a + N_j(\epsilon_i - \epsilon_e)}}{1 - \dfrac{f}{3} \sum\limits_{j=x,y,z} \dfrac{N_j(\epsilon_i - \epsilon_e)}{\epsilon_a + N_j(\epsilon_i - \epsilon_e)}} \tag{9.10}$$

Of course, if the apparent permittivity is that of the environment ($\epsilon_a = \epsilon_e$), the Maxwell Garnett rule (9.9) is recovered. But the choice $\epsilon_a = \epsilon_{\text{eff}}$ gives the Coherent potential formula, and the Bruggeman symmetric formula can be written from (9.10) by setting $\epsilon_a = \epsilon_{\text{eff}} - N_i(\epsilon_{\text{eff}} - \epsilon_e)$.

9.4   Other mixing models

Of the very large set of the remaining mixing rules that are being used in the random medium theories and practical applications, the following ones deserve to be introduced.

9.4.1   Power-law models

A widely used class of mixing models is formed by the "power-law" approximations:

$$\epsilon_{\text{eff}}^{\beta} = f\epsilon_i^{\beta} + (1-f)\epsilon_e^{\beta} \tag{9.11}$$

This is a simple principle: a certain power of the permittivity is averaged by volume weights. For example, in the Birchak formula[4] [13] the parameter is $\beta = 1/2$, which means that the square roots of the component permittivities add up to the square root of the mixture permittivity. This model also means that the refractive index of a mixture of nonmagnetic gases is the volume average of the indices of the components, and it is often, among microwave remote sensing investigators, referred to as the "refractive mixing model." This averaging law has existed for a long time in the folklore of optical physics [14]. And for the case of liquid properties, the careful measurements by Gladstone and Dale [15] gave confirmation to this law a long time ago.

Another famous formula is the Looyenga formula [16] for which $\beta = 1/3$, although this formula can be found elsewhere, too. For example, it is given as the "permittivity of a mixture" in Section 9 of the authoritative text by Landau and Lifshitz [17].

One can also find in the literature (see for example [18, p. 1080]) the linear law, *Silberstein* formula [12],

$$\epsilon_{\text{eff}} = f\epsilon_i + (1-f)\epsilon_e \tag{9.12}$$

which corresponds to $\beta = 1$ in (9.11). This mixing rule can be given theoretical confirmation if the mixture is formed of plates or other inclusions for which no depolarisation is induced. If the depolarisation factor is $N_x = 0$, one can obtain formula (9.12) from (4.25).[5]

---

[4] Also called the Beer formula [12].
[5] This mixing rule is claimed (Lichtenecker [19, p. 117]) to have a history going back to Isaac Newton!

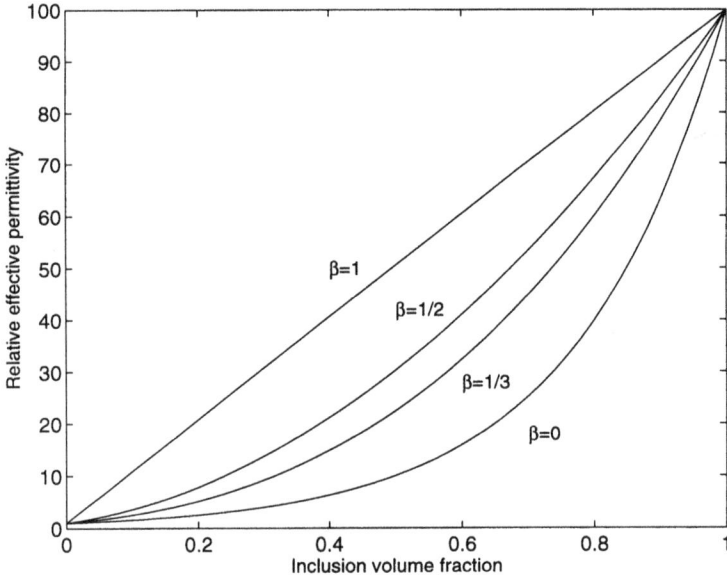

Figure 9.2: *The relative effective permittivity of a high-contrast mixture $\epsilon_i/\epsilon_e = 100$ according to power-law models. The highest curve corresponds to $\beta = 1$, then $1/2$, $1/3$, and the lowest one is the Lichtenecker model ($\beta = 0$).*

The *Lichtenecker* formula, which averages the logarithms of the permittivities, could also be seen as a special case of the power-law mixtures:

$$\epsilon_{\text{eff}} = \epsilon_i^f \epsilon_e^{1-f} \tag{9.13}$$

The prediction for various exponential rules is illustrated by Figure 9.2.

9.4.2 Differential mixing models

Because the geometrical description of a mixture deals with its internal microstructure, quite often in the literature mixing rules are based on differential analysis. The resulting mixing rules can often be recognised by the fact that powers of one-third appear in the formulas. One example of such a rule is the Looyenga formula

$$\epsilon_{\text{eff}}^{1/3} = f\epsilon_i^{1/3} + (1 - f)\epsilon_e^{1/3} \tag{9.14}$$

which can be seen to be "Bruggeman-like" because of the symmetry with respect to the inclusion and environment. Other mixing rules resulting from differential analysis are the "Bruggeman nonsymmetric" formula [1, 20]

$$\frac{\epsilon_i - \epsilon_{\text{eff}}}{\epsilon_i - \epsilon_e} = (1 - f)\left(\frac{\epsilon_{\text{eff}}}{\epsilon_e}\right)^{1/3} \tag{9.15}$$

and its "complement," by Sen, Scala, and Cohen [21]:[6]

$$\frac{\epsilon_{\text{eff}} - \epsilon_e}{\epsilon_i - \epsilon_e} = f\left(\frac{\epsilon_{\text{eff}}}{\epsilon_i}\right)^{1/3} \tag{9.16}$$

which contain the unknown effective permittivity implicitly, requiring a third-order equation to be solved to enumerate ϵ_{eff}. For raisin-pudding mixture (meaning $\epsilon_i > \epsilon_e$), the Looyenga formula predicts the smallest effective permittivity, Sen-formula the largest, and the Bruggeman nonsymmetric formula falls in between. This fact is perhaps more clearly visible from the Taylor expansions around $f = 0$ which are useful for dilute mixtures. For Looyenga:

$$\epsilon_{\text{eff}} = \epsilon_e + f3\epsilon_e^{2/3}(\epsilon_i^{1/3} - \epsilon_e^{1/3}) + f^2 3\epsilon_e^{1/3}(\epsilon_i^{1/3} - \epsilon_e^{1/3})^2 + \cdots \tag{9.17}$$

Bruggeman nonsymmetric:

$$\epsilon_{\text{eff}} = \epsilon_e + f3\epsilon_e\frac{\epsilon_i - \epsilon_e}{\epsilon_i + 2\epsilon_e} + f^2 3\epsilon_e\left(\frac{\epsilon_i - \epsilon_e}{\epsilon_i + 2\epsilon_e}\right)^2\frac{2\epsilon_i + \epsilon_e}{\epsilon_i + 2\epsilon_e} + \cdots \tag{9.18}$$

and Sen, Scala, and Cohen:

$$\epsilon_{\text{eff}} = \epsilon_e + f(\epsilon_i - \epsilon_e)\left(\frac{\epsilon_e}{\epsilon_i}\right)^{1/3} + f^2\frac{(\epsilon_i - \epsilon_e)^2}{3\epsilon_e^{1/3}\epsilon_i^{2/3}} + \cdots \tag{9.19}$$

Note that the Bruggeman nonsymmetric formula carries the same linear term as the mixing rules from the unified formula (9.7). A comparison of (9.8) and (9.18) shows that up to the second order the Bruggeman nonsymmetric agrees with the unified rule for the choice $\nu = 1$.

9.4.3   Periodical lattice models

A great amount of literature also exists for effective properties of mixtures where spherical inclusions are arranged in a cubic array in a background matrix. The first such analysis was given in the classical paper by Rayleigh [22]. The Rayleigh result (3.26) for the spherical two-phase mixture reads in the accurate form

$$\epsilon_{\text{eff}} = \epsilon_e + \frac{3f\epsilon_e}{\dfrac{\epsilon_i + 2\epsilon_e}{\epsilon_i - \epsilon_e} - f - 1.305\dfrac{\epsilon_i - \epsilon_e}{\epsilon_i + 4\epsilon_e/3}f^{10/3}} \tag{9.20}$$

Successive improvements for this Rayleigh result have been presented by Runge [23], Meredith and Tobias [24], McPhedran, McKenzie, and Derrick [25, 26], Doyle [27], and by Lam [28]. However, these formulas are derived for ordered mixtures, though not all necessarily for cubic-centred lattices, and from the point of view of application

---

[6]The complementary mixture means the mixture where the phases and their volume fractions are interchanged: $\epsilon_e \to \epsilon_i, \epsilon_i \to \epsilon_e, f \to 1 - f$.

to random media, they suffer from the disadvantage of predicting infinite effective permittivities as the inclusions come into contact with each other.

The deviations of the effective permittivity of a mixture from the plain Maxwell Garnett model can be estimated also by including the interactions between the inclusion spheres in the mixture. Quadrupoles and even higher-order multipoles are included in the studies by Felderhof (see, for example, [29]).

Many comparisons and tests of the predictions of various mixing rules against measured results have been made. The results are not conclusive in the sense that a certain mixing rule would be universally closest to mixtures of the real world. This is natural because the nature and character of a mixture changes drastically with the volume fraction and dielectric contrast. The reader interested in the comparison of mixing model results for microwave permittivities may find the references [30] and [31] interesting.

9.4.4   Random medium model

If the mixture is not formed of discrete phases but rather its permittivity is a continuous dielectric function of the spatial co-ordinates $\epsilon = \epsilon(\mathbf{r})$, the previous models are difficult to apply. A perturbational model for the effective permittivity for this case can be written if the statistical variation of the dielectric function is known [17]. If the function is separated into the volume-average part $< \epsilon(\mathbf{r}) > = \epsilon_{\mathrm{ave}}$ and a zero-mean varying part $\Delta\epsilon(\mathbf{r})$:

$$\epsilon(\mathbf{r}) = \epsilon_{\mathrm{ave}} + \Delta\epsilon \tag{9.21}$$

the effective permittivity can be written as a function of the average of the squared variation $< (\Delta\epsilon)^2 >$:

$$\epsilon_{\mathrm{eff}} = \epsilon_{\mathrm{ave}} - \frac{< (\Delta\epsilon)^2 >}{3\epsilon_{\mathrm{ave}}} \tag{9.22}$$

This result can be written explicitly for a discrete two-phase mixture for which

$$\epsilon_{\mathrm{eff}} = \epsilon_{\mathrm{ave}} - f(\epsilon_i - \epsilon_e)\frac{\epsilon_i - \epsilon_{\mathrm{ave}}}{3\epsilon_{\mathrm{ave}}} \tag{9.23}$$

where $\epsilon_{\mathrm{ave}} = \epsilon_e + f(\epsilon_i - \epsilon_e)$ and $< (\Delta\epsilon)^2 > = f(1-f)(\epsilon_i - \epsilon_e)^2$. The result is only valid for weak fluctuations, and predicts clearly unphysical results if the contrast ϵ_i/ϵ_e is much larger than the environment permittivity.

9.5   Chiral and bi-anisotropic mixtures

In terms of more complex materials than plain dielectrics, the various mixing rules can be certainly written to describe the effective parameters in a generalised manner. Here the dyadic and six-vector algebra are of valuable assistance. Consider, for example, the case of bi-isotropic mixtures. Let bi-isotropic spheres with material six-dyadic M_i be located within another bi-isotropic material M_e.

Regarding the exponential models which are straightforward weighted averages of the components, the mixing rule reads naturally

$$\mathsf{M}_{\text{eff}}^{\beta} = f\mathsf{M}_e^{\beta} + (1-f)\mathsf{M}_e^{\beta} \tag{9.24}$$

where β is the exponent into which power the six-matrices have to be raised.

Also mixing formulas with a more analytical background can be generalised. This is perhaps easiest to represent using the unified mixing rule (9.7). Its obvious generalisation reads, using the material six-dyadic of the effective medium M_{eff}:

$$(\mathsf{M}_{\text{eff}}-\mathsf{M}_e)\cdot[\mathsf{M}_{\text{eff}}+2\mathsf{M}_e+\nu(\mathsf{M}_i-\mathsf{M}_e)]^{-1} = f(\mathsf{M}_i-\mathsf{M}_e)\cdot[\mathsf{M}_i+2\mathsf{M}_e+\nu(\mathsf{M}_i-\mathsf{M}_e)]^{-1} \tag{9.25}$$

Here, again, the various special cases emerge: Maxwell Garnett ($\nu = 0$), Bruggeman symmetric ($\nu = 2$), and Coherent potential ($\nu = 3$).

One problem for the numerical evaluation of the effective parameters is the nonlinearity of the six-dyadic equation (9.25). The solution for M_{eff} can be found for example iteratively [32], [33], or by a diagonalisation procedure [34].

The example of a mixture with mirror-image materials that was discussed in Section 6.3.4 can be used as an application of the formula (9.25). There chiral spheres were embedded into another chiral background, and the two components only differed from each other by their handedness, in other words the sign of their chirality parameter. The material parameters of the components, relative to a vacuum are $\epsilon_r = 2$, $\mu_r = 1.5$, and $\kappa = 1$ for the inclusions, and $\epsilon_r = 2$, $\mu_r = 1.5$, and $\kappa = -1$ for the background medium. Use of Equation (9.25) gives easily the results in Figure 9.3.

The results show that the macroscopic chirality parameter is different depending on the used mixing model. The symmetry inherent in the Bruggeman model can be seen in the fact that the mixture becomes racemic ($\kappa_{\text{eff}} = 0$) exactly for the mixing ratio $f = 0.5$. Despite this, the Bruggeman curve is not a straight line. Maxwell Garnett favours the chirality of the environment, staying below the Bruggeman prediction, and Coherent potential gives more weight for the properties of the inclusions in the effective chirality.

9.6   Numerical approaches for homogenisation

Because the effective modelling problem of any sample of a random medium has no exact, analytical solution, an appealing approach is to solve the problem numerically. Typically, in the numerical approaches the sample is sliced into small cells and the fields are solved in a finite number of points, instead of looking for continuous functions. The limitation of this way of solving the problem is that the size of the volume to be discretised cannot exceed the capacity of the computer.

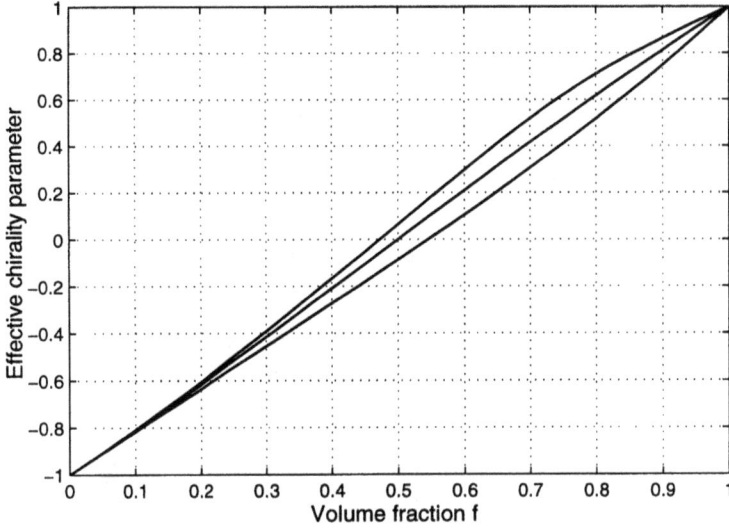

Figure 9.3: *The macroscopic chirality parameter κ_{eff} of a chiral-in-chiral mirror-image mixture, according to different mixing models. The three curves are, starting from the lowest one, Maxwell Garnett, Bruggeman, and Coherent potential.*

In this section, one example is discussed in which a numerical method shows its power in predicting the effective permittivity of a mixture. The approach is based on the Finite Difference Time Domain (FDTD) principle for solving the fields in a finite region. Currently FDTD enjoys popularity among the electromagnetics community and is a much used method to solve field problems [35].

FDTD is a dynamical method to solve the full set of Maxwell equations, in other words, the time dependence of the fields is respected and taken correctly into account. Despite this emphasis on time dependence, it turns out that the method is also suitable in the study of quasistatic problems. In [36] a numerical study is reported where the effective permittivity of randomly inhomogeneous mixtures is being evaluated using the FDTD method. For computational restrictions, a two-dimensional mixture is treated instead of the full three-dimensional case. Two-dimensional spheres are circles, and one example of such mixture is shown in Figure 9.4, where the positions of the circular inclusions are randomly chosen. Clustering is allowed which means that the spheres can touch each other and overlap.

A cut of such mixture is in the numerical simulation put into a parallel-plate waveguide, and the reflection coefficient from that slab is then calculated by the FDTD method. Then, the effective permittivity is defined to the permittivity of such a homogeneous sample from which the reflection coefficient is the same as from the mixture.

The results are compared in Figure 9.5 to the most common mixing rule predic-

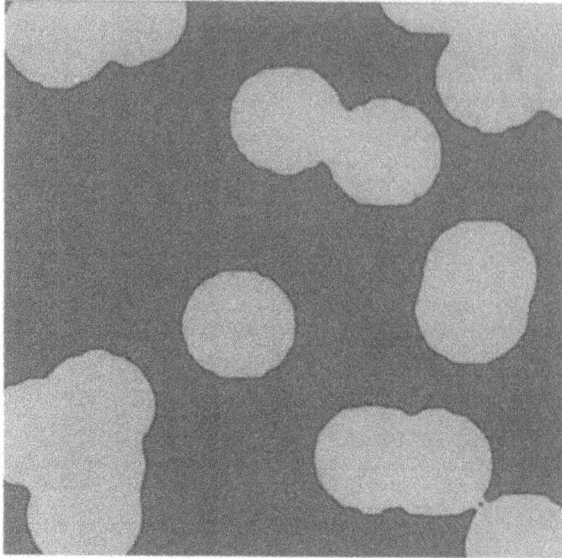

Figure 9.4: *A simulated mixture in two dimensions. The circle positions are randomly chosen.*

tions. The Maxwell Garnett rule, Bruggeman formula, and the Coherent potential formula can be written in two dimensions as

$$\frac{\epsilon_{\text{eff}} - \epsilon_e}{\epsilon_{\text{eff}} + \epsilon_e + \nu(\epsilon_{\text{eff}} - \epsilon_e)} = f \frac{\epsilon_i - \epsilon_e}{\epsilon_i + \epsilon_e + \nu(\epsilon_{\text{eff}} - \epsilon_e)} \tag{9.26}$$

where now $\nu = 0$ gives the Maxwell Garnett prediction, $\nu = 1$ corresponds to Bruggeman, and $\nu = 2$ to the Coherent potential formula.

From the simulations one may conclude that the Bruggeman formula is a quite good prediction for the effective permittivity, although it gives rather an overestimate.

Problems

9.1 Prove that the form (9.4) for spherical inclusions follows from the general expression (9.5) for the effective permittivity of a mixture according to the Coherent potential approximation.

9.2 Let us study the accuracy of perturbation expansion of mixing formulas in the prediction of the effective permittivity of dilute mixtures. Take into consideration the three mixing principles:

(a) Maxwell Garnett formula

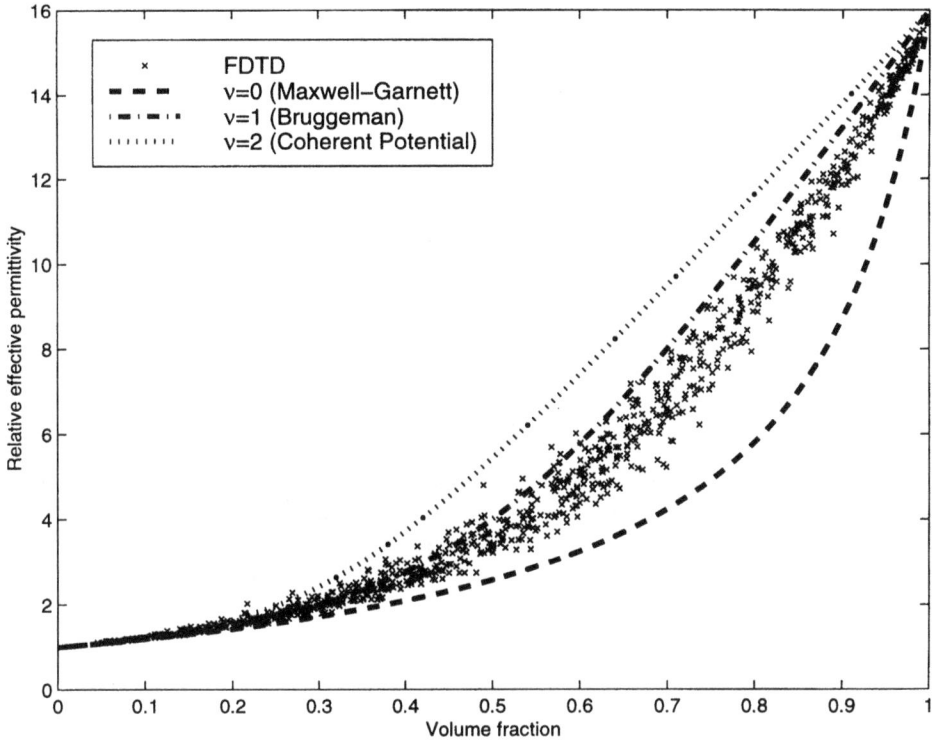

Figure 9.5: *The effective permittivity of simulated two-dimensional mixtures calculated with the FDTD method. The dielectric contrast between the inclusions and the environment is $\epsilon_i/\epsilon_e = 16$.*

(b) Bruggeman symmetric rule

(c) Coherent potential case.

Use the expansion (9.8) and, for each of the mixing rules, calculate what is the error, compared to the exact rule, when one uses a two-term Taylor series (containing the constant term and the one linear in f), and also when one uses the three-term expansion (the quadratic term is also included). Take all the following cases: the volume fraction is $f = 0.1$ and 0.5, and the inclusion to background contrast is $\epsilon_i/\epsilon_e = 2$ and 10.

9.3 Repeat the previous problem for the differential mixing models:

(a) Looyenga formula

(b) Bruggeman nonsymmetric rule

(c) Sen, Scala, and Cohen formula.

Use the Taylor expansions (9.14)–(9.16) and compare with the "exact" predictions given by (9.17)–(9.19).

9.4 Show that the logarithmic mixing law, the Lichtenecker formula (9.13), which can be written as

$$\ln \epsilon_{eff} = f \ln \epsilon_i + (1 - f) \ln \epsilon_e \qquad (9.27)$$

belongs to the family of power-law models,

$$\epsilon_{eff}^{\beta} = f \epsilon_i^{\beta} + (1 - f) \epsilon_e^{\beta} \qquad (9.28)$$

by taking the limit $\beta \to 0$.

9.5 Show that the unified six-vector mixing formula (9.25) is symmetric between the interchange of the roles of inclusion and environment for $\nu = 2$, in other words, when we follow the Bruggeman mixing philosophy.

9.6 Consider the case of a Swiss cheese mixture ($\epsilon_i < \epsilon_e$). The macroscopic permittivity of such a mixture with (three-dimensional) spheres can be calculated using the unified mixing rule (9.7), and is a function of ν. Try to see if possible problems emerge in the evaluation of ϵ_{eff} for certain values of ν when the dielectric contrast between the phases increases. Plot the Coherent potential prediction for the effective permittivity in the case $\epsilon_e/\epsilon_i = 10$.

References

[1] BRUGGEMAN, D.A.G.: 'Berechnung verschiedener physikalischer Konstanten von heterogenen Substanzen, I. Dielektrizitätskonstanten und Leitfähigkeiten der Mischkörper aus isotropen Substanzen', *Annalen der Physik*, 1935, Series 5, **24**, pp. 636-679

[2] POLDER, D., and VAN SANTEN, J.H.: 'The effective permeability of mixtures of solids', *Physica*, 1946, **XII**, (5), pp. 257-271

[3] DE LOOR, G.P.: 'Dielectric properties of heterogeneous mixtures containing water', *The Journal of Microwave Power*, 1968, **3**, pp. 67-73

[4] BÖTTCHER, C.J.F.: 'Theory of electric polarization' (Elsevier, Amsterdam, 1952)

[5] GRIMVALL, G: 'Thermophysical properties of materials' (North-Holland, Amsterdam, 1986)

[6] TSANG, L., KONG, J.A., and SHIN, R.T.: 'Theory of microwave remote sensing' (Wiley, New York, 1985)

[7] KOHLER, W.E., and PAPANICOLAOU, G.C.: 'Some applications of the coherent potential approximation', *in* CHOW,P.L., KOHLER, W.E., and PAPANICOLAOU, G.C. (Eds.): 'Multiple scattering and waves' (North Holland, New York, 1981), pp. 199-223

[8] ELLIOTT, R.J., KRUMHANSL, J.A., and LEATH, P.L.: 'The theory and properties of randomly disordered crystals and related physical systems', *Reviews of Modern Physics*, July 1974, **46**, (3), pp. 465-543

[9] SIHVOLA, A., and KONG, J.A.: 'Effective permittivity of dielectric mixtures', *IEEE Transactions on Geoscience and Remote Sensing*, 1988, **26**, (4), pp. 420-429. Correction, *ibid.*, 1989,**27**, (1), pp. 101-102

[10] SHENG, P.: 'Introduction to wave scattering, localization, and mesoscopic phenomena' (Academic Press, San Diego, 1995)

[11] SIHVOLA, A.: 'Self-consistency aspects of dielectric mixing theories', *IEEE Transactions on Geoscience and Remote Sensing*, 1989, **27**, (4), pp. 403-415

[12] LICHTENECKER, K., and ROTHER, K.: 'Die Herleitung des logarithmischen Mischungsgesetzes aus allgemeinen Prinzipien der stationären Strömung', *Physik. Zeitschr.*, 1931, **XXXII**, pp. 255-260

[13] BIRCHAK, J.R., GARDNER, L.G., HIPP, J.W., and VICTOR, J.M.: 'High dielectric constant microwave probes for sensing soil moisture', *Proceedings of the IEEE*, 1974, **62**, (1), pp. 93-98

[14] BEER, A.: 'Einleitung in die höhere Optik' (Friedrich Vieweg und Sohn, Braunschweig, 1853)

[15] GLADSTONE, J.H., and DALE, T.P.: 'Researches on the refraction, dispersion, and sensitiveness of liquids', *Philosophical Transactions of the Royal Society of London*, 1863, **153**, pp. 317-343

[16] LOOYENGA, H.: 'Dielectric constants of mixtures', *Physica*, 1965, **31**, pp. 401-406

[17] LANDAU, L.D., and LIFSHITZ, E.M.: 'Electrodynamics of continuous media', Second Edition (Pergamon Press, Oxford, 1984)

[18] ULABY, F.T., MOORE, R.K., and FUNG, A.K: 'Microwave remote sensing – Active and passive', Vol. III (Artech House, Norwood, Mass., 1986)

[19] LICHTENECKER, K.: 'Die Dielektrizitätskonstante natürlicher und künstlicher Mischkörper', *Physikalische Zeitschrift*, 1926, **27**, (4/5), pp. 115-158

[20] MERRILL, W.M., DIAZ, R.E., LORE, M. M., SQUIRES, M.C., and ALEXOPOULOS, N.G.: 'Effective medium theories for artificial materials composed of multiple sizes of spherical inclusions in a host continuum', *IEEE Transactions on Antennas and Propagation*, 1999, **47**, (1), pp. 142-148

[21] SEN, P.N., SCALA, C., and COHEN, M.H.: 'A self-similar model for sedimentary rocks with application to the dielectric constant of fused glass beads', *Geophysics*, 1981, **46**, (5), pp. 781-795

[22] RAYLEIGH, Lord: 'On the influence of obstacles arranged in rectangular order upon the properties of the medium', *Philosophical Magazine*, 1892, **34**, pp. 481-502.

[23] RUNGE, I.: 'Zur elektrischer Leitfähigkeit metallischer Aggregate', *Zeitschrift für technische Physik*, 1925, 6. Jahrgang, (2), pp. 61-68

[24] MEREDITH, R.E., and TOBIAS, C.W.: 'Resistance to potential flow through a cubical array of spheres', *J. Applied Physics*, 1960, **31**, (7), pp. 1270-1273

[25] McPHEDRAN, R.C., and McKENZIE, D.R.: 'The conductivity of lattices of spheres. I. The simple cubic lattice', *Proceedings of the Royal Society of London*, 1978, **A359**, pp. 45-63

[26] McKENZIE, D.R., McPHEDRAN, R.C., and DERRICK, G.H.: 'The conductivity of lattices of spheres. II. The body centred and face centred cubic lattices', *Proceedings of the Royal Society of London*, 1978, **A362**, pp. 211-232

[27] DOYLE W.T.: 'The Clausius-Mossotti problem for cubic array of spheres', *J. Applied Physics*, 1978, **49**, (2), pp. 795-797

[28] LAM, J.: 'Magnetic permeability of a simple cubic lattice of conducting magnetic spheres', *J. Applied Physics*, 1986, **60**, (12), pp. 4230-4235

[29] HINSEN, K., and FELDERHOF, B.U.: 'Dielectric constant of a suspension of uniform spheres', *Physical Review B*, 1992, **46**, (10), pp. 12955-12963

[30] NELSON, S.O., and YOU, T-S.: 'Relationships betweeen microwave permittivities of solid and pulverised plastics', *Journal of Physics D: Applied Physics*, 1990, **23**, pp. 346-353

[31] SIHVOLA, A.H., and LINDELL, I.V.: 'Polarizability modeling of heterogeneous media', *in* PRIOU, A. (Ed.): Dielectric Properties of Heterogeneous Materials, *Progress in Electromagnetics Research*, **6**, (Elsevier, New York, 1992), pp. 101-151

[32] SIHVOLA, A.H., and PEKONEN, O.P.M.: 'Effective medium formulas for bi-anisotropic mixtures', *Journal of Physics D: Applied Physics*, 1996, **29**, pp. 514-521

[33] SIHVOLA, A.H., and PEKONEN, O.P.M.: 'Six-vector mixing formulae defended', *Journal of Physics D: Applied Physics*, 1997, **30**, pp. 291-292

[34] SIHVOLA, A., and OLYSLAGER, F.: 'Eigenvector approach for solving bi-anisotropic mixing formulas', 1996, *Radio Science*, **31**, (6), pp. 1399-1405

[35] TAFLOVE, A.: 'Computational electrodynamics: the finite difference time domain method' (Artech House, Norwood, Mass., 1995)

[36] PEKONEN, O., KÄRKKÄINEN, K., SIHVOLA, A., and NIKOSKINEN, K.: 'Numerical testing of dielectric mixing rules by FDTD method', *Journal of Electromagnetic Waves and Applications*, 1999, **13**, (1), pp. 67-87

Chapter 10

Towards higher frequencies

The homogenisation principles in the present book have been very much based on the quasi-static treatment of the fields. To be frank, the analysis was truly static. The effective properties of the mixtures under study have been defined to be relations between a divergence-free flow and a curl-free field, although the results turn out to be useful also for time-varying excitations, provided that the time dependence is slow enough. When the frequency of the driving field increases, the waves become more powerful in their resolution of the details of the mixture. Consequently, a quasi-static homogenisation in this "Laplacian" sense starts to lose its validity.

In this chapter possibilities are sought how the effective-medium models could be hardened; in other words, to generalise them so that their domain of applicability would extend to frequencies for which the inclusions and inhomogeneities can no longer be considered very small. Certainly, the scattering of single bodies in free space has been studied thoroughly for various canonical shapes. See the extensive collection by Bowman et al. [1] for these results. However, the aim here is not to look for full solutions of Maxwell equations in random media but rather, how the most important correction terms could be added to mixing rules if we wish to try to describe the dynamic response of the dielectric inhomogeneity and randomness of the material.

Because of the connectedness of electric and magnetic fields and responses for time-varying fields, an exact treatment of high-frequency material modelling needs to account for the magnetic macroscopic polarisation due to the time-dependent electric field and vice versa. One example of such a material behaviour is the diamagnetic character of mixtures which contain conducting inclusions. A time-dependent excitation creates eddy currents which, by Lenz' law, oppose change, and are equivalent to magnetic dipoles. Such effects are, however, left aside in the present chapter.

10.1   Rayleigh scattering contribution

Scattering means that an inclusion reradiates part of the incident electromagnetic energy. And because the radiation by an object that is limited in spatial extension is a spherical wave this means that the energy is lost from the original wave field. The situation is illustrated in Figure 10.1.

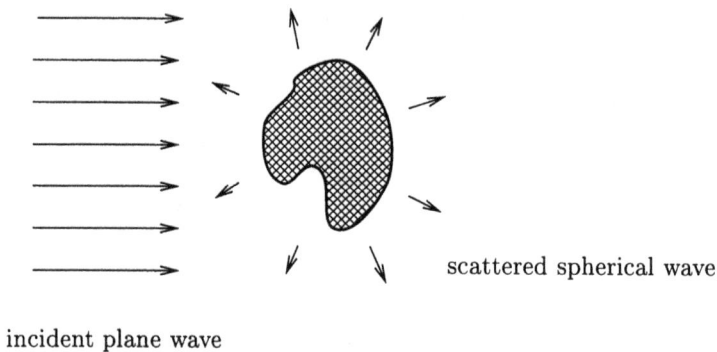

incident plane wave

scattered spherical wave

Figure 10.1: *An inclusion in free space acts as a scatterer if the incident field varies with time. An electromagnetic wave is radiated by the polarisation current within the inclusion.*

The exact solution for the scattering problem requires solving the Maxwell equations both inside and outside the scatterer and binding the two solutions together through the interface conditions.

However, as has been emphasised many times in the earlier chapters, for small scatterers the field distribution inside the scatterer does not vary greatly,[1] and the scattering efficiency can be calculated from the knowledge of the static field solution. This is the Rayleigh scattering regime [2, 3].

In the following, let us restrict our analysis to the case that a dielectric scatterer is located in free space (ϵ_0, μ_0). Of course also other environments can be handled, but then the extra parameters in the equations may make the text harder to follow and to extract the salient points in Rayleigh scattering.

10.1.1   Rayleigh scattering of a single inclusion

Assume that a scatterer is exposed to a plane wave with amplitude $\mathbf{E}_i$. A plane wave varies sinusoidally both in space (along the propagation axis) and in time, but if the scatterer is very small compared to the wavelength, the spatial variance can be neglected over the volume of the scatterer. The instantaneous internal electric field

---

[1]Depending on the shape of the scatterer, also the static field inside it may be inhomogeneous; however, the wave character with the consequent spatial variation is absent.

has the same relation to the external field as in the electrostatic case. Of course, looking at the time-variance, the internal field has the same sinusoidal behaviour as the external field. This means that the dipole moment vector induced in the inclusion vibrates in time. This oscillating dipole produces electromagnetic radiation. The radiation forms the scattered wave, and its energy is lost from the incident wave. From the point of view of the propagating wave, this scattering means losses just like the loss contribution due to ordinary absorption.

Radiated field

The field radiated by a current source $\mathbf{J}(\mathbf{r})$, in the far field, is [4]

$$\mathbf{E}_s(\mathbf{r}) = j\omega\mu_0 \frac{e^{-jk_0 r}}{4\pi r}\mathbf{u}_r \times \left(\mathbf{u}_r \times \int_{source} \mathbf{J}(\mathbf{r}')e^{jk_0\mathbf{u}_r\cdot\mathbf{r}'}dV'\right) \tag{10.1}$$

where the integration is over the the current source $\mathbf{J}$, the wave number is $k_0 = \omega\sqrt{\mu_0\epsilon_0}$, and the source is vibrating with the angular frequency ω. The vector $\mathbf{u}_r$ is the unit vector pointing to the field position, at a distance r.

What is the current source in the case of a dielectric scatterer with permittivity ϵ_i and internal field $\mathbf{E}_i$? A displacement current in the Maxwell–Ampère equation is $j\omega\epsilon_i\mathbf{E}_i$. If this is juxtaposed with the corresponding equation for a free current in free space, $\mathbf{J} + j\omega\epsilon_0\mathbf{E}_i$, the polarisation current can be identified as

$$\mathbf{J} = j\omega(\epsilon_i - \epsilon_0)\mathbf{E}_i \tag{10.2}$$

Consider the radiation from a small scatterer. The smallness is measured relative to the wavelength, which means that the phase term in the integral of (10.1) is constant, and only the current term is left to be integrated:

$$\int_{source} \mathbf{J}(\mathbf{r})dV' = j\omega V(\epsilon_i - \epsilon_0)\mathbf{E}_i = j\omega\mathbf{p} \tag{10.3}$$

with $\mathbf{p}$ being the dipole moment of the scatterer. The radiated field is, then

$$\mathbf{E}_s(\mathbf{r}) = -\omega^2\mu_0\frac{e^{-jk_0 r}}{4\pi r}\mathbf{u}_r \times (\mathbf{u}_r \times \mathbf{p}) \tag{10.4}$$

The effect of the operation $-\mathbf{u}_r \times (\mathbf{u}_r \times \mathbf{p})$ on the dipole moment vector $\mathbf{p}$ is to take its component perpendicular to the scattering direction $\mathbf{u}_r$, which is the obvious direction of the polarisation vector of a scattered electric field. The angular dependence, due to the cross-product, is $\sin\theta$, where θ is the angle between the dipole moment vector and the scattering direction according to Figure 10.2.

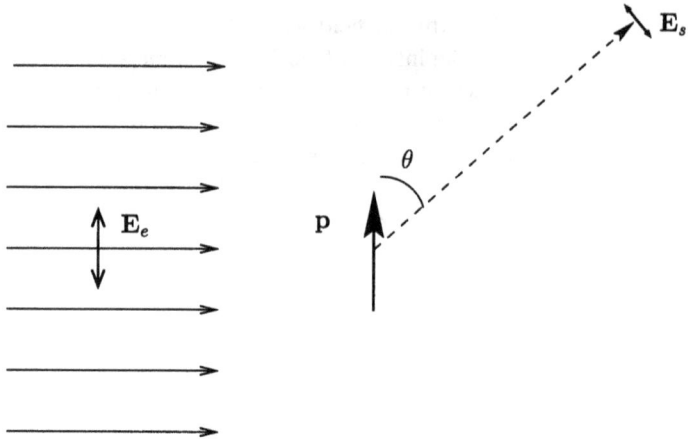

Figure 10.2: *The field $\mathbf{E}_e$ of the incident plane wave induces a dipole moment $\mathbf{p}$, which creates a spherical scattered wave with field $\mathbf{E}_s$.*

Scattering cross-section

The bistatic scattering cross-section σ of an inclusion is defined by [5, 6]

$$\sigma = 4\pi r^2 \frac{|\mathbf{E}_s|^2}{|\mathbf{E}_e|^2} \tag{10.5}$$

where $\mathbf{E}_s$ is the scattered field at the (far-field) distance r from the scatterer, which is exposed to the incident wave with the field $\mathbf{E}_e$. We can write the scattering cross-section for a small scatterer with polarisability α:

$$\sigma = \frac{k_0^4}{4\pi} \left(\frac{\alpha}{\epsilon_0}\right)^2 \sin^2\theta \tag{10.6}$$

where $k_0^2 = \omega^2 \mu_0 \epsilon_0$ and the relation of the dipole moment to the polarisability is again $\mathbf{p} = \alpha \mathbf{E}_e$.[2]

If the radar cross-section is to be calculated, the scattering direction is perpendicular to the induced dipole moment, and the angular dependence term is $\sin^2\theta = 1$.

For example, for a dielectric sphere with radius a and permittivity ϵ_i, this (low-frequency) scattering cross-section is

$$\sigma = 4\pi a^2 \sin^2\theta (k_0 a)^4 \left|\frac{\epsilon_i - \epsilon_0}{\epsilon_i + 2\epsilon_0}\right|^2 \tag{10.7}$$

where again the scattering angle is θ.

---

[2] For a complex object, the scattering cross-section σ depends on both the incidence angle direction and the scattering angle direction.

The total scattering cross-section σ_s is a quantity with which one can calculate the scattering efficiency of the inclusion. It takes into account the scattering power into all directions:

$$\sigma_s = \frac{1}{4\pi} \int \sigma \, d\Omega \tag{10.8}$$

where the integration has to be performed over the whole solid angle 4π. The angular dependence of the dipole scattering is $\sin^2 \theta$, which is easy to integrate:

$$\iint \sin^2 \theta \, d\Omega = \int_0^\pi \sin^2 \theta \, \sin \theta \, d\theta \int_0^{2\pi} d\varphi = \frac{8\pi}{3} \tag{10.9}$$

This gives us the total scattering cross-section as a function of the polarisability

$$\sigma_s = \frac{k_0^4}{6\pi} \left(\frac{\alpha}{\epsilon_0} \right)^2 \tag{10.10}$$

The corresponding quantity for a small dielectric sphere reads

$$\sigma_s = \frac{8}{3} \pi a^2 (k_0 a)^4 \left| \frac{\epsilon_i - \epsilon_0}{\epsilon_i + 2\epsilon_0} \right|^2 \tag{10.11}$$

This relation shows the strong dependence on frequency that small scatterers have, as already noted by Lord Rayleigh in 1871. Sky is blue because shorter wavelengths of the spectrum of light from the Sun are scattered much more strongly than the longer ones which are located in the red part of the spectrum.

10.1.2   Rayleigh attenuation

Let us consider next a medium consisting of randomly distributed scatterers. If each of the inclusions acts as a Rayleigh scatterer, we can estimate the loss factor due to this process by considering the situation in Figure 10.3. A plane wave is incident on the scattering medium, and attenuated.

As the plane wave propagates through the scattering medium, the scattered energy is taken away from the incident wave, and there is exponential attenuation. The electric field variation is determined by the effective wave number $k_{\text{eff}} = k'_{\text{eff}} - jk''_{\text{eff}}$:

$$E(z) \sim e^{-jk_{\text{eff}}z} = e^{-jk'_{\text{eff}}z} \cdot e^{-k''_{\text{eff}}z} \tag{10.12}$$

where k'_{eff} and k''_{eff} are the phase and attenuation constants for the field in the scattering medium. The power density of the plane wave $S(z)$ (the magnitude of the real part of the Poynting vector) is proportional to the square of the electric field, which means that the differential condition for the power density is

$$\frac{dS(z)}{dz} = -2k''_{\text{eff}} S(z) \tag{10.13}$$

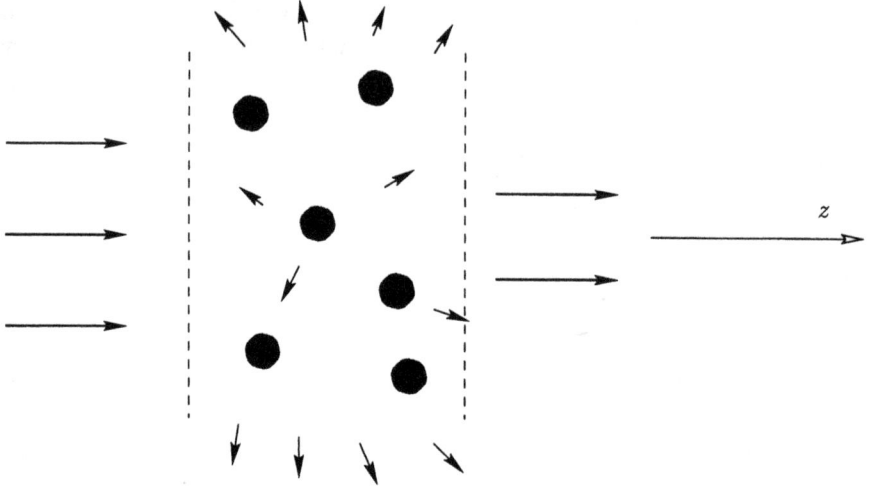

Figure 10.3: *The intensity of a propagating plane wave decreases as a result of the Rayleigh scattering of the individual inclusions.*

Consider a volume with cross-section A in the plane orthogonal to the z axis, and of depth dz. The power loss in such a volume $V = A\,dz$ is NP_s, where N is the number of scatterers in that volume and P_s is the scattered power by a single inclusion. The connection of the scattering cross-section to the scattered power and the incident power density, for a single inclusion, is

$$P_s = \sigma_s S \tag{10.14}$$

By power balance, then the lost power is $NP_s = -A\,dS$. We then have

$$k''_{\text{eff}} = \frac{N\sigma_s}{2V} = \frac{n\sigma_s}{2} \tag{10.15}$$

where $n = N/V$ is the number density of the scatterers. It is important to remember that all scatterers were assumed to be of the same size.[3]

This relation (10.15) gives us the connection between the scattering of small particles and the attenuation of the effective medium. One must, however, remember the assumption of independent scattering which was implicitly made. No multiple scattering or interference effects were taken into account.

Because now scattering contributes to the attenuation of the effective medium, it also gives rise to a term in the imaginary part of the effective permittivity. The connection of the effective permittivity and the attenuation constant of the plane wave was discussed in Section 2.2.2, and we can write

$$k''_{\text{eff}} = -\omega\sqrt{\mu_0\epsilon_0}\,\text{Im}\sqrt{\epsilon_{\text{eff}}/\epsilon_0} \tag{10.16}$$

---

[3]For a size distribution, the multiplication by the number density n has to be replaced by an integral.

A particularly simple relation between the scattering and the imaginary part of the effective permittivity occurs for very dilute mixtures. Then the real part of the effective permittivity is nearly that of background: $\epsilon'_{\text{eff}} \approx \epsilon_0$, and the square root can be simplified

$$\sqrt{\epsilon_{\text{eff},r}} = \sqrt{\frac{\epsilon_{\text{eff}}}{\epsilon_0}} \approx \sqrt{1 - j\frac{\epsilon''_{\text{eff}}}{\epsilon_0}} \approx 1 - j\frac{\epsilon''_{\text{eff}}}{2\epsilon_0} \tag{10.17}$$

and we have the connection

$$\frac{\epsilon''_{\text{eff}}}{\epsilon_0} = \frac{n\sigma_s}{k_0} \tag{10.18}$$

where $k_0 = \omega\sqrt{\mu_0\epsilon_0}$ is the free-space wave number. Note that the background in which the scatterers were sparsely immersed here was assumed to be a vacuum.

Absorption cross-section

One has to remember that in addition to scattering, absorption is the other contributing mechanism to attenuation. Let us define, analogously to the scattering cross-section σ_s, the corresponding cross-section for the absorption loss: let σ_a be the area from which the inclusion "captures" energy from the incident wave and transfers to dielectric losses. To calculate the absorption cross-section, neglect now the scattering, and make the connection corresponding to (10.15)

$$\sigma_a = \frac{2}{n}k''_{\text{eff}} \tag{10.19}$$

where now k''_{eff} is connected to the imaginary part of the effective permittivity ϵ_{eff}, which for a dilute mixture is (see Equation (3.31))

$$\epsilon_{\text{eff}} = \epsilon_0\left(1 + 3f\frac{\epsilon_r - 1}{\epsilon_r + 2}\right) \tag{10.20}$$

Here $\epsilon_r = \epsilon_i/\epsilon_0$ is the relative (complex) permittivity of the inclusions and $f = nV$ their volume fraction, again assumed to be small: $f \ll 1$. The inclusions are assumed to be spherical.

This gives us the absorption cross-section for a dielectric sphere:

$$\sigma_a = -3k_0 V \, \text{Im}\left\{\frac{\epsilon_r - 1}{\epsilon_r + 2}\right\} \tag{10.21}$$

which can also be written in the form

$$\sigma_a = -4\pi a^2 (k_0 a) \, \text{Im}\left\{\frac{\epsilon_r - 1}{\epsilon_r + 2}\right\} = 12\pi a^2 (k_0 a)\frac{\epsilon''_r}{|\epsilon_r + 2|^2} \tag{10.22}$$

where a is the radius of the sphere. Here the fact that $k = k_0$ appears in the relations is a reminder that the coherent wave propagates as in free space. For a low volume fraction of the scatterers this is to be expected.[4]

---

[4]The inconvenient play here with the minus signs is a result of the definitions $\epsilon_r = \epsilon'_r - j\epsilon''_r$ and correspondingly $k''_{\text{eff}} = -\text{Im}\{k_{\text{eff}}\}$.

10.2   Mie scattering

The problem of electromagnetic scattering by a particle becomes much more difficult when the frequency of the field variation increases. Then the size of the inclusion is comparable or large compared to the wavelength, and the polarisation current is not in the same phase throughout the particle. The scattering problem requires a numerical approach for an arbitrary-shaped inclusion. However, if the particle is a homogeneous sphere, a series solution can be written for the scattered field and power. This series is known in the literature as the *Mie solution*, after the work of Gustav Mie in 1908 [7].[5]

Construction principles of the Mie series

The idea in the Mie solution is to expand the electric and magnetic fields of the problem in vector spherical harmonics. The field functions then become linear combinations of terms that are products of separable functions of the three spherical co-ordinates: spherical Bessel functions of the distance from the sphere, associated Legendre polynomials of the zenith angle, and sinusoidal functions of the azimuth angle [10,11]. The unknown coefficients in these series are then determined by the boundary conditions at the surface of the sphere. The result is the scattered field as an infinite sum of the partial waves. The sum converges. The number of terms that needs to be calculated to get a reasonable estimate for the scattering of a given sphere depends for the most part on the size of the sphere relative to wavelength. A criterion much used in practical calculations is the one by Wiscombe [12], according to which one should take N terms, with

$$N = x + 4\sqrt[3]{x} + 2 \qquad (10.23)$$

Here the important quantity is the *size parameter*, defined as $x = k_0 a$, where k_0 is the free-space wave number of the field[6] and a is the radius of the sphere. The scattering coefficients are determined by the size parameter and the refractive index m of the sphere. For a dielectric sphere, the connection of the relative permittivity and the refractive index is

$$m = \sqrt{\epsilon_r} \qquad (10.24)$$

For an arbitrary sphere, the scattering cross-section σ_s can be calculated from the square amplitudes of the Mie coefficients. Another important parameter that comes from the same coefficients is the *extinction cross-section*, σ_e. It takes into account all attenuation that the sphere causes to the propagating field: both scattering

---

[5]The paternity of the sphere scattering solution can be debated; other good candidates are L. Lorenz and P. Debye. See [8,9].

[6]The theory applies as well to inclusions embedded in another material, not necessarily a vacuum; then the size parameter is defined as ka, where the relative permittivity of the environment $\epsilon_{e,r}$ determines the wave number $k = \sqrt{\epsilon_{e,r}}k_0$.

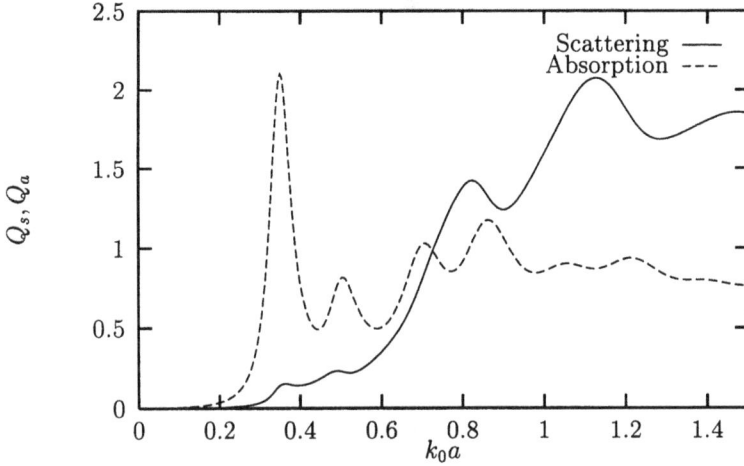

Figure 10.4: *The scattering and absorption efficiencies* Q_s *and* Q_a *of a water droplet as a function of the size parameter* k_0a *at the frequency of 3 GHz. The radius of the sphere varies between 0 and 2.4 cm.*

and absorption. Finally, the absorption cross-section can be calculated by a simple subtraction

$$\sigma_a = \sigma_e - \sigma_s \tag{10.25}$$

The low-frequency absorption and scattering cross-sections (10.22) and (10.11) emerge from the first terms of the Mie coefficients.

The economical enumeration of the terms in the Mie series is a science in itself; see [8, 13, 14] for principles and examples. It is also very rewarding to study the classic text by van de Hulst [11], where much effort has been put into solving the problem of how to approach infinite Mie sums with limited computational capacity.

Quite often in scattering studies the *efficiencies* are discussed. These are the cross-sections relative to the geometrical cross-section:

$$Q_s = \frac{\sigma_s}{\pi a^2} \tag{10.26}$$

$$Q_a = \frac{\sigma_a}{\pi a^2} \tag{10.27}$$

$$Q_e = \frac{\sigma_e}{\pi a^2} \tag{10.28}$$

Example: extinction by a water droplet

As an example of the behaviour of the scattering and absorption due to a lossy dielectric sphere, Figures 10.4–10.5 show the scattering and absorption efficiencies

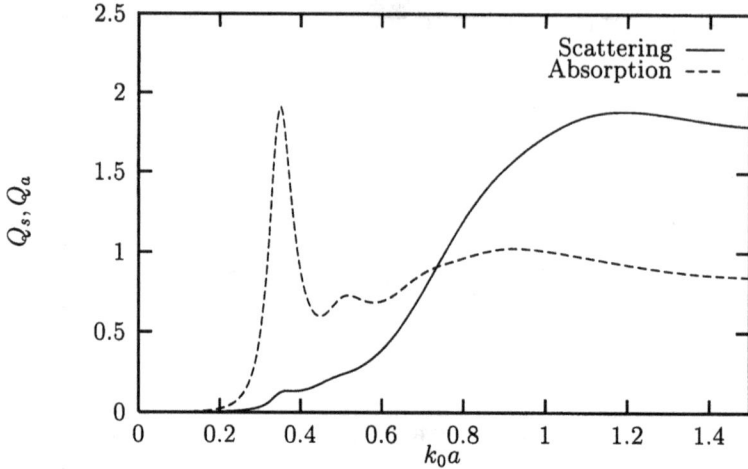

Figure 10.5: *The same as in Figure 10.4, but the radius a of the sphere is kept constant (a = 5 mm), and the frequency varies, between 0 and 14.3 GHz. (The corresponding free-space wavelengths are from infinity to 2 cm.)*

of a water droplet at optical frequencies.[7]

The efficiencies are shown as functions of the size parameter $k_0 a = 2\pi a/\lambda$, where $\lambda = 2\pi c/\omega$ is the wavelength of the field in free space. The difference in the figures is that the size parameter is changed by varying either the radius (Figure 10.4) or the frequency (Figure 10.5). The reason behind the difference of the corresponding curves in the two figures is the dispersion of the permittivity of water, which means that in Figure 10.4, the complex permittivity is constant but in Figure 10.5 it changes along with $k_0 a$. In the calculations, the Debye dispersion has been used for the relative permittivity of water [16]:

$$\epsilon_r(\omega) = 4.9 + \frac{75.2}{1 + j\omega\tau} \tag{10.29}$$

where the parameter $\tau = 10.1 \times 10^{-12}$ s is the relaxation time for water molecules at the temperature of 20° C.

The curves show clearly how both scattering and absorption increase as the electrical size of the scattering sphere increases. But the increase is not monotonical: there are maxima and minima due to interference effects. The physics of the various ripple structure details are discussed in [8] in a very elucidative way.

Towards the low end of the size parameter values, the scattering and absorption cross-sections are very small, as is to be expected from the relations (10.11) and (10.22). One can also observe that in the low-frequency regime, scattering becomes

---

[7]The results are calculated with the *Indiascat* software [15].

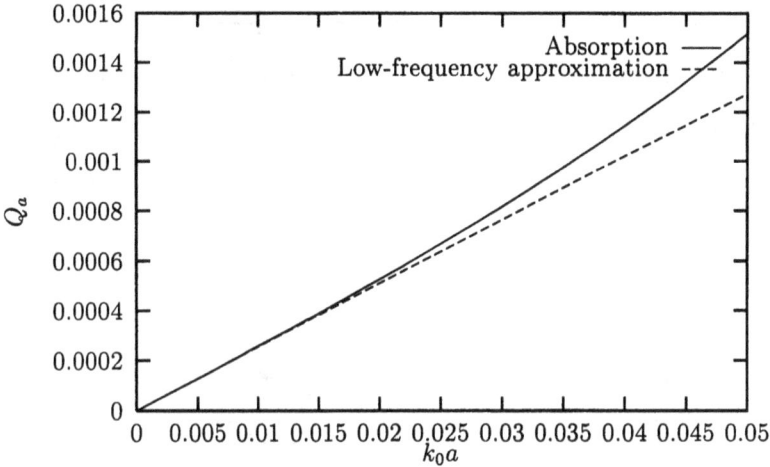

Figure 10.6: *The absorption efficiency of the dielectric sphere (water) in the low end of the size parameter scale. The frequency is 3 GHz. The low-frequency approximation (10.22) grows linearly with the size parameter.*

negligible compared with absorption, which can be explained by the strong fourth-power wavelength-dependence of Rayleigh scattering.

In Figures 10.6 and 10.7, the closer look at the long-wavelength regime shows that the absorption starts off with a linear dependence on the size parameter (10.6), whereas the fourth-power dependence of scattering (10.7) makes Q_s asymptotically vanishingly small.

In the figures, also the low-frequency approximations (10.11) and (10.22) are depicted. A comparison of the horizontal scales of the figures shows that the validity range of the low-frequency approximation is much broader for scattering than for absorption for water droplets at this frequency range.

Between Rayleigh and Mie

In the literature about scattering by complex objects and their radar reflection characteristics, usually the approach is fully dynamical. Wave effects and interference are important mechanisms that determine the response of the object. But it is probably valuable to emphasise the connection of static solutions with electromagnetic scattering as a natural low-frequency limiting case. Today's computational electromagnetics is capable of solving large problems with high accuracy and quite often problems are treated with a broad dynamic range in the time domain, not least because various FDTD-based software is available. One may save, however, a lot of effort in solving low-frequency scattering problems if the quasistatic solution is taken as the starting point, and the dynamical results are calculated with an asymptotic

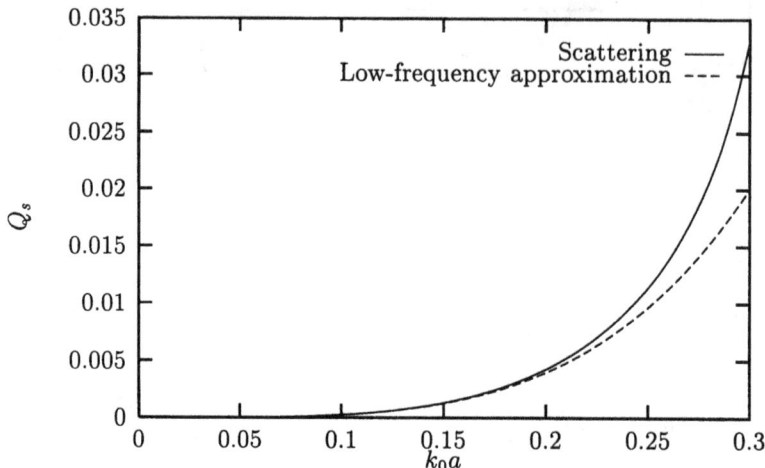

Figure 10.7: *The scattering efficiency of the dielectric sphere (water) in the low end of the size parameter scale. The frequency is 3 GHz. The low-frequency approximation (10.11) grows with the fourth power of the size parameter.*

correction to the Rayleigh scattering. The classic contrubution in this vein is the pair of articles by Stevenson in the 1950s [17].

10.3 Scattering in random media

The previous analysis has probably proved that it is very difficult to include scattering effects into the effective permittivity of random materials. The scattering analysis in the Rayleigh regime in Section 10.1.2 gave us the basic term of scattering loss in sparse-mixture conditions. It can be added as the imaginary part of the effective permittivity of a mixture:

$$\epsilon''_{\text{eff}} = \frac{n\sigma_s\epsilon_0}{k_0} \tag{10.30}$$

with Equation (10.11). Then the Maxwell Garnett formula reads as follows:

$$\epsilon_{\text{eff}} = \epsilon_0 + 3f\epsilon_0 \frac{\epsilon_i - \epsilon_0}{\epsilon_i + 2\epsilon_0}\left[1 - j\frac{2}{3}(k_0a)^3 \frac{\epsilon_i - \epsilon_0}{\epsilon_i + 2\epsilon_0}\right] \tag{10.31}$$

where the inclusions have been assumed lossless (ϵ_i is real).

10.3.1 Quasi-crystalline approximation

The sparse-mixture assumption under which Equation (10.31) works can be relaxed using the quasicrystalline approximation for the multiple scattering equations [6,

Chapter 6]. This approach takes into account the positions of neighbouring scatter-ers by the so-called pair-distribution function $g(\mathbf{r})$. This is a normalised probability that a particle occupies the position in the neighbourhood of the particle under consideration. With that, the improvement over (10.31) can be written as follows:

$$
\epsilon_{\text{eff}} = \epsilon_0 + 3f\epsilon_0 \frac{\dfrac{\epsilon_i - \epsilon_0}{\epsilon_i + 2\epsilon_0}}{1 - f\dfrac{\epsilon_i - \epsilon_0}{\epsilon_i + 2\epsilon_0}}
$$

$$
\times \left\{ 1 - j\frac{2}{3}(k_0 a)^3 \frac{\dfrac{\epsilon_i - \epsilon_0}{\epsilon_i + 2\epsilon_0}}{1 - f\dfrac{\epsilon_i - \epsilon_0}{\epsilon_i + 2\epsilon_0}} \left[1 + 4\pi n \int_0^\infty r^2[g(r) - 1]dr \right] \right\} \tag{10.32}
$$

where n is the number of scatterers per unit volume and here the pair distribution function is radially symmetric: $g(\mathbf{r}) = g(r)$.

10.3.2 Size-dependent polarisability approach

Another possibility to add frequency dependence to the mixing rules is to study the behaviour of the polarisability as the size of the scatterer grows. The familiar relation between the external and internal fields for a dielectric sphere, Equation (3.8),

$$
\mathbf{E}_i = \frac{3}{\epsilon_r + 2} \mathbf{E}_e \tag{10.33}
$$

cannot preserve its credibility when the frequency increases. To find a more accurate relation, it is good to start with the integral equation that governs the exact field relations [4]

$$
\mathbf{E}_i(\mathbf{r}) = \mathbf{E}_e(\mathbf{r}) + k_0^2 \int_V \overline{\overline{G}}(\mathbf{r}, \mathbf{r}') \cdot (\epsilon_r - 1)\mathbf{E}_i(\mathbf{r}')dV' \tag{10.34}
$$

where the ordinary free-space Green dyadic is

$$
\overline{\overline{G}}(\mathbf{r}, \mathbf{r}') = \left(\overline{\overline{I}} + \frac{\nabla\nabla}{k_0^2} \right) \frac{e^{-jk_0 R}}{4\pi R} \tag{10.35}
$$

with $R = |\mathbf{r} - \mathbf{r}'|$ and $\overline{\overline{I}}$ is the unit dyadic.

Given that the external field is uniform across the homogeneous sphere, and the sphere is small with respect to the wavelength, the singularity of the Green dyadic gives the difference of the external and internal fields according to the second term of the right-hand side of (10.34), and we have

$$
k_0^2 \int_V \overline{\overline{G}}(\mathbf{r}, \mathbf{r}') \cdot (\epsilon_r - 1)\mathbf{E}_i(\mathbf{r}')dV' \approx -\frac{1}{3}(\epsilon_r - 1)\mathbf{E}_i \tag{10.36}
$$

This result for the (uniform) internal field can be directly written down with the knowledge that the depolarisation dyadic of a spherical volume is $\overline{\overline{I}}/3$.

Peltoniemi expansion

Size-dependent corrections to (10.36) have been presented [18, 19]. Let us, however, take advantage of a recent and more detailed study by Peltoniemi [20] in the careful evaluation of the volume integral equation. By extracting a spherical subvolume V_b around the singularity point $\mathbf{r}' = \mathbf{r}$, the integral can be split into two parts: over V_b and over $V - V_b$, of which the latter is nonsingular. With the Taylor expansion of the Green dyadic around the singularity, the integral can be analytically integrated, and the second-order expansion for the field relation can be seen to read as follows [20]:

$$\mathbf{E}_i(\mathbf{r}) \approx \mathbf{E}_e(\mathbf{r}) + \left[-\frac{1}{3} + G_1(y) + \epsilon_r G_2(y) \right] (\epsilon_r - 1)\mathbf{E}_i$$
$$+ \int_{V-V_b} \overline{\overline{G}}(\mathbf{r}, \mathbf{r}') \cdot (\epsilon_r - 1)\mathbf{E}_i(\mathbf{r}')dV' \tag{10.37}$$

where the scalar functions

$$G_1(y) = \frac{2}{3}\left[(1 + jy)e^{-jy} - 1\right] \tag{10.38}$$

$$G_2(y) = \left(1 + jy - \frac{7}{15}y^2 - j\frac{2}{15}y^3\right)e^{-jy} - 1 \tag{10.39}$$

are functions of $y = k_0 b$, where b is the radius of the extracted sphere V_b.

Generalised polarisability

Instead of full scattering calculations, the Peltoniemi result can be exploited for asymptotic analysis of the polarisability of a finite sphere. To that end, we need the relations between the internal and the external fields. Let us choose $V = V_b$ in (10.37), in other words, the excluded volume occupies the whole spherical inclusion. Then the integral in (10.37) vanishes because the support of the integrand is V. We then have

$$\mathbf{E}_e(\mathbf{r}) = \left\{ 1 - (\epsilon_r - 1)\left[-\frac{1}{3} + G_1(x) + \epsilon_r G_2(x) \right] \right\} \mathbf{E}_i(\mathbf{r}) \tag{10.40}$$

Here now $x = k_0 a$ is the size parameter of the dielectric sphere with radius a. Note that in the limit of vanishing size parameter ($x \to 0$), also $G_1(x) \to 0$ and $G_2(x) \to 0$, and (10.40) becomes (10.33).

Proceeding in the enumeration of the dipole moment as was done in Chapter 3, the size-dependent polarisability can be calculated:

$$\alpha = 3V\epsilon_0 \frac{\epsilon_r - 1}{\epsilon_r + 2}\left(1 - 3\frac{\epsilon_r - 1}{\epsilon_r + 2}[G_1(x) + \epsilon_r G_2(x)]\right)^{-1} \tag{10.41}$$

Of course, this result is only the first step beyond the purely static Lorenz–Lorentz polarisability

$$\alpha = 3V\epsilon_0 \frac{\epsilon_r - 1}{\epsilon_r + 2} \tag{10.42}$$

and therefore breaks down for very large size parameter values x. It is instructive to write down the first terms of the Taylor series of the polarisability (10.41):

$$\alpha \approx 3V\epsilon_0 \frac{\epsilon_r - 1}{\epsilon_r + 2}\left(1 + \frac{\epsilon_r - 1}{\epsilon_r + 2}\cdot\frac{\epsilon_r + 10}{10}x^2 - j\frac{2}{3}\frac{\epsilon_r - 1}{\epsilon_r + 2}x^3 + \cdots\right) \tag{10.43}$$

Note that no term appears which is linear in x.

Generalised Maxwell Garnett formula

Once the polarisability is known, one can write the Maxwell Garnett mixing rule with size dependence of the inclusions. Assume that the inclusions in the mixture are all spheres with the same radius a and they occupy the volume fraction $f = nV$ in the whole mixture, with n being the number density of the spheres. In the Maxwell Garnett relation, let us assume that the volume fraction is small: $f \ll 1$. Then we can write for the effective permittivity: $\epsilon_{\text{eff}} \approx \epsilon_0 + n\alpha$. We have

$$\epsilon_{\text{eff}} \approx \epsilon_0 + 3f\epsilon_0\frac{\epsilon_r - 1}{\epsilon_r + 2}\left(1 + \frac{\epsilon_r - 1}{\epsilon_r + 2}\frac{\epsilon_r + 10}{10}x^2 - j\frac{2}{3}\frac{\epsilon_r - 1}{\epsilon_r + 2}x^3 + \cdots\right) \tag{10.44}$$

It is interesting to see that the first term corresponding to losses in this equation, proportional to the cube of the size parameter, is exactly the same as the one in Equation (10.31) which was due to the incoherent scattering by a collection of Rayleigh-scattering spheres.

Problems

10.1 Show that for lossless spheres of permittivity ϵ_i occupying volume fraction f in the free-space environment, the scattering loss is equivalent to the imaginary part of the effective permittivity in Equation (10.31):

$$\epsilon_{\text{eff}}'' = 2f\epsilon_0(k_0 a)^3 \left(\frac{\epsilon_i - \epsilon_0}{\epsilon_i + 2\epsilon_0}\right)^2$$

where $k_0 a$ is the size parameter of the spheres (all assumed of equal size). Use Equations (10.11), (10.15), and (10.16).

10.2 Calculate the scattering and absorption cross-sections for a dielectric ellipsoid with depolarisation factors N_x, N_y, and N_z. Use Rayleigh scattering principles.

10.3 Consider the polarisability according to the result (10.41). Using Taylor expansions for the functions $G_1(x)$ and $G_2(x)$ for values of x close to zero, calculate the term of the polarisability that has x^4 dependence of the size parameter. Does this term contribute to the phase or the attenuation of the coherent wave if a wave propagated through a random mixture containing such dielectric spheres?

References

[1] BOWMAN, J.J., SENIOR T.B.A., and USLENGHI, P.L.E. (Eds.): 'Electromagnetic and acoustic scattering by simple shapes' (North-Holland, Amsterdam, 1969)

[2] KLEINMAN, R.E.: 'The Rayleigh region', *Proceedings of the IEEE*, 1965, **53**, pp. 848-856

[3] SENIOR, T.B.A.: 'Low-frequency scattering by a dielectric body', *Radio Science*, 1976, **11**, (5), pp. 477-482

[4] LINDELL, I.V.: 'Methods for electromagnetic field analysis' (IEEE Press and Oxford University Press, 1995)

[5] ULABY, F.T., MOORE, R.K., and FUNG, A.K.: 'Microwave remote sensing: Active and passive', Vol. I (Addison-Wesley, Reading, Massachusetts, 1981)

[6] TSANG, L., KONG, J.A., and SHIN, R.T.: 'Theory of microwave remote sensing' (Wiley, New York, 1985)

[7] MIE, G.: 'Beiträge zur Optik trüber Medien, speziell kolloidaler Metallösungen', *Annalen der Physik*, 1908, **25**, pp. 377-445.

[8] BOHREN, C.F., and HUFFMAN, D.R.: 'Absorption and scattering of light by small particles' (Wiley, New York, 1983)

[9] KERKER, M.: 'The scattering of light and other electromagnetic radiation' (Academic Press, New York, 1969)

[10] STRATTON, J.A.: 'Electromagnetic theory' (McGraw Hill, New York, 1941)

[11] VAN DE HULST, H.C.: 'Light scattering by small particles' (Wiley, New York, 1957; Dover, New York, 1981)

[12] WISCOMBE, W.J.: 'Improved Mie scattering algorithms', *Applied Optics*, 1980, **19**, (9), pp. 1505-1509

[13] BARBER, P.W., and HILL, S.C.: 'Light scattering by particles: Computational methods' (Advanced Series in Applied Physics, World Scientific, 1990)

[14] DEIRMENDJIAN, D.: 'Electromagnetic scattering on spherical polydispersions' (Elsevier, New York, 1969)

[15] SHARMA, R., and SIHVOLA, A.: 'Mie scattering code for dielectric sphere', *Helsinki Univ. Tech. Electromagnetics Laboratory Report Ser.*, December 1998, **281**.

[16] HASTED, J.B.: 'Aqueous dielectrics' (Chapman & Hall, London, 1973)

[17] STEVENSON, A.F.: 'Solution of electromagnetic scattering problems as power series in the ratio (dimension of scatterer)/wavelength', *Journal of Applied Physics*, 1953, **24**, (9), pp. 1134-1142; 'Electromagnetic scattering by an ellipsoid in the third approximation', *ibid.*, pp. 1143-1151

[18] FIKIORIS, J.G.: 'Electromagnetic field inside a current-carrying region', *Journal of Mathematical Physics*, November 1965, **6**, (11), pp. 1617-1620.

[19] WANG, J.J.H.: 'A unified and consistent view on the singularity of the electric dyadic Green's function in the source region', *IEEE Transactions on Antennas and Propagation*, May 1982, **30**, (3), pp. 463-468

[20] PELTONIEMI, J.I.: 'Variational volume integral equation method for electromagnetic scattering by irregular grains', *J. Quant. Spectrosc. Radiat. Transfer*, 1996, **55**, (5), pp. 637-647

Chapter 11
Dispersion and time-domain analysis

Electrical engineering textbooks are predominantly preaching frequency-domain ideology. This means that the time variation of the electromagnetic phenomena under consideration is very simple. It is sinusoidal, and the frequency of the wave variation is a measure for the time derivatives that are needed in Maxwell equations. Use of complex vectors helps to eliminate totally the time-dependence in the field quantities, and admittedly the remaining equations are easier to solve than the original time-varying system.

However, the single-frequency description of an electromagnetic problem is an idealisation. A sinusoidal wave is eternal, and it cannot have a beginning or an end, and hence it does not exist. Also, in ordinary applications of electromagnetic waves, the signal needs to be modulated in order to be capable of information transmission. And modulation means that in the frequency domain, the field quantities require a broader band than a single peak.

This is not to say that the frequency-domain approach to electromagnetic field problems could not be used for broadband applications. Certainly our understanding of the response of complex systems is assisted by the concepts like bandpass filters and dispersion curves that describe the behaviour of a system as a function of the frequency of the excitation. But it is important to bear in mind that the basic measurable quantities are real and varying in time and space rather than complex combinations of components in temporal and spatial frequency.

The juxtaposition of time-domain and frequency-domain approaches, and argumentation in favour of one or the other may seem foreign in the context of the present book because of the emphasis here, which is on the static and quasi-static limit of the spectrum. Nonetheless, frequency and time are there. The implicit message in the earlier chapters has been that the results of various mixing rules could be used at different frequencies. The user should use inclusion permittivity data valid for that frequency for which she needs to calculate the mixture permittivity.

And because many materials are strongly dispersive one needs to pay attention to the way these frequency-dependent properties of media are affected if materials are mixed. It is well-known and was observed even a hundred years ago that the chromatic optical properties of materials may be very different if the material is in bulk form compared to the particulate state [1, 2].

Let us therefore in the present chapter concentrate on the question how the mixing process affects the temporally dispersive properties of materials. The discussion shall be partially continued in Chapter 12 where particular dispersive phenomena in heterogeneous materials are given special attention. A deeper insight into this question is gleaned if the analysis is done in both frequency and time domains.

11.1   Constitutive relations as operators

Susceptibility kernel

The previously introduced constitutive relations were algebraic: the relation between the displacement and the electric field in a point within the material was a scalar multiplication (isotropic), dyadic multiplication (anisotropic), or six-dyadic multiplication (bi-anisotropic materials). Although the relation could be dyadic, it still was local in time and space: the response at a certain point in a certain time instant only depended on the simultaneous excitation at that very point.

For dispersive media, the relation is instead an integral operator. For the sake of simplicity, let us consider dispersion in media which display neither anisotropy, magnetic response, nor magnetoelectric coupling. The dielectric constitutive relation has a general form for a temporally dispersive medium that looks like

$$\mathbf{D}(\mathbf{r}, t) = \epsilon_0(\mathbf{r})\mathbf{E}(\mathbf{r}, t) + (\chi * \mathbf{E})(\mathbf{r}, t) \qquad (11.1)$$

Here, the electric field $\mathbf{E}$ and displacement $\mathbf{D}$ appearing in the constitutive relation have an explicit space and time dependence.

The dispersive part is described by the susceptibility kernel $\chi(\mathbf{r}, t)$ as a function of time.[1] The operation denoted by $*$ stands for the temporal convolution:

$$(\chi * \mathbf{E})(t) = \int\limits_{-\infty}^{t} \chi(t - t')\mathbf{E}(t')\, dt' \qquad (11.2)$$

Due to causality, the dielectric response of the medium cannot be affected by field values in future times. This is the reason that the upper limit in the integral is t. Another way of confirming the same causal state of the response is that the susceptibility kernel $\chi(t)$ has to vanish for $t < 0$, for negative values of its temporal argument. The dependence $\chi(t)$ contains the information about the memory effects

---

[1]Note that this is a different quantity from the nonreciprocity parameter which appeared in Chapter 6. Also, the susceptibility kernel is not dimensionless.

of the medium. The values for small t represent the quickest polarisation response. One may include also a discontinuity in the origin, i.e. it can happen that

$$\chi(t = 0^+) \neq 0$$

Although this option of noncontinuity across the origin is sometimes doubted in the literature [3, p. 310], it does not violate causality.

The optical response

Because of the integral in the constitutive relation (11.1), a certain time is needed for the material response to emerge. In engineering applications, however, the time scales of the field variation are limited, and therefore the material response relation can be simplified by introducing the so-called optical response.

The optical response can be connected with the susceptibility kernel $\chi(t)$ as follows [4]. Assume that the susceptibility kernel is a sum of two terms, a rapidly varying function χ_r and a slowly varying term χ_s. For microwave applications, the electronic polarisability terms that show their dispersion in optical and ultraviolet regimes could be such rapidly varying terms. We therefore can write

$$
\begin{aligned}
\mathbf{D}(t) &= \epsilon_0 \mathbf{E}(t) + (\chi_r * \mathbf{E})(t) + (\chi_s * \mathbf{E})(t) \\
&= \epsilon_\infty \mathbf{E}(t) + (\chi_s * \mathbf{E})(t)
\end{aligned}
\tag{11.3}
$$

where now the optical response is defined as

$$\epsilon_\infty = \epsilon_0 + \int_0^\infty \chi_r(t) dt \tag{11.4}$$

The separation could be made[2] because the field variation $\mathbf{E}(t)$ is much slower than the variation of $\chi_r(t)$.

Now we can see that the practical time-domain constitutive relation (11.3) consists of two parts: the optical response and a dispersive part. The function ϵ_∞ describes the instantaneous response of the material, because the time-dependence mediated by this part is simultaneous in $\mathbf{E}$ and $\mathbf{D}$.

Connection to frequency domain

The frequency-domain counterpart of the dielectric constitutive relation in (11.1) is

$$\mathbf{D}(\omega) = \epsilon(\omega)\mathbf{E}(\omega) \tag{11.5}$$

where the Fourier transformations of the quantities are defined by the following convention:

$$\mathbf{D}(\omega) = \int_{-\infty}^{\infty} \mathbf{D}(t)e^{-j\omega t}\, dt \tag{11.6}$$

---

[2] Of course, if the field has infinite bandwidth, the optical response is $\epsilon_\infty = \epsilon_0$.

$$\mathbf{E}(\omega) = \int_{-\infty}^{\infty} \mathbf{E}(t) e^{-j\omega t} \, dt \tag{11.7}$$

$$\epsilon(\omega) - \epsilon_0 = \int_{-\infty}^{\infty} \chi(t) e^{-j\omega t} \, dt \tag{11.8}$$

The relation (11.1) can also be written for anisotropic and bi-anisotropic materials [5] and then the susceptibility kernel is a dyadic or even a six-dyadic.

The connection between the low-frequency behaviour of the permittivity to the kernels can be seen from

$$\epsilon(\omega = 0) = \epsilon_{\infty} + \int_{0}^{\infty} \chi_s(t) dt = \epsilon_0 + \int_{0}^{\infty} \chi(t) dt \tag{11.9}$$

which can be interpreted such that for sufficiently slow fields ($\omega \to 0$), everything happening in the material is quasi-instantaneous and can be collected in the optical response.

In the following analysis in the present chapter, the practical definition for the constitutive relation will be used where the optical response can be different from the vacuum response. The susceptibility kernel $\chi_s(t)$ will be denoted by $\chi(t)$; the rapid terms are included in ϵ_{∞}.

The subsequent analysis makes use of reference [4].

11.2   Susceptibility models

In this section, let us introduce some important polarisation response models that natural materials display. In any sample of a real material, many different polarisation mechanisms may be found. But in a limited frequency range, one of the following models perhaps can provide a satisfactory approximation for the dielectric behaviour of the medium.

11.2.1   Debye model

The model labelled after Debye describes well the dielectric response of fluids with permanent electric dipole moments. These dipole moments feel a torque in the electric field which means that the polarisation requires time to reach its equilibrium state. This means that the susceptibility kernel reads

$$\chi(t) = \Theta(t) \beta e^{-t/\tau} \tag{11.10}$$

where τ is the so-called relaxation time. The Heaviside function $\Theta(t)$, which vanishes for negative values of t and is unity for positive ones, guarantees causality: the material should not develop polarisation before the field is applied.

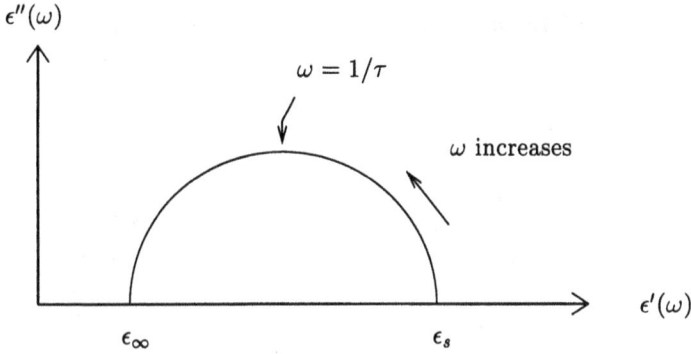

Figure 11.1: *The Cole–Cole plot of the complex permittivity of a Debye material. Both axes have linear scales.*

The Fourier transform of the Debye relation gives us the permittivity in the frequency domain:

$$\epsilon(\omega) = \epsilon_\infty + \frac{\epsilon_s - \epsilon_\infty}{1 + j\omega\tau} \qquad (11.11)$$

where ϵ_s and ϵ_∞ are the low-frequency and optical permittivities of the material. The connection between the parameters is $\beta\tau = \epsilon_s - \epsilon_\infty$. The material is lossy, and the dissipativity of the material is seen from the fact that the imaginary part of $\epsilon(\omega)$ is negative for all frequencies. The imaginary part has its minimum[3] at the relaxation frequency $f_r = 1/(2\pi\tau)$. Figure 11.1 shows the locus of the permittivity in the complex plane. This is the so-called Cole–Cole plot and it shows how the permittivity follows a semicircle as the frequency increases.

Consider next a mixture where spherical inclusions with Debye-type permittivity are in a dispersionless background with permittivity ϵ_e. Then the Maxwell Garnett prediction (3.27) for the effective permittivity of the mixture can be calculated, and it is

$$\epsilon_{\text{eff}}(\omega) = \epsilon_{\infty,\text{eff}} + \frac{\epsilon_{s,\text{eff}} - \epsilon_{\infty,\text{eff}}}{1 + j\omega\tau_{\text{eff}}} \qquad (11.12)$$

In other words, the mixture also obeys the Debye law. The parameters of the mixture are

$$\epsilon_{\infty,\text{eff}} = \epsilon_e + 3f\epsilon_e \frac{\epsilon_\infty - \epsilon_e}{\epsilon_\infty + 2\epsilon_e - f(\epsilon_\infty - \epsilon_e)} \qquad (11.13)$$

$$\epsilon_{s,\text{eff}} = \epsilon_e + 3f\epsilon_e \frac{\epsilon_s - \epsilon_e}{\epsilon_s + 2\epsilon_e - f(\epsilon_s - \epsilon_e)} \qquad (11.14)$$

$$\tau_{\text{eff}} = \tau \frac{(1-f)\epsilon_\infty + (2+f)\epsilon_e}{(1-f)\epsilon_s + (2+f)\epsilon_e} \qquad (11.15)$$

---

[3]Since Im$\{\epsilon\}$ is negative, its absolute value has a maximum.

The result shows that the static and high-frequency permittivities of the mixture are, as expected, Maxwell Garnett averages of the component values. But an interesting fact is that the relaxation frequency of the mixture is different from that of water. As an example of how drastic the change can be, let us consider a mixture of water droplets in air. Water is a good example of a Debye material [6]. At the temperature of +20°C, the values are [7]: $\epsilon_\infty/\epsilon_0 \approx 4.90$, $\epsilon_s/\epsilon_0 \approx 80.1$, and the relaxation time $\tau \approx 10.1$ ps. This means that the relaxation frequency (the frequency at which the imaginary part attains its maximum absolute value) is $f_r \approx 15.8$ GHz.[4]

But now, as water is in the form of droplets, and its volume fraction is small, as is the case in rain and fog, the relaxation time becomes smaller, as can be seen from (11.15). For mixtures with $f \ll 1$, we have $\tau_{\text{eff}} \approx 0.87$ ps. This means that the relaxation frequency for dilute water–air mixtures is around 188 GHz at 20° C. In the next chapter, we shall return to this question in more detail.

11.2.2   Lorentz model

The Lorentz model is a widely used model in solid state physics, and it predicts the frequency dependence of the permittivity function as

$$\epsilon(\omega) = \epsilon_\infty + \epsilon_0 \frac{\omega_p^2}{\omega_0^2 - \omega^2 + j\omega\nu} \tag{11.16}$$

where again ϵ_∞ is the high-frequency permittivity of the material. The other parameters are the plasma frequency ω_p, the resonance frequency ω_0, and the damping amplitude ν, also with dimensions of frequency.

The time-domain counterpart of the Lorentz model shows the susceptibility kernel

$$\chi(t) = \epsilon_0 \, \Theta(t) \frac{\omega_p^2}{\nu_0} \sin(\nu_0 t) e^{-\nu t/2} \tag{11.17}$$

where $\nu_0^2 = \omega_0^2 - (\nu/2)^2$. Note that, unlike in the Debye model, the kernel of the Lorentz model vanishes for $t = 0$.

Use of the Maxwell Garnett rule (3.27) shows that a mixture with Lorentz material in dispersionless background medium is also a Lorentz material, but the model parameters are transformed to

$$\epsilon_{\infty,\text{eff}} = \epsilon_e + 3f\epsilon_e \frac{\epsilon_\infty - \epsilon_e}{\epsilon_\infty + 2\epsilon_e - f(\epsilon_\infty - \epsilon_e)} \tag{11.18}$$

$$\omega_{p,\text{eff}} = \sqrt{f} \frac{3\epsilon_e}{(1-f)\epsilon_\infty + (2+f)\epsilon_e} \omega_p \tag{11.19}$$

$$\omega_{0,\text{eff}}^2 = \omega_0^2 + \frac{1-f}{(1-f)\epsilon_\infty/\epsilon_0 + (2+f)\epsilon_e/\epsilon_0} \omega_p^2 \tag{11.20}$$

---

[4]To treat water as a Debye material is valid only within a limited spectral range. When frequency increases sufficiently, the simple orientational dispersion model fails. For example, water is very well transparent at optical wavelengths. But according to the Debye model (11.12), if extended into the visible window, the attenuation remains considerable.

$$\nu_{\text{eff}} = \nu \tag{11.21}$$

These results show that a mixture has a higher resonance frequency than the inclusion phase, and the shift of $\omega_{0,\text{eff}}$ from ω_0 is largest for dilute mixtures. Another observation is that the plasma frequency of a mixture increases for higher concentrations. This is natural because ω_p in fact is a measure for the permittivity magnitude, as can be seen from Equation (11.16). The damping factor ν is not affected by mixing.

11.2.3   Drude model

The resonant response of Lorentz can be modelled by an electric RLC-circuit [8]. It is interesting to note that if in this model the loss term dominates over the inductance, the result is an overdamped resonance, and the resulting RC-circuit behaves as a Debye model.

Another special case of the Lorentz model is the Drude model, used to describe the optical properties of metals. It comes from (11.16) by letting the electrons be free and setting the resonance frequency to zero, $\omega_0 = 0$:

$$\epsilon(\omega) = \epsilon_\infty - \epsilon_0 \frac{\omega_p^2}{\omega^2 - j\omega\nu} \tag{11.22}$$

The typical conductivity behaviour for low frequencies

$$\epsilon(\omega) \to -\frac{j\sigma}{\omega} \tag{11.23}$$

is apparent in this model, where the conductivity is $\sigma = \omega_p^2 \epsilon_0/\nu$.

The susceptibility kernel of the Drude material is

$$\chi(t) = \epsilon_0 \, \Theta(t) \frac{\omega_p^2}{\nu}(1 - e^{-\nu t}) \tag{11.24}$$

Consider next a mixture where Drude spheres are embedded in dispersionless environment ϵ_e. The Maxwell Garnett prediction for the permittivity of this mixture is a Lorentz model, where $\epsilon_{\infty,\text{eff}}, \omega_{p,\text{eff}}$, and ν are as in the Lorentz mixture (11.18)–(11.21) but $\omega_{0,\text{eff}}$ is different. This means that the mixture does not follow a metal-type Drude model but an insulator-type resonator model (11.16). This is understandable because separate metal particles do not form a conducting lattice and therefore there is no low-frequency divergence of the permittivity. The mixture permittivity has a resonance frequency:

$$\omega_{0,\text{eff}} = \omega_p \sqrt{\frac{1 - f}{(1 - f)\epsilon_\infty/\epsilon_0 + (2 + f)\epsilon_e/\epsilon_0}} \tag{11.25}$$

For dilute mixtures and vanishing high-frequency response of the metal ($\epsilon_\infty = 1$) this condition reads

$$\omega_{0,\text{eff}} = \frac{\omega_p}{\sqrt{1 + 2\epsilon_e/\epsilon_0}} \tag{11.26}$$

Metal colloids have been studied extensively and indeed the condition (11.26) has been observed to hold for many metals like sodium, aluminium, and gold [2, 9]. The experiments show that a strong peak occurs for the absorption cross-section of small spherical metal particles at this frequency $\omega_{0,\text{eff}}$.

The condition that determines the point of strong absorption for the small spherical metal inclusions carries also other names in the literature: terms like surface mode, surface plasmon, or Fröhlich frequency are associated with it. The studies dealing with plasmons and Fröhlich modes do not consider mixing rules but rather the scattering and absorption coefficients of inclusions. Looking at the polarisability of a sphere,

$$\alpha = V(\epsilon_i - \epsilon_e) \frac{3\epsilon_e}{\epsilon_i + 2\epsilon_e} \tag{11.27}$$

it is then obvious that for a frequency at which $\epsilon_i = -2\epsilon_e$ something catastrophic happens, and the amplitude of the resulting absorption peak is determined by the imaginary part of the inclusion permittivity, assuming the background to be real. This condition corresponds to the sparse-mixture limit of the more general resonance condition (11.25). In Chapter 12, the Fröhlich resonances will be considered in more detail.

11.2.4   Modified Debye model

The susceptibility kernel of the Debye model (11.10) was seen to have a discontinuity at the origin $t = 0$. A modification of the Debye kernel has been suggested to smooth out the discontinuity. This kernel looks like

$$\chi(t) = \epsilon_0 \, \Theta(t) \, \omega_p^2 \, t e^{-\nu t/2} \tag{11.28}$$

which still has the same exponential damping character as the original Debye model. The frequency function of the permittivity of the modified Debye model reads

$$\epsilon(\omega) = \epsilon_\infty + \epsilon_0 \frac{\omega_p^2}{(\omega_0 + j\omega)^2} \tag{11.29}$$

and here ω_p and ω_0 are the plasma and resonance frequencies.

A mixture with modified Debye material embedded in a nondispersive host (ϵ_e) obeys the following frequency-dependent permittivity function:

$$\epsilon_{\text{eff}}(\omega) = \epsilon_{\infty,\text{eff}} + \epsilon_0 \frac{\omega_{p,\text{eff}}^2}{\omega_{0,\text{eff}}^2 - \omega^2 + j\omega\nu_{\text{eff}}} \tag{11.30}$$

where

$$\epsilon_{\infty,\text{eff}} = \epsilon_e + 3f\epsilon_e \frac{\epsilon_\infty - \epsilon_e}{\epsilon_\infty + 2\epsilon_e - f(\epsilon_\infty - \epsilon_e)} \tag{11.31}$$

$$\omega_{p,\text{eff}} = \sqrt{f} \frac{3\epsilon_e}{(1-f)\epsilon_\infty + (2+f)\epsilon_e} \omega_p \tag{11.32}$$

$$\omega_{0,\text{eff}}^2 = \omega_0^2 + \frac{1-f}{(1-f)\epsilon_\infty/\epsilon_0 + (2+f)\epsilon_e/\epsilon_0} \omega_p^2 \tag{11.33}$$

$$\nu_{\text{eff}} = 2\omega_0 \tag{11.34}$$

It is seen that the mixture is no longer a modified Debye material but a more general Lorentz material. The effective parameters $\epsilon_{\infty,\text{eff}}, \omega_{p,\text{eff}}, \omega_{0,\text{eff}}$ obey the same rules as a mixture with Lorentz inclusions. A difference in the mixing process is the appearance of a damping factor (absent in the inclusion permittivity) which in the present case is twice the resonance frequency.

Figure 11.2 shows the schematic behaviour of the susceptibility kernels as functions of time. The details of the curves are determined by their model parameters, but particular features of the various models can be recognised from the figure. For example, one might pay attention to the damped resonance of the Lorentz model, the nonzero initial value of the Debye model, and the fact that the response according to the Drude model does not vanish even if one waits for a long time.

11.2.5 Other susceptibility models

Along with the previous basic polarisation models in natural materials, also others exist. The following susceptibility kernel is interesting:

$$\chi(t) = \Theta(t) \frac{\epsilon_s - \epsilon_\infty}{\tau \cos \psi} e^{-t/\tau} \cos(\omega_0 t + \psi) \tag{11.35}$$

where $\tan \psi = -\omega_0 \tau$. The permittivity in the frequency domain reads for this model

$$\epsilon(\omega) = \epsilon_\infty + \frac{1}{2}(\epsilon_s - \epsilon_\infty) \left[\frac{1 + j\omega_0\tau}{1 + j(\omega + \omega_0)\tau} + \frac{1 - j\omega_0\tau}{1 + j(\omega - \omega_0)\tau} \right] \tag{11.36}$$

which was suggested by Van Vleck and Weisskopf [10] for a collection of dipolar molecules whose orientations change rapidly. If the frequency is much higher than the resonant frequency ω_0, this model reduces to the Debye model.

The reader is referred to the monograph by Scaife for more information of material behaviour types [11].

11.3 Mixing in time domain

In the previous section, the dispersive characteristics of mixtures were calculated by their frequency-domain description of the permittivities. This was easy because in the frequency domain, the convolutions with the susceptibility kernels are replaced by algebraic products. And as long as the resulting effective permittivity could be seen to fall into one of the analysed dispersion models, the susceptibility kernel could also be straightforwardly written down.

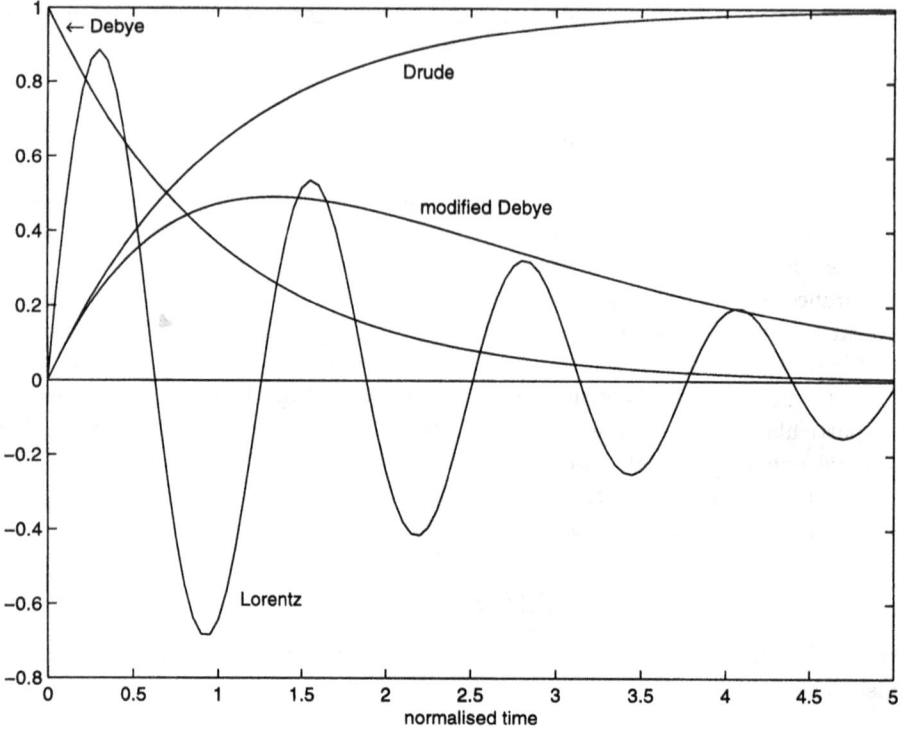

Figure 11.2: *Normalised susceptibility kernels $\chi(t)$ of basic dielectric response models: Debye, Lorentz, Drude, and modified Debye.*

But if two dispersive materials are mixed with each other, more often than not it happens that the effective permittivity of the mixture does not follow any of the models treated before. If one wishes to find out the susceptibility kernel of the effective medium, the mixing analysis is perhaps best performed in the time domain [4].

11.3.1   Quasi-static time-domain fields

The time-domain treatment of the mixing problem requires now that the permittivity ϵ is an operator. To distinguish the operator from an algebraic constant, let us use the symbol ε (rather than ϵ) for the operator of (11.3):

$$[\varepsilon \mathbf{E}](t) \equiv [(\epsilon_\infty + \chi*)\mathbf{E}](t) = [(\epsilon_\infty \delta + \chi) * \mathbf{E}](t) \tag{11.37}$$

where $\delta(t)$ is the Dirac delta function which gives the effect of the simultaneous response via convolution. Using now the permittivity operators, the problem of inclusions (with permittivity operator ε_i) in the environment (operator ε_e) can be solved using the Maxwell Garnett principles.

In the frequency domain, the fundamental quasi-static limitation is often expressed by saying that the size of the scatterer should be small in comparison with the wavelength of the field. However, in the time domain formulation there are no wavelengths. Instead, the corresponding limitation relates the general time variation of the source with the propagation velocity v and the size d of the inclusions. When the "temporal size" $\Delta t = d/v$ of the inclusion is small compared with the inverse of the relative time derivative of the source, we can consider the field inside an inclusion homogeneous even if the field varies temporally. Time only appears as a parameter in the field expressions.

Although the retardation effects are excluded within the homogenisation analysis, memory effects are allowed in the dielectric materials. These are contained in the susceptibility kernels. The high-frequency components of the material response are integrated into the optical response (delta-function) part of the representation (11.37).

For the polarisability of a spherical inclusion, the internal field needs to be solved. The relation between the internal field $\mathbf{E}_i(t)$ and the external field $\mathbf{E}_e(t)$ comes as

$$(\varepsilon_i + 2\varepsilon_e)\mathbf{E}_i(t) = 3\varepsilon_e\mathbf{E}_e(t) \tag{11.38}$$

This equation looks formally similar to the purely static result but attention must be paid to the fact that the permittivities are here convolution operators. Hence, to solve for the internal field, a deconvolution has to be performed:

$$\mathbf{E}_i(t) = (\varepsilon_i + 2\varepsilon_e)^{-1}3\varepsilon_e\mathbf{E}_e(t) \tag{11.39}$$

The polarisability operator reads accordingly

$$\alpha = V(\varepsilon_i - \varepsilon_e)(\varepsilon_i + 2\varepsilon_e)^{-1}3\varepsilon_e \tag{11.40}$$

And finally, when the local field is taken into account in a mixture, the time-domain Maxwell Garnett formula can be written in the form

$$\varepsilon_{\text{eff}} = \varepsilon_e + f(\varepsilon_i - \varepsilon_e)[\varepsilon_i + 2\varepsilon_e - f(\varepsilon_i - \varepsilon_e)]^{-1}3\varepsilon_e \tag{11.41}$$

where f is the volume fraction of the inclusions. One must again note that the inverses here are inverses of convolution operators. The Maxwell Garnett result also requires use of the commutativity property of convolution kernels. The result

$$[(a + A*)(b + B*)]^{-1} = (a + A*)^{-1}(b + B*)^{-1} \tag{11.42}$$

is valid for nonzero constants a and b.

11.3.2 Deconvolution of kernels

A comparison of the results (3.27) and (11.41) gives us the impression that the time-domain and frequency-domain Maxwell Garnett mixing rules are very similar. One must not, however, forget that in the time domain, the inverses mean deconvolutions. Fortunately, very often the deconvolution is a well-behaving operation [4].

When inverting convolutions one can make use of a so-called *resolvent* operator which is another integral operator [12]. Consider an operator with the "optical response" as unity: $\mathcal{A} = 1 + A*$, the inverse of which is defined with the resolvent kernel A_{res} in the following manner:

$$\mathcal{A}^{-1} = (1 + A*)^{-1} \equiv 1 + A_{\text{res}}* = (\delta + A_{\text{res}})* \tag{11.43}$$

This operator satisfies $\mathcal{A}\mathcal{A}^{-1} = \mathcal{A}^{-1}\mathcal{A} = 1$. The resolvent kernel satisfies the linear Volterra equation of the second kind

$$A_{\text{res}}(t) + A(t) + (A_{\text{res}} * A)(t) = 0 \tag{11.44}$$

This is numerically a well-behaving problem and has a unique solution. A series solution for the resolvent reads

$$A_{\text{res}}(t) = \sum_{k=1}^{\infty} (-1)^k \left[(A*)^{k-1} A \right](t) \tag{11.45}$$

11.3.3   A mixture of two Debye materials

As an example where the dispersive behaviour of a mixture is quite complex, consider a mixture where both the spherical inclusions and the background are Debye materials obeying the relation (11.11). The effective permittivity is a complicated function in both frequency and time-domain descriptions [4]. Take the example of the Debye liquids water and alcohol (ethanol). We take the values at 20° C, whence the parameters are for water

$$\epsilon_s/\epsilon_0 = 80.1, \qquad \epsilon_\infty/\epsilon_0 = 4.9, \qquad \tau = 1.01 \times 10^{-11}\,\text{s} \tag{11.46}$$

and those of ethanol

$$\epsilon_s/\epsilon_0 = 25.1, \qquad \epsilon_\infty/\epsilon_0 = 4.4, \qquad \tau = 1.2 \times 10^{-10}\,\text{s} \tag{11.47}$$

Note the 12 times lower relaxation frequency for alcohol compared to liquid water.

The Cole–Cole diagrams display the frequency behaviour of the complex permittivity. Consider two mixtures: alcohol-in-water and water-in-alcohol, and for both mixtures the inclusion volume fraction is 50%. Figure 11.3 shows the first case (water as background, spherical ethanol inclusions) and Figure 11.4 shows the second case (ethanol as background, inclusions as water drops).

The two figures display the dispersion of a mixture with the same amount components. But as expected from the character of the Maxwell Garnett mixing, the effective permittivity behaviour is not the same if the roles of host and guest are reversed. One might observe that the mixture permittivity is "favouring" the host

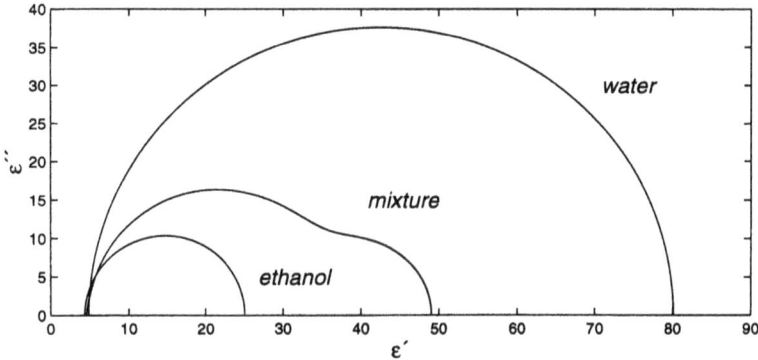

Figure 11.3: *The Cole–Cole plot (the real and imaginary parts of the relative permittivity with frequency as parameter) of the mixture of spherical ethanol droplets in water (50% ethanol). Also shown are the Cole–Cole diagrams of pure water and pure ethanol.*

permittivity: looking at the low-frequency part of the curve (the region of large ϵ'_{eff}), the ethanol-in-water curve (Figure 11.3) is slightly closer to water than the water-in-ethanol curve (Figure 11.4), despite the same alcohol percentage.

But the frequency behaviour of the mixture permittivity is certainly very interesting. It is not a Debye-type semicircle but there exists a bend in the curve, at different frequencies for the water-in-alcohol mixture and alcohol-in-water mixture. This bend is reminiscent of the curve shape of a Davidson–Cole formula for the frequency dependence of the permittivity of polar materials. The Davidson–Cole plot is one of the modifications of Debye behaviour of isotropic media with permanent dipole moments, and in the complex plane the effect is to distort the pure semicircle of the basic Cole–Cole plot [11, Sec. 3.7].

To emphasise that in these examples the mixing results were calculated using the Maxwell Garnett principles, let us repeat the water–alcohol mixture calculation using the Bruggeman model. Figure 11.5 shows the permittivity dispersion of a 50–50 water and ethanol mixture, according to the relation (9.2). The Bruggeman model is symmetric which means that no distinction can be made between the "guest" and "host" phase, and therefore a mixture with equal amounts of both constituents has the same permittivity regardless of whether water forms inclusions and ethanol the environment or vice versa.

A similar bend can be observed in the Bruggeman permittivity for the mixture as appeared in Figures 11.3 and 11.4. Anyway, it is clear that the Bruggeman Cole–Cole plot is different in shape from both Maxwell Garnett mixing cases. The bend appears at the mid-frequencies against the low- and high-frequency bends of the Maxwell Garnett mixings.

If the reader is interested in the correspondence of the frequency dispersion of

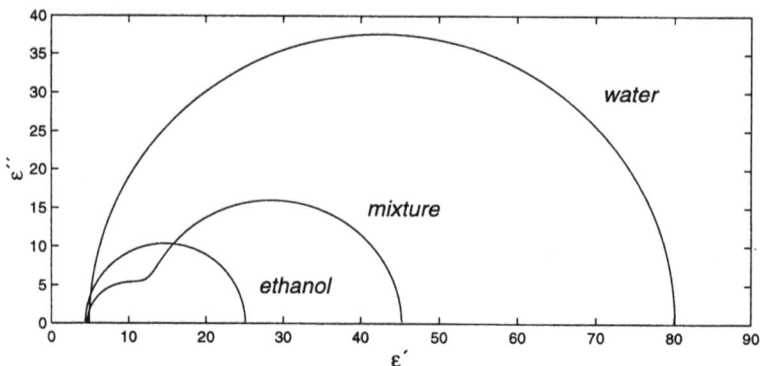

Figure 11.4: *The same as the Cole–Cole plot in Figure 11.3, for a mixture of water drops in ethanol (50% ethanol).*

Debye–in–Debye mixtures in the time domain, a look at the effective susceptibility kernels may be enlightening. These are studied in [4].

11.4   Temporal dispersion in anisotropic and chiral materials

Anisotropic materials

The dielectric response of the models above was assumed to be isotropic. Temporal dispersion meant that the induced electric polarisation had to be calculated as a convolution in the time domain if the Fourier transform is not available. Then the susceptibility (or permittivity) is an integral operator.

This treatment of dispersion can very well be extended to anisotropic materials. Then the susceptibility kernel is a dyadic, or a matrix: $\overline{\overline{\chi}}(t)$, and the components of this dyadic are causal functions of time. Correspondingly, the constitutive relation between the electric field and the displacement has to be written down as

$$
\begin{aligned}
\mathbf{D}(\mathbf{r}, t) &= \overline{\overline{\epsilon}}_\infty \cdot \mathbf{E}(\mathbf{r}, t) + (\overline{\overline{\chi}} * \mathbf{E})(\mathbf{r}, t) \\
&= \overline{\overline{\epsilon}}_\infty \cdot \mathbf{E}(\mathbf{r}, t) + \int_{-\infty}^{t} \overline{\overline{\chi}}(\mathbf{r}, t - t') \cdot \mathbf{E}(\mathbf{r}, t') \, dt'
\end{aligned} \tag{11.48}
$$

This relation in the frequency domain is more familiar:

$$
\mathbf{D}(\mathbf{r}, \omega) = \overline{\overline{\epsilon}}(\mathbf{r}, \omega) \cdot \mathbf{E}(\mathbf{r}, \omega) \tag{11.49}
$$

In this relation the various components of the dyadic may follow different frequency dependencies. An anisotropic solid is a good example. Quartz, for example, has a

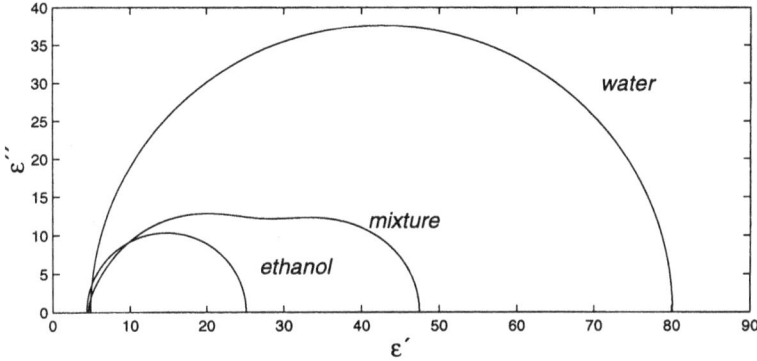

Figure 11.5: *The same as the Cole–Cole plot in Figure 11.3, but calculated according to the Bruggeman mixing rule instead of Maxwell Garnett. The mixture consists of 50% water and 50% ethanol in volume.*

hexagonal crystal structure which makes its optical and dielectric response uniaxially anisotropic. The model for such a medium with electrons and ions as harmonic oscillators leads to the Lorentz dispersion with resonance properties. But the anisotopy means that the spring constants and attenuation are different between excitations along the optical axis and perpendicular to it. Therefore a description of the dispersive properties of quartz needs different parameters in the two directions for the oscillator strengths and resonance and damping frequencies.[5]

Of course, when the materials are anisotropic, the mixing principles have to be applied using dyadic algebra and the relations in Chapter 5. Then more complicated frequency-dependent mixture behaviours can be expected than in the plain isotropic case because the dispersive responses in various directions may be interacting by the mixing process.

Bi-anisotropic materials

Similarly to the anisotropic generalisation of the susceptibility kernels also bi-anisotropic materials can be represented in time and frequency domains. The convolution integral can be kept formally as simple as in (11.48) but now the susceptibility kernel is neither scalar, nor three-dyadic, but a six-dyadic which is a function of time (and possible, space). We have to write

$$\mathsf{X}(\mathbf{r}, t) = \begin{pmatrix} \overline{\overline{\chi}}_{e,e}(\mathbf{r}, t) & \overline{\overline{\chi}}_{e,m}(\mathbf{r}, t) \\ \overline{\overline{\chi}}_{m,e}(\mathbf{r}, t) & \overline{\overline{\chi}}_{m,m}(\mathbf{r}, t) \end{pmatrix} \tag{11.50}$$

---

[5]In reality, a broadband model for the dispersion of a solid needs several oscillators each with their own plasma, resonance, and damping frequencies. In the anisotropic case, then, the multiple oscillators have varying parameters along the principal axis directions.

and play with the rules of six-dyadic algebra (Section 6.2). Likewise to the susceptibility six-dyadic, the frequency-domain permittivity relations have to be taken with magnetic polarisation included, and the material six-dyadic is a function of frequency: $M(\omega)$.

Chiral materials and Condon model

Of bi-anisotropic and bi-isotropic materials, the famous particular type is chiral medium. Chiral media, which were discussed in Section 6.3.1, obey in the bi-isotropic case the following constitutive relations:

$$\mathbf{D}(\omega) \;=\; \epsilon(\omega)\mathbf{E}(\omega) - j\kappa(\omega)\sqrt{\mu_0\epsilon_0}\,\mathbf{H}(\omega) \tag{11.51}$$

$$\mathbf{B}(\omega) \;=\; \mu(\omega)\mathbf{H}(\omega) + j\kappa(\omega)\sqrt{\mu_0\epsilon_0}\,\mathbf{E}(\omega) \tag{11.52}$$

with the dimensionless chirality parameter $\kappa(\omega)$. Chiral materials are microscopically formed of handed elements, like helices. Therefore it is obvious that such a material is dispersive because if the magnetoelectric effect is due to the resonating character of these helices, they respond strongly to exciting fields that have wavelengths of the size scale of the elements. One might also talk about spatial dispersion in connection with these materials because in fact the polarisation response is primarily dependent on the space variation of the field, in other words the magnitude of the wavevector $\mathbf{k}$. But for time-harmonic fields this fact is connected to the frequency variation, and the magnetoelectric couplings are therefore dependent on the temporal variation of the wave.[6]

In the early studies of chiral media and optical activity, dispersion played a central role. In the early part of the 19th century, after French scientists Arago and Biot discovered the rotation of the plane of polarisation of propagating light, the phenomenon was called "optical rotatory dispersion" (ORD). The activity was observed both in solid state materials, like quartz and gypsum, and also in isotropic fluids, like the exhaust gas of boiling turpentine. The rotation is proportional to the real part of the chirality parameter in the constitutive relations. On the other hand, the imaginary part causes the transmitted radiation to become elliptically polarised. This "circular dichroism" (CD) was observed in crystals by Haidinger (1847) and in liquids by Cotton (1895) [14]. Both these terms describing optical activity, ORD and CD, contain the observation that there is spectral structure in the chiral behaviour of these materials, and optical activity is fundamentally a dispersive phenomenon.

Various types of resonance models were proposed for the optical activity of chiral materials starting from Drude at the end of the 19th century. One of the basic models for the dispersion of optical activity was presented by Edward U. Condon in 1937 [15]. His quantum-mechanical theory gave a multiresonant expression that takes into account the transitions between states of optically active molecules and rotational strengths of the absorption lines.

---

[6]Cf. the concise treatment of optical activity as weak spatial dispersion in [13, Chapter 104].

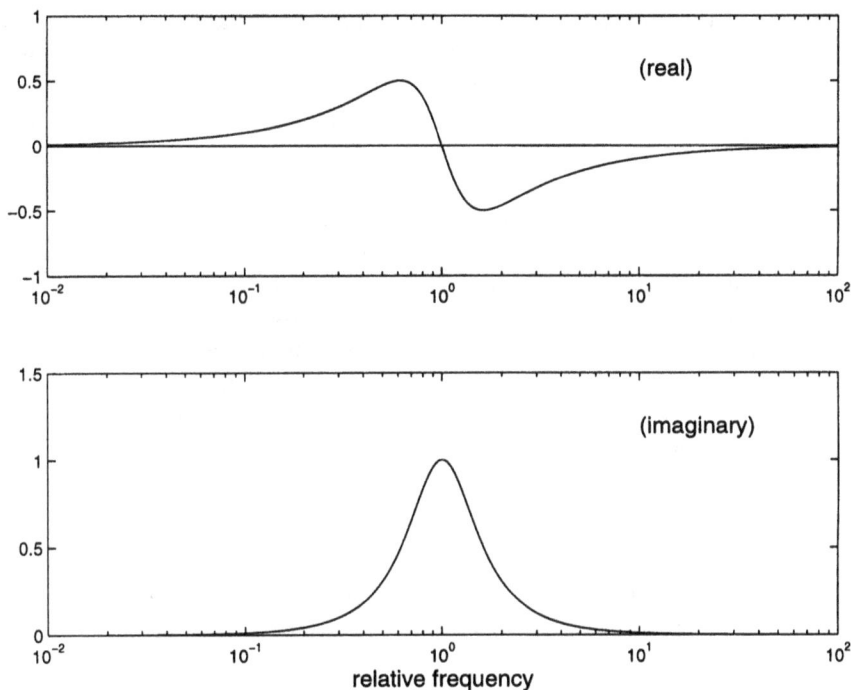

Figure 11.6: *The dispersion of the chirality parameter $\kappa'(\omega) - j\kappa''(\omega)$ according to the one-resonance Condon model. The frequency is normalised to the resonance frequency ω_0.*

If the Condon model is rewritten for the chirality parameter $\kappa(\omega)$ instead of the optical rotation, it looks like

$$\kappa(\omega) = \sum_i \frac{\omega R_i}{\omega_{o,i}^2 - \omega^2 + j\omega\Gamma_i} \tag{11.53}$$

where each term in the summation corresponds to one molecular transition with index i. Here R_i measures the rotational strength of the transition, $\omega_{o,i}$ its resonant frequency, and Γ_i is the associated damping, similarly to the attenuation being included in the Lorentz model (11.16) for the dielectric response of the material. The basic characteristics of the Condon model are illustrated in Figure 11.6 where the real and imaginary parts of the chirality parameter are shown in the case that only one molecular transition is accounted for in the sum of (11.53).

Condon model is capable of explaining certain dispersive properties of chiral materials. For example, the model predicts that the chirality parameter vanishes for DC signals, i.e. $\omega = 0$. This is certainly true: we need to have spatial variation in the electromagnetic wave so that it would feel handedness in matter. Another interesting observation is the fact that the real part of the chirality parameter $\kappa'(\omega)$

changes sign at the resonance frequency. Therefore also the angle of optical rotation that a sample of this material can produce also changes sign as frequency varies. In other words, a material with, say, left-handed microstructure can cause both left- and right-handed optical rotation depending on the frequency of the wave. This is also a known phenomenon, called the Cotton effect.

What can we say about the dispersive properties of mixtures with chiral constituent materials? Although it is perhaps not totally foolish to connect the dispersive properties of chiral media with quasi-static mixing rules,[7] the results of such an endeavour may lack meaning. In order to have considerable optical activity effects, the wavelength of the operating field needs to be close to the size of the handed elements. But the elements themselves constitute an inhomogeneity in the medium, and therefore we violate quasi-static assumptions on which the mixing principles are based if we try to straightforwardly homogenise such a mixture. The order of size scales has to be appreciated.

Problems

11.1 The frequency dependence of the permittivity according to the Debye model is given in Equation (11.12). Calculate the attenuation of an electromagnetic plane wave propagating through such a medium. Plot the attenuation in decibels per metre for water at a temperature of 20° C, when the parameters are $\epsilon_s/\epsilon_0 = 80.1$, $\epsilon_\infty/\epsilon_0 = 4.90$, and $\tau = 10.1$ ps. How large is—in this Debye model—the attenuation in the high-frequency limit, into which the optical window certainly belongs? Compare the result with your everyday experience of the opacity of water.

11.2 Consider a mixture with two dispersive materials. Assume that both obey the Lorentz model (11.16), with parameters $\epsilon_{i,\infty}$, $\omega_{i,p}$, $\omega_{i,0}$, and ν_i for the inclusions, and $\epsilon_{e,\infty}$, $\omega_{e,p}$, $\omega_{e,0}$, and ν_e for the environment. Plot the behaviour of the real and imaginary parts of ϵ_{eff} as a function of frequency.

11.3 Solve the previous problem of a mixture with two Lorentz materials in the time domain. In other words, solve the susceptibility kernel of the mixture. Plot the time-dependence behaviour of $\chi_{\text{eff}}(t)$.

11.4 Prove that the dispersion models

   (a) Debye model, (11.11)

   (b) Lorentz model, (11.16)

   (c) Drude model, (11.22)

   (d) modified Debye model, (11.29)

---

[7]For an attempt in this direction, see [16].

satisfy the Kramers–Kronig relations (2.21)–(2.22).

11.5 Show that the susceptibility kernel and the frequency-dependent permittivity are Fourier-transform pairs for all the four basic models of Section 11.2. The transform between the two domains is given in (11.8).

11.6 Calculate the resolvent kernels $\chi_{res}(t)$ for the four susceptibility models of Section 11.2.

11.7 Plot the behaviour of a Lorentz medium in dispersion curves:

(a) the real and imaginary parts of the permittivity $\epsilon(\omega) = \epsilon'(\omega) - j\epsilon''(\omega)$ as functions of frequency ω

(b) the susceptibility kernel $\chi(t)$ as a function of time

(c) The locus of the permittivity in the complex plane, with $\epsilon'(\omega)$ and $\epsilon''(\omega)$ as the axes.

11.8 Repeat the previous problem for the Drude model.

11.9 Repeat the previous problem for the modified Debye model.

11.10 Repeat the previous problem for the Van Vleck–Weisskopf model.

11.11 For a causal susceptibility kernel $\chi(t)$, the following is evident:

$$\chi(t) = \Theta(t)\chi(t)$$

where $\Theta(t)$ is the Heaviside step function, which vanishes for negative arguments. Take the Fourier transform of this relation, and, using the Fourier transform convolution theorem, derive the Kramers–Kronig relations "in two lines" [17].

References

[1] MAXWELL GARNETT, J.C.: 'Colours in metal glasses and metal films', *Trans. of the Royal Society*, (London), Vol. CCIII, 1904, pp. 385-420

[2] BOHREN, C.F., and HUFFMAN, D.R.: 'Absorption and scattering of light by small particles' (Wiley, New York, 1983)

[3] JACKSON, J.D.: 'Classical electrodynamics' (Second Edition, John Wiley & Sons, New York, 1975)

[4] KRISTENSSON, G., RIKTE, S., and SIHVOLA, A.: 'Mixing formulas in time domain', *Journal of the Optical Society of America*, A, 1998, **15**, (5), pp. 1411-1422

[5] KARLSSON, A., and KRISTENSSON, G.: 'Constitutive relations, dissipation and reciprocity for the Maxwell equations in the time domain', *Journal of Electromagnetic Waves and Applications*, 1992, **6**, (5/6), pp. 537-551

[6] HASTED, J.B.: 'Aqueous dielectrics' (Chapman & Hall, London, 1973)

[7] CHANG, A.T.C., and WILHEIT, T.T.: 'Remote sensing of atmospheric water vapor, liquid water, and wind speed at the ocean surface by passive microwave techniques from the Nimbus 5 satellite', *Radio Science*, 1979, **14**, (5), pp. 793-802

[8] HIPPEL, A. VON: 'Dielectrics and waves' (Artech House, Boston, 1995)

[9] KREIBIG, U., and VOLLMER, M.: 'Optical properties of metal clusters', Materials Science Series, **25** (Springer, Berlin, 1995)

[10] VAN VLECK, J.H., and WEISSKOPF, V.F.: 'On the shape of collision-broadened lines', *Reviews of Modern Physics*, 1945, **17**, pp. 227-236

[11] SCAIFE, B.K.P.: 'Principles of dielectrics' (Clarendon Press, Oxford, 1989)

[12] KRISTENSSON, G., and RIKTE, S.: 'Transient wave propagation in reciprocal bi-isotropic media at oblique incidence', *Journal of Mathematical Physics*, 1993, **34**, (4), pp. 1339-1359

[13] LANDAU, L.D., and LIFSHITZ, E.M.: 'Electrodynamics of continuous media', Second Edition (Pergamon Press, Oxford, 1984)

[14] LINDELL, I.V., SIHVOLA, A.H., TRETYAKOV, S.A., and VIITANEN, A.J.: 'Electromagnetic waves in chiral and bi-isotropic media' (Artech House, Boston and London, 1994)

[15] CONDON, E.U.: 'Theories of optical rotatory power', *Reviews of Modern Physics*, 1937, **9**, pp. 432-457

[16] SIHVOLA, A.H.: 'Temporal dispersion in chiral composite materials: a theoretical study', *Journal of Electromagnetic Waves and Applications*, 1992, **6**, (9), pp. 1177-1196

[17] HU, B.Y-K.: 'Kramers–Kronig in two lines', *American Journal of Physics*, 1989, **57**, (9), p. 821

Chapter 12

Special phenomena caused by mixing

The evident message of mixing formulas is that a mixture has properties that are dependent and determined by those of its constituents but different from them—just like the phenotype of a child may differ in some respects from parents' appearance. Although the dielectric properties of a mixture are a certain average of the component permittivities we have observed in previous chapters that sometimes the whole character of the dielectric behaviour of the mixture is changed by the mixing process. The effective medium can possess properties that are totally absent from the inclusions and the environment. Let us concentrate on description and analysis of such effects in this chapter.

12.1 Dispersion of the permittivity of mixtures

Temporal dispersion, in other words the frequency dependence of the permittivity of various types materials, was analysed in the previous chapter. Frequency dispersion is certainly a character of heterogeneous materials which is crucially dependent on their geometric structure and material parameters of the components. For example, the mixture where inclusions follow a Debye-type dispersion while the background is nondispersive, also obeys Debye-type polarisation response, but with different model parameters. And in Section 11.3 there was discussion on the change of the dispersive character of a mixture when not only the inclusions but both the components were dispersive. Let us now return to the question how the model parameters of a mixture depend on the mixing geometry and materials.

12.1.1 Water and polar molecules

To remind ourselves of the nature of a simple dispersive mixture, take the case of spherical inclusions with Debye-type relaxation behaviour in a nondispersive environment. The result was an effective permittivity according to Equations (11.13)–(11.15). The effect of mixing is that the relaxation time of the dielectric response of the mixture is smaller than that of the inclusions. A good example of such a mixture is the water–air mixture: rain, fog, and clouds. According to the mixing theory, the relaxation frequency is higher for these mixtures than that of water. For water with the Debye dispersion formula

$$\epsilon_w(\omega) = \epsilon_\infty + \frac{\epsilon_s - \epsilon_\infty}{1 + j\omega\tau} \tag{12.1}$$

the following parameters (low- and high-frequency permittivities ϵ_s and ϵ_∞, and the relaxation time τ) can be written as functions of the temperature T in kelvins [1–3]:

$$\epsilon_s/\epsilon_0 = 190.0 - 0.375T \tag{12.2}$$

$$\epsilon_\infty/\epsilon_0 = 4.90 \tag{12.3}$$

$$\tau = \frac{1.99}{T} e^{2140/T} \times 10^{-12} \text{ s} \tag{12.4}$$

Using the result (11.15) for the relaxation time τ_{eff} of the permittivity of the mixture, the relaxation behaviour is illustrated in Figure 12.1. There the relaxation frequency $1/(2\pi\tau_{\text{eff}})$, which for bulk water is around $1/(2\pi\tau) \approx 17\,\text{GHz}$ at room temperature (cf. (12.4)), is shown as a function of the volume fraction of the water phase which is composed of spheres.[1]

As the figure illustrates, the relaxation frequency may be much higher for the mixture compared with bulk water. A look at the real and imaginary parts of the Debye-dispersive permittivity (12.1) tells that the frequency region within the relaxation regime corresponds to high attenuation for waves that propagate through such a dispersive medium. For rain, it is experimentally difficult to study the absorption effect at these millimetre wave frequencies because the scattering effect of raindrops is much higher than the absorption attenuation. However, fog and clouds consist of water droplets that are of micrometre size, which ensures that below $300\,\text{GHz}$ the scattering effects are negligible compared with the absorption phenomena. In fact, broadband microwave and millimetre-wave propagation experiments have been performed to determine the attenuation of fogs and clouds [4, 5], and indeed the maximum attenuation has been observed to be near $200\,\text{GHz}$, although the measurements are often impeded by additional attenuation factors in the atmosphere, such as the absorption peaks of water vapour and other molecules present in the troposphere.

What, then, if the water inclusions in the mixture are not spherical? Consider the case of an anisotropic setting: ellipsoidal inclusions that are all aligned. If the inclusions have the Debye permittivity function $\epsilon_w(\omega)$ according to (12.1) and the constant permittivity of the background is ϵ_e, the Maxwell Garnett prediction for the effective permittivity of the mixture is also a Debye mixture

---

[1] At the relaxation frequency the imaginary part ϵ''_{eff} attains its maximum.

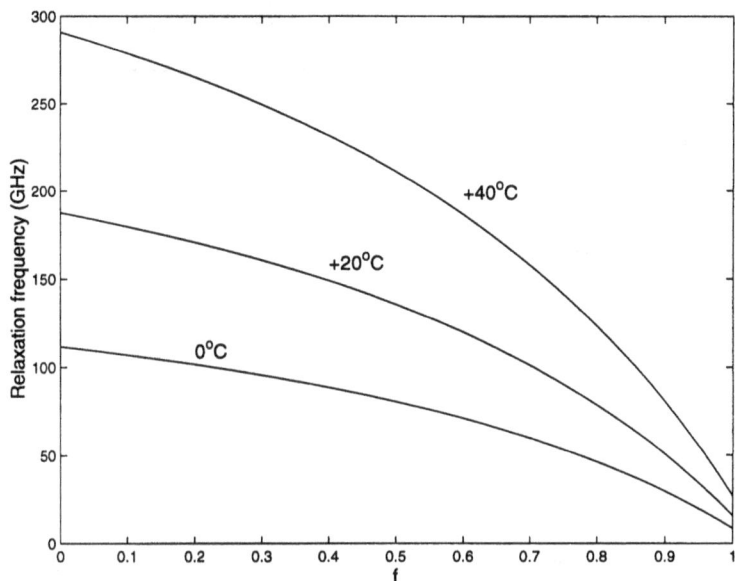

Figure 12.1: *Relaxation frequency of a water–air mixture (spherical water droplets in air) as a function of the volume fraction of the water component. Three different temperature curves are shown.*

$$\epsilon_{\text{eff}}(\omega) = \epsilon_{\infty,\text{eff}} + \frac{\epsilon_{s,\text{eff}} - \epsilon_{\infty,\text{eff}}}{1 + j\omega\tau_{\text{eff}}} \tag{12.5}$$

with parameters

$$\epsilon_{\infty,\text{eff}} = \epsilon_e + f\epsilon_e \frac{\epsilon_\infty - \epsilon_e}{\epsilon_e + (1 - f)N(\epsilon_\infty - \epsilon_e)} \tag{12.6}$$

$$\epsilon_{s,\text{eff}} = \epsilon_e + f\epsilon_e \frac{\epsilon_s - \epsilon_e}{\epsilon_e + (1 - f)N(\epsilon_s - \epsilon_e)} \tag{12.7}$$

$$\tau_{\text{eff}} = \tau \frac{\epsilon_e + (1 - f)N(\epsilon_\infty - \epsilon_e)}{\epsilon_e + (1 - f)N(\epsilon_s - \epsilon_e)} \tag{12.8}$$

where N is the depolarisation factor of the ellipsoids in the direction of the electric field. The drastic change that the mixing causes on the relaxation time is shown in Figure 12.2. There the susceptibility parameters for water at room temperature are used: $\epsilon_s/\epsilon_e \approx 80.1$ and $\epsilon_\infty/\epsilon_e \approx 4.9$, and the environment is assumed to be air. As was observed from Figure 12.1, in fog or rain where the inclusions are spherical, the shift of the relaxation frequency (which is proportional to the inverse of the relaxation time) was over tenfold which corresponds here to the starting point $(f = 0)$ of the curve $N = 1/3$. But now we can see that the shift is still larger if the depolarisation factor is larger than $1/3$. This amount of depolarisation is induced

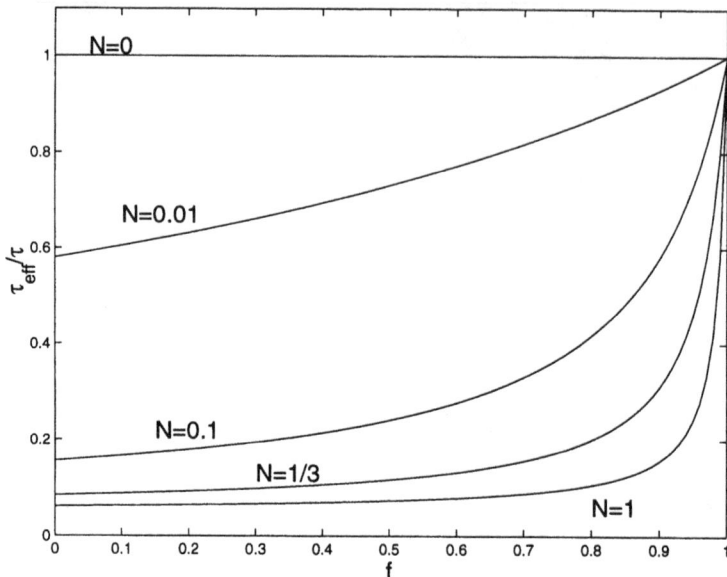

Figure 12.2: *The relaxation time* τ_{eff} *of a mixture relative to the relaxation time of the inclusion material as a function of the volume fraction and the depolarisation factor N of the inclusions in the direction of the field. Here all the Debye-type ellipsoidal inclusions are aligned in the mixture.*

in the case that the ellipsoids are flattened and the electric field vector is parallel to the shorter axis. On the other hand, if the field vector points tangentially to the flat plane of an oblate ellipsoid there is little depolarisation ($N \approx 0$) and the relaxation time of the mixture is approximately the same as that of the inclusions.

The permittivity of the mixture follows the Debye dispersion only if the ellipsoids are aligned. If there is an orientational distribution of the inclusions, the averaging makes the frequency dependence of the effective permittivity a more complicated function.

12.1.2　Metals and Drude dispersion

In the above cases of polar, Debye-inclusions in a nondispersive environment the mixture and the inclusions had the same type of dispersion behaviour. The same is the case if a Lorentz-responding medium[2] forms separated inclusions in a nondispersive matrix; then the plasma frequency and the resonance frequencies change but the dispersion characteristics remain the same.

But as was noted in Chapter 11, dispersive materials exist for which the effective

---

[2]The Lorentz model (Section 11.2.2) describes the dielectric response of many solids, and takes into account the resonance behaviour.

susceptibility kernel may be of different type compared to the inclusion and the environment. This happens for example when metal particles are distributed into a nonconducting host matrix. Then the mixture is a Lorentz-type medium typical of a nonconducting solid rather than metal.

To illustrate this effect, recall that the classical characterisation of metal permittivity is the Drude dispersion (11.22)

$$\epsilon(\omega) = \epsilon_\infty - \frac{\omega_p^2}{\omega^2 - \mathrm{j}\omega\nu} \epsilon_0 \tag{12.9}$$

with the plasma frequency ω_p and the damping amplitude ν. If we now mix small spheres of this type into a nondispersive environment, the mixture response displays a resonant behaviour. In addition to plasma frequency and damping, there is a resonance phenomenon. This situation is depicted in Figure 12.3. There the metallic inclusions display a dispersion for which the imaginary part of permittivity diverges strongly in the low-frequency regime. However, this divergence is absent in the mixture dispersion which only shows a resonance. The imaginary part of the mixture permittivity is small everywhere else except around this resonance region. Of course, the result can be expected because the separated metal spheres only contribute to polarisation and do not make the sample conducting.

The resonance frequency $\omega_{0,\mathrm{eff}}$ of the mixture with spherical Drude inclusions in nondispersive environment (permittivity ϵ_e), according to the Maxwell Garnett model, is

$$\frac{\omega_{0,\mathrm{eff}}}{\omega_p} = \sqrt{\frac{1 - f}{(1 - f)\epsilon_\infty/\epsilon_0 + (2 + f)\epsilon_e/\epsilon_0}} \tag{12.10}$$

The behaviour of the resonance frequency of the mixture is shown in Figure 12.4. The resonance frequency depends on the permittivity contrast $\epsilon_\infty/\epsilon_e$ and the volume fraction of inclusions. Obviously, the resonance frequency disappears for $f \to 1$.

12.2 Polarisation enhancement

Mixing rules can predict sometimes unexpectedly strong polarisation and effective response for inclusions and homogenised materials. Furthermore, there are several possible mechanisms for such enhancement built in the mixing formulas. For the first, this amplification effect can be connected with the coupling of the dipole moments surrounding an inclusion to its local field. On the other hand, it can be due to the interaction between materials with strongly contrasting dielectric properties. And finally, an analysis of the dipole moment in a single scatterer can reveal the existence of distinguished field modes in a case when the material permittivity is dispersive in a particular manner. In this section, the various polarisation enhancement mechanisms are introduced and discussed to some detail.

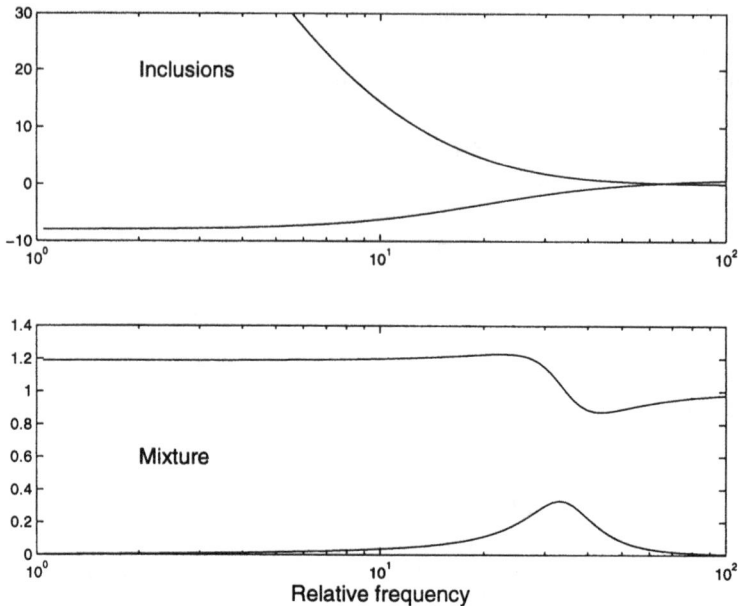

Figure 12.3: *The dispersion of the relative permittivity of Drude inclusions (12.9) with increasing imaginary part and a negative real part towards low frequencies (upper figure). The mixture permittivity (lower figure) shows a resonant behaviour with vanishing imaginary part for both high and low frequencies.*

12.2.1   Mossotti catastrophe

In the basic analysis of the effective dielectric response of a mixture the local field played a very important role. As a reminder from Chapter 3, the field exciting a single particle is dependent on the average field and average polarisation

$$\mathbf{E}_L = <\mathbf{E}> + \frac{1}{3\epsilon_0} <\mathbf{P}> \tag{12.11}$$

provided that the inclusions were spherical. This gave us the effective permittivity as

$$\epsilon_{\text{eff}} = \epsilon_0 + \frac{n\alpha}{1 - \dfrac{n\alpha}{3\epsilon_0}} \tag{12.12}$$

where α is the polarisability of a single inclusion, and n is the number density of the inclusions, assumed here to be located in the free-space environment, permittivity ϵ_0.

The expression (12.12) contains a singular behaviour for the case that the denominator vanishes. For that to happen, the polarisability of an inclusion has to be sufficiently large. There exist models for the polarisability of certain materials

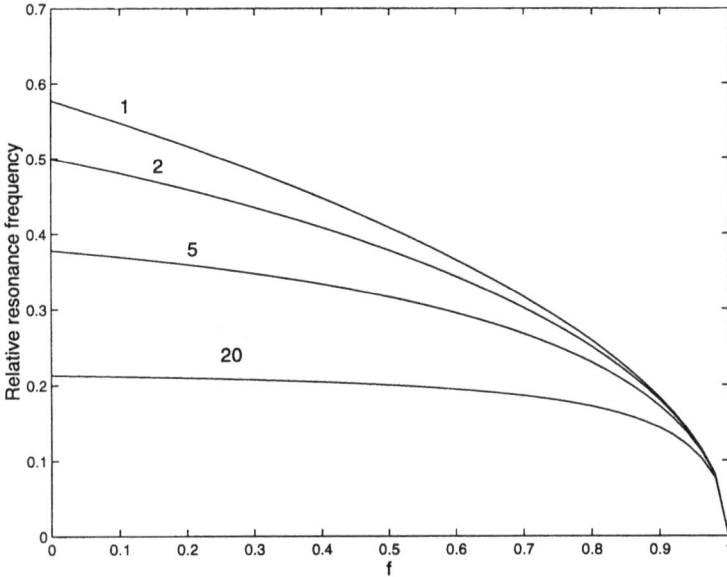

Figure 12.4: *The resonance frequency $\omega_{0,\text{eff}}$ of a mixture relative to the plasma frequency ω_p of the inclusion spheres (Drude dispersion) as a function of the volume fraction and the permittivity ratio $\epsilon_\infty/\epsilon_0$. Note that both the inclusions and the environment (which is here assumed free space: $\epsilon_e = \epsilon_0$) are nonresonant.*

that predict such counterintuitive things. Consider, for example, fluids that contain polar molecules with permanent electric dipole moments. A model for such a fluid is analogous to Langevin-type paramagnetism and in the electric regime known by the studies of Debye.

The random orientation of the moments is maintained by their thermal movement. This causes the average electric polarisation to be zero in the absence of an external electric field. When an external field is applied, the dipole moments tend to align themselves with it. The alignment, however, is only partial because of the thermal fluctuations. Therefore the polarisability is dependent on the temperature T of the fluid. It is inversely proportional to the thermal equilibrium energy kT, where $k \approx 1.3807 \times 10^{-23}$ J/K is the Boltzmann constant.

The classical Debye model for the average orientational polarisability per dipole molecule is

$$\alpha_{\text{or}} = \frac{\mu^2}{3kT} \tag{12.13}$$

where μ is the magnitude of the permanent dipole moment of a single molecule.

This basic model predicts increasing polarisabilities for falling temperatures owing to the inverse proportionality to the temperature. Connecting (12.13) with the basic Clausius–Mossotti relation (12.12) we can observe that at the critical temperature T_c the effective permittivity of the material approaches infinity:

$$T_c = \frac{n\mu^2}{9\epsilon_0 k} \tag{12.14}$$

Naturally, other polarisation mechanisms than orientational are here neglected. Using the critical temperature, the effective permittivity (12.12) can be rewritten as

$$\epsilon_{\text{eff}} = \epsilon_0 + 3\epsilon_0 \frac{T_c}{T - T_c} \tag{12.15}$$

and here the "Mossotti catastrophe" is visible as the temperature decreases to T_c. A similar temperature dependence of the polarisation response is known in the nature of ferromagnetism. There the linear dependence of the inverse magnetic susceptibility on the temperature is known as the Curie–Weiss law.

Ferromagnetism is a commonly observed phenomenon. In the electric regime, one is tempted—using the simple reasoning that led to (12.15)—to predict an analogous effect, a ferroelectric behaviour for polar substances. How would this take place? As the temperature decreases, the feedback mechanisms of the surrounding polarisation to the Lorentzian field grow. The growth of the excitation of a single molecule exceeds all limits and a spontaneous polarisation would take place. However, in the everyday world, ferroelectrics are rare, and this type of Mossotti catastrophe does not occur in ordinary polar materials, like water. According to this simple ferroelectric model [6], water should solidify when the temperature falls below 1520 K !

12.2.2   Onsager model

Liberation from the Mossotti catastrophe can be sought from a closer analysis of the exciting field. The classical way to account for this fictitious field is too simple in the case of permanent dipole moments. The neighbouring polarisation, in the Lorentzian picture which was followed in Chapter 3, is allowed to have a too strong effect on the exciting field. Lars Onsager in 1936 gave the first model where this excitation was analysed in more detail [7].

Onsager made the distinction between the so-called *cavity* field and the *reaction* field which both are present in the surroundings of a given permanent dipole in the fluid. The cavity field is the static field that can be calculated in a cavity cut within the continuum. This is the field component that contributes to orienting the permanent dipole. But it is important to note that this field is only caused by the external sources.

On the other hand, in the Onsager model there is an additional field in the internal field: the reaction field. In the reaction field, the effects of the oriented permanent dipoles in the surrounding polarisation are allowed. But this field is parallel to the permanent dipole moment in the cavity, and hence does not contribute to orienting it. Therefore, because of the lack of feedback mechanism, the "boot-strapping" effect of the orientational polarisation through the local field, and the subsequent Mossotti catastrophe, is avoided.

The response of the orientational polarisation affects the low-frequency behaviour of the effective permittivity. Therefore the Onsager model makes the distinction between the effective permittivity ϵ_{eff} and the internal permittivity of the mixture which is the frequency-independent effective permittivity ϵ_{np}, caused by the nonpolar components in the material. The final result, according to Onsager theory, for the effective permittivity reads

$$\frac{(\epsilon_{\text{eff}} - \epsilon_{\text{np}})(2\epsilon_{\text{eff}} + \epsilon_{\text{np}})}{\epsilon_{\text{eff}}(\epsilon_{\text{np}} - \epsilon_0)} = \frac{n\mu^2}{9\epsilon_0 kT} \tag{12.16}$$

and here again, ϵ_{np} is the effective permittivity of the mixture with no polar molecules present. Obviously, for $n = 0$, the effective permittivity is the same as the ordinary Mossotti prediction: $\epsilon_{\text{eff}} = \epsilon_{\text{np}}$.

This theory, while good for nonassociated polar liquids, fails to take into account the short-range order of the molecules due to bonding. Hindered relative rotation of neighbouring molecules produces a correlation between their orientations which requires a more detailed analysis. Such analysis has been attempted by Kirkwood [8].

12.2.3 Single scattering and Fröhlich modes

In the study of the optical and dielectric response of materials, the character of metal colloids and solutions has historically been very important in advancing the understanding of how the mixing process affects the dispersion in materials. In medieval and ancient times, glassware recipes were very valuable with which craftsmen could produce vessels of preciously purple colour [9]. After the birth of electricity as science, Faraday, Clerk Maxwell, Mie, and Maxwell Garnett paid much attention to the colours of the solutions containing solid metal particles, like fine spherical grains of silver, gold, sodium, and potassium. For example, Maxwell Garnett [10] was able to explain the colours of thin films and glasses that were composed of minute spheres of various metals although the colours displayed by the light trasmitted through the glass were seemingly unrelated to the colour of the corresponding metal in bulk form.

Surface modes

The absorption characteristics of particles in solutions are quite often analysed with the concept of *surface modes*. These modes correspond to certain conditions for the Mie scattering coefficients. The electromagnetic scattering by a sphere is analysed by Mie scattering as was qualitatively described in Section 10.2, and the surface mode conditions correspond to the situation that these coefficients grow large. Mie scattering is presented in series form with these coefficients, and if the limiting value of the coefficients for low frequencies is studied, the condition that the denominator of the nth coefficient in the series vanish is [11]

$$\frac{\epsilon_i}{\epsilon_e} = -\frac{n+1}{n} \tag{12.17}$$

where a nonmagnetic sphere with permittivity ϵ_i is assumed to be located in a dielectric environment of permittivity ϵ_e. When this condition is satisfied, the nth term dominates scattering. The electric energy density for a given mode is the more concentrated to the surface of the sphere the larger n is. For the lowest term, $n = 1$, the internal field is constant. This mode of uniform polarisation is responsible for the ordinary Rayleigh scattering phenomena.

For small spheres, it is sufficient to concentrate on the mode $n = 1$. This uniform internal field corresponds to the quasi-static situation which has been the basis of mixing formulas earlier in the present text. Then the condition for the surface mode is

$$\epsilon_i = -2\epsilon_e \tag{12.18}$$

and it is easy to understand that this corresponds to anomalously strong polarisation in the inclusion which has the polarisability

$$\alpha = 3V\epsilon_e \frac{\epsilon_i - \epsilon_e}{\epsilon_i + 2\epsilon_e} \tag{12.19}$$

The condition (12.18) makes the denominator vanish and α grows without limits. If the polarisability is extremely large, the sphere scatters very strongly (even in this case that it is electrically small). A consequence is that the mixture extinguishes the propagating wave.

Fröhlich frequencies

But the condition (12.18) means that the real part of the inclusion permittivity relative to the environment has to be negative. This is well possible for metals with Drude dispersion and so also for solid materials in the region of anomalous dispersion as was discussed in Section 11.2. Of course the permittivity is then strongly dependent on the frequency. For a particular frequency the real parts of the two permittivities in the denominator of (12.19) cancel each other; then the polarisability amplitude is determined by the imaginary part of the inclusion permittivity. If that is small at the frequency where the cancellation takes place, a very strong extinction is to be expected. Such a frequency is called the *Fröhlich frequency*, and it is evident that a mixture with small metal spheres has clear spectral lines in its transmission and scattering spectra at these Fröhlich frequencies. The bulk metal is naturally opaque around these frequencies because the negative permittivity means that all radiation is reflected. But since the bulk metal may be very weakly dispersive in the vicinity of the Fröhlich frequencies, it is understandable that the colour of the bulk metal clearly differs from the mixture colour.

Among the messages conveyed by Section 4.2 was that the polarisability of a small dielectric inclusion is dependent on the depolarisation factor of the sphere. Hence the polarisability of an ellipsoid, for example, is determined by its shape. And for an ellipsoid the relation corresponding to the sphere relation (12.19) reads

$$\alpha = V\epsilon_e \frac{\epsilon_i - \epsilon_e}{\epsilon_e + N(\epsilon_i - \epsilon_e)} \tag{12.20}$$

where N is the depolarisation factor of the ellipsoid in the field direction. Considering now the condition for Fröhlich frequencies, the polarisability is seen to vanish for such a frequency where the inclusion permittivity relative to the environment is

$$\frac{\epsilon_i}{\epsilon_e} = -\frac{1 - N}{N} \tag{12.21}$$

This relative permittivity can vary from zero to negative infinity between discs and needles. Given the spectral characteristic of a material with negative values for the permittivity it is easy to find the required depolarisation factor N that would yield a desired Fröhlich frequency according to the relation (12.21). Micromachining small metal inclusions into the required ellipsoidal shape, and mixing those in dispersionless matrix, composites with sharp and predetermined colours can be designed. In this connection a reference to the appendices of the extensive textbook by Kreibig and Vollmer [9] is in place, where very detailed tabulations and illustrations of the optical properties of various metal clusters can be found.

Although here the terms "surface modes" and "Fröhlich frequencies" have been dominantly used in connection with the polarisation enhancement process, it is not uncommon to find the mechanism analysed with other concepts and quantities. Sometimes the surface mode is treated as dipolar *surface-plasmon resonance*, and analysed with the Mie scattering theory. The sharp Fröhlich frequencies according to the resonance conditions above may of course be red-shifted or blue-shifted owing to size-dependent effects for inclusions which are no longer very small compared to the effective wavelength of the radiation.[3]

The same surface-mediated plasmon is also responsible for the enhancement of other optical and dielectric effects in composites. In the study of nonlinear responses of gold, aluminium, and polystyrene inclusions in water, a very strong increase of the third-order nonlinear susceptibility has been observed. Enhancement factors of 10^8 have been reported for the phase-conjugate reflectivity [12].

---

[3]Red-shift is more common because the real part of the permittivity is often an increasing function around the Fröhlich regime [11].

12.3   Percolation

Percolation and phenomena associated with it have attracted certain attention within several different fields of science, after the early studies of Hammersley [13] some decades ago. Percolation theory has indeed been successful and very applicable to several phenomena in different fields of physics and even other disciplines [14,15]. Among these applications one may mention ferromagnetism, soil moisture studies, oil penetration in rocks, the spread of epidemics, forest fires, and wafer-scale integration in the manufacture of microchips.

Percolation is inherently connected with heterogeneous and random media. *Percolate* means to flow through. Essential in percolation is the very process of flowing and the manner in which the associated flow finds its way. This, naturally, depends on the microstructure of the material. What flows need not be a fluid: it can even be a rumour in the network formed by a group of humans, or, like in the case of dielectric mixtures, it can be the electric flux. Percolation is a nonlinear phenomenon in the sense that a very abrupt change in the behaviour of certain parameters of the percolating material suddenly occurs. The geometry of the matter where percolation takes place is very special: be there even small changes in the fractions of the components forming the material, the structure behaves totally differently. This fact is characteristic to percolation processes. In percolation studies one often encounters terms like site and bond percolation, mixed and oriented percolation, lattice animals, and clustering. Often the parameters characterising percolation behavior in the critical region have been enumerated through Monte Carlo simulations [16]. In this section, let us study how parameters associated with percolation—especially the percolation threshold—are connected with mixing rules [17].

12.3.1   Generalised mixing rule

Let us return to the familiar isotropic two-phase mixture with spherical inclusions (permittivity ϵ_i) occupying the volume fraction f in the environment of permittivity ϵ_e. Then the generalised mixing rule (9.7)

$$\frac{\epsilon_{\text{eff}} - \epsilon_e}{\epsilon_{\text{eff}} + 2\epsilon_e + \nu(\epsilon_{\text{eff}} - \epsilon_e)} = f\frac{\epsilon_i - \epsilon_e}{\epsilon_i + 2\epsilon_e + \nu(\epsilon_{\text{eff}} - \epsilon_e)} \tag{12.22}$$

gives us the effective permittivity according to several mixing models, including Maxwell Garnett ($\nu = 0$) and Bruggeman ($\nu = 2$), as was shown in Chapter 9.

Consider this general mixing formula from the percolation point of view. Does a percolation phenomenon exist, and, if this is the case, how does one extract the percolation threshold dependence on ν? Percolation threshold is an important parameter. It is the volume fraction around which the effective permittivity changes strongly.

One approach is to let the dielectric contrast ϵ_i/ϵ_e in Equation (12.22) become large. In that case, the equation is approximately

$$\frac{\epsilon_{\text{eff}} - \epsilon_e}{\epsilon_{\text{eff}} + 2\epsilon_e + \nu(\epsilon_{\text{eff}} - \epsilon_e)} = f \qquad (12.23)$$

However, this relation is only valid as long as $\epsilon_{\text{eff}}/\epsilon_e$ is small, i.e. only up to those volume fraction values where percolation appears. The effective permittivity is in this region

$$\epsilon_{\text{eff}} = \epsilon_e \frac{1 + f(2 - \nu)}{1 - f(1 + \nu)} \qquad (12.24)$$

Clearly this result breaks down as the denominator reaches the value zero. That very point is interpreted as the percolation threshold point $f = f_c$:

$$f_c = \frac{1}{1 + \nu} \qquad (12.25)$$

Depending on ν, the threshold varies between $f_c = 1$ (Maxwell Garnett) and $f_c = 0$ ($\nu = \infty$). The simple result and the values for the percolation threshold f_c in different mixing formulas are collected in Table 12.1.

Table 12.1: *Percolation threshold for different mixing models*

ν	Mixing rule	Percolation threshold f_c
0	Maxwell-Garnett	1
1		0.5
2	Bruggeman	0.333
3	Coherent potential	0.25

In addition to permittivity percolation, it is instructive to take a look at conductivity percolation. In fact, the analysis is only a little more complicated for the conductivity behaviour. The imaginary part of the permittivity ϵ''_{eff} is proportional to conductivity.[4] Provided that the conductivity of the inclusion phase is the dominant phenomenon of the mixing process, the following result applies in the subpercolation regime ($f < f_c$)

$$\epsilon''_{\text{eff}} = \epsilon_e \sqrt{\frac{2 - \nu}{1 + \nu} \cdot \frac{1 + f(2 - \nu)}{1 - f(1 + \nu)}} \qquad (12.26)$$

Here, the assumption $\epsilon''_{\text{eff}} \gg \epsilon'_{\text{eff}}$ has been made.

The earlier reasoning that was applied to the real part of the effective permittivity ϵ'_{eff} is also valid here. Looking for the point where the denominator vanishes,

---

[4]The connection is $\epsilon''_{\text{eff}} = \sigma_{\text{eff}}/\omega$.

the same percolation thresholds (of Equation (12.25)) apply for the conductivity as for the permittivity mixing.

Numerically, the effects are clearly seen in Figure 12.5. There, the real and imaginary parts of the effective permittivity are shown as a function of the volume fraction of the inclusions. A strong contrast is chosen between the phases: $\epsilon_i/\epsilon_e = 500 - j500$. An obvious observation is that there exists a threshold volume fraction at which the global character of the permittivity changes. The percolation threshold is clearly in agreement with the results of Table 12.1.

And also the percolation threshold for a given ν is the same for both ϵ'_{eff} and ϵ''_{eff}. An interesting fact also appears from Figure 12.5 which is not obvious from the analytical reasoning of the previous paragraphs. This is the fact that the percolation behaviour is more pronounced for conductivity: for dielectric and conductivity contrasts of similar magnitude, the percolation bends are much sharper in the conductivity curves compared to permittivity curves.

12.3.2   Effect of spatial dimension

Percolation behaviour is also dependent on the dimension d of the space which permeates the medium. The analysis of the previous section is valid for three-dimensional space, $d = 3$. But the fact that the electric polarisability of a two-dimensional sphere is different from a three-dimensional one makes mixing formulas dimension-dependent. A two-dimensional sphere is a circle. The polarisability of a circle is the same as that of a circular cylinder when the electric field is limited to the plane perpendicular to the cylinder axis. The polarisability of a circle with permittivity ϵ_i is

$$\alpha = 2V\epsilon_e \frac{\epsilon_i - \epsilon_e}{\epsilon_i + \epsilon_e} \tag{12.27}$$

corresponding to an ellipsoid with depolarisation factor $N = 1/2$.[5] In general, the polarisability of a d-dimensional sphere is

$$\alpha = d\,V\epsilon_e \frac{\epsilon_i - \epsilon_e}{\epsilon_i + (d-1)\epsilon_e} \tag{12.28}$$

The appearance of Equation (12.22) in the case where d-dimensional spheres are embedded in a background medium is

$$\frac{\epsilon_{\text{eff}} - \epsilon_e}{\epsilon_{\text{eff}} + (d-1)\epsilon_e + \nu(\epsilon_{\text{eff}} - \epsilon_e)} = f\frac{\epsilon_i - \epsilon_e}{\epsilon_i + (d-1)\epsilon_e + \nu(\epsilon_{\text{eff}} - \epsilon_e)} \tag{12.29}$$

This means that Equation (12.24), which gives the effective permittivity below percolation, looks in d dimensions like

---

[5]Here, α and V are the two-dimensional quantities for the polarisability and volume of the circle.

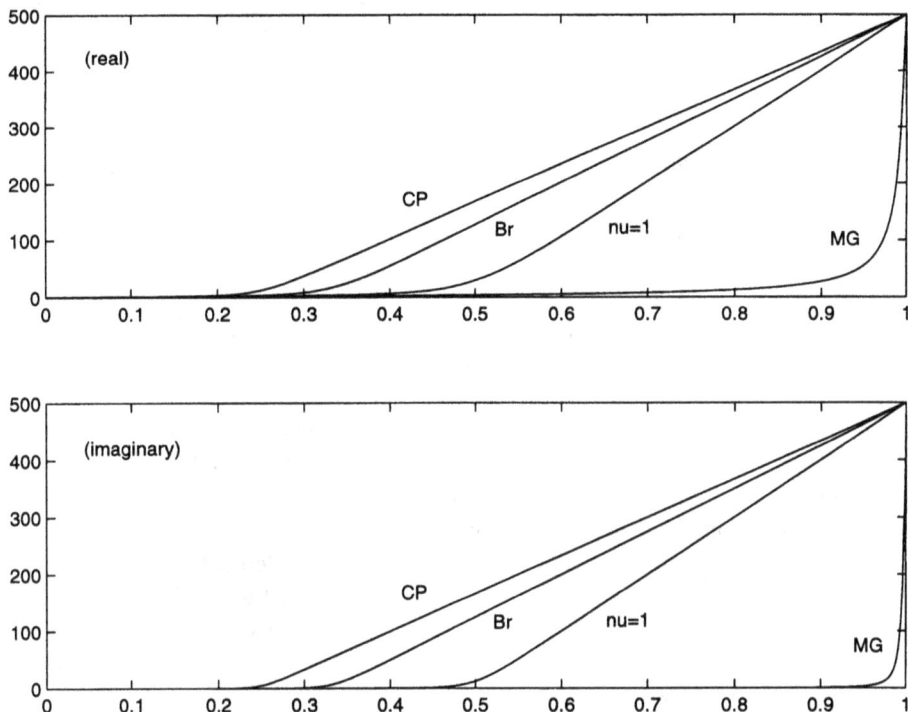

Figure 12.5: *The effective permittivity of a high-contrast mixture: the inclusions (three-dimensional spheres) relative to the environment have permittivity $\epsilon_i/\epsilon_e = 500 - j500$. The effective permittivity relative to the environment $\epsilon_{eff}/\epsilon_e$ is shown as a function of the volume fraction of the inclusion phase. The unified mixing rule (12.22) is used with the four values for the parameter ν: 0 (Maxwell Garnett), 1, 2 (Bruggeman), and 3 (Coherent potential). Upper figure: the real part; lower figure: the imaginary part of the relative effective permittivity.*

$$\epsilon_{\text{eff}} = \epsilon_e \, \frac{1 + f[d - (1 + \nu)]}{1 - f(1 + \nu)} \qquad (12.30)$$

allowing us to readily calculate the percolation threshold in d dimensions.

The result clearly is again the same, $f_c = 1/(1 + \nu)$, as in the three-dimensional case and the conclusion is that the percolation threshold of a given model does not depend on the dimension of the space of this mixing problem.[6]

The conductivity behaviour in d dimensions can be accordingly seen from the

---

[6]An exception, however, is the case $d = \infty$. In that case Equation (12.29) gives $\epsilon_{\text{eff}} = \epsilon_e + f(\epsilon_i - \epsilon_e)$, i.e. the mixing rule becomes a volume average of the permittivities. Furthermore, the percolation behaviour degenerates into threshold at $f = 0$, independent of ν. This kind of a mixture has sometimes been termed "Bethe lattice" [14].

formula (valid again below percolation, $f < f_c$)

$$\epsilon''_{\text{eff}} = \epsilon_e \sqrt{\frac{d - 1 - \nu}{1 + \nu} \cdot \frac{1 + f(d - 1 - \nu)}{1 - f(1 + \nu)}} \tag{12.31}$$

This simplifies to (12.26) for $d = 3$. And since here the denominator does not depend on d, the percolation threshold is the same in all cases regardless of the number of dimensions of the problem, just as in the case of permittivity.

Do the effective permittivity curves, then, look identical in two and three dimensions, if now the beginning, end, and threshold points are the same? The global character is certainly similar but differences can be observed in the sharpness of the effective behaviour around the percolation point. Let us illustrate this effect using Figure 12.6.

In Figure 12.6, a vertical cut is taken of the ordinary effective permittivity curves. There, ϵ_{eff} is shown at a given volume fraction ($f = 0.333$), and as a function of the dimension d of the spheres (with permittivity $\epsilon_i/\epsilon_e = 500 - j500$). The volume fraction is intentionally taken to be around the percolation threshold of the $\nu = 2$ model. Then the curves show that indeed, ϵ_{eff} does not depend too much on the spatial dimension of the spheres in the mixture, expect in the particular case if we are close to percolation. Then we can observe that a decrease of the dimension lowers the effective permittivity. From this it can be inferred that in lower dimensions the percolation effect is sharper. Another observation that is worth mentioning from the curves is that again, the effect is different between the real and imaginary parts of the effective permittivity. Concerning ϵ''_{eff}, the dielectric contrast used in the calculations is so strong that there is practically no variation with d unless we are at the percolation threshold (which happens for $\nu = 2$).

Now it is safe to conclude that the dependence of the percolation threshold on ν (Equation (12.25)) is independent of the spatial dimension d of the problem. But the threshold dependence on d of a given mixing model needs to be looked at more carefully. For example, the important Bruggeman mixing rule corresponds to the case $\nu = 2$ for the three-dimensional mixing, according to the rule (12.22). But if the inclusions are two-dimensional spheres, the Bruggeman model appears from (12.29) with the choice $\nu = 1$.[7] This means that the percolation threshold for the Bruggeman mixing rule is $f_c = 0.333$ for three-dimensional mixtures but $f_c = 0.5$ for the two-dimensional case.

Another observation is that the Maxwell Garnett rule is an extreme model as far as percolation behaviour is concerned. The percolation threshold is pushed to the limit $f = 1$, to such a situation that the guest phase reclaims all space within the mixture. This is, however, not forbidden: experimental studies show that the

---

[7]Reminder from Equation (9.1): the mixing according to the Bruggeman philosophy would read in d dimensions as:

$$\sum_j f_j \frac{\epsilon_j - \epsilon_e}{\epsilon_j + (d - 1)\epsilon_e}$$

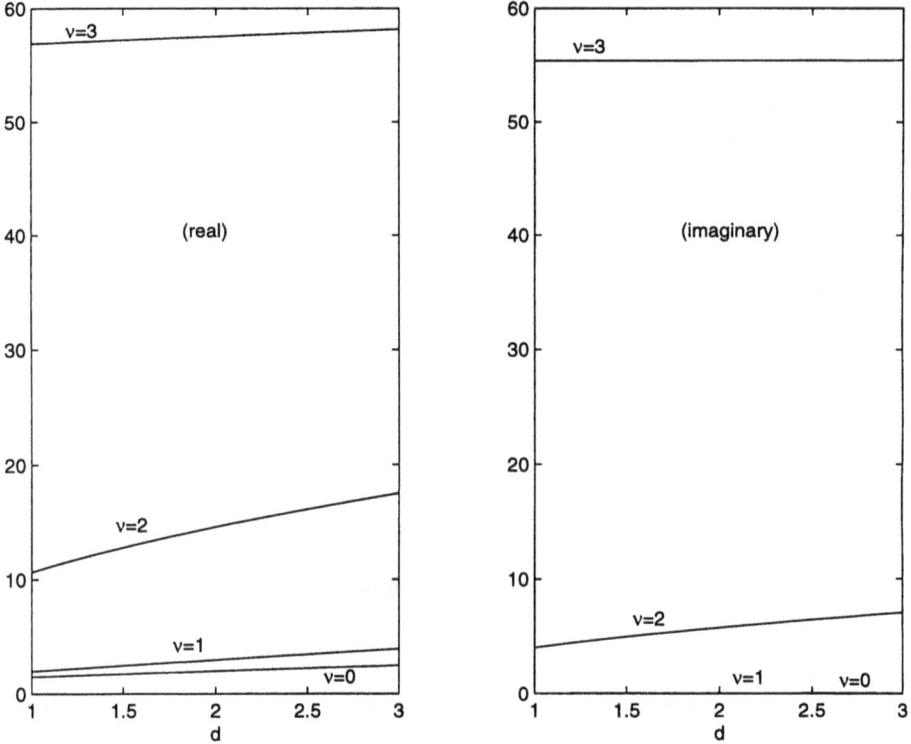

Figure 12.6: *The effect of spatial dimension d on ϵ'_{eff} and ϵ''_{eff} for various mixing models ($\nu = 0, 1, 2, 3$). The mixture consists of d-dimensional spheres with permittivity contrast $\epsilon_i/\epsilon_e = 500 - j500$ to the environment. The volume fraction is fixed: $f = 0.333$, corresponding to the percolation threshold for $\nu = 2$. Note that in the right-hand figure (ϵ''_{eff}), the values for $\nu = 0$ and $\nu = 1$ are so small that the curves cannot be distinguished from the $\epsilon''_{\text{eff}} = 0$ axis.*

percolation threshold can have a broad variation. Landauer [18] holds that the figure of percolation threshold can be anything between 0% and 100%. It is true that real-life f_c values are usually towards the lower end of inclusion volume fraction, and this is an often-mentioned weakness of Maxwell Garnett mixing principles. But in a larger perspective, this fact that the percolation does not clearly emerge from the MG mixing rule does not need to discourage supporters of the Maxwell Garnett theory in the competition against Bruggeman mixing theories because on the other hand, MG beautifully predicts the Mossotti catastrophe and the existence of Fröhlich frequencies. These real effects are much more difficult to explain using Bruggeman effective-medium models, as is often pointed out in the literature [19–21].

Problems

12.1 Figure 12.1 shows the relaxation frequency of a mixture where spherical water drops occupy the volume fraction f in dispersionless air. For $f = 1$ it is clear from the curves that the relaxation frequency of the mixture is the same as that of bulk water. However, for the other, dilute mixing limit ($f = 0$), the mixture possesses another relaxation frequency. But for that limit, the mixture is equivalent to air, which was assumed to display no relaxation behaviour.

Explain this paradox.

12.2 Show that the anisotropic mixture with aligned Debye-type ellipsoids in a nondispersive environment also follows the Debye law (derive equations (12.5)–(12.8)). Use (4.25).

12.3 Derive relation (12.26) from which the percolation threshold can be determined for the conductivity behaviour of the mixture.

12.4 Consider the mixture that was termed "Bethe lattice" in Section 12.3.2. What kind of geometrical structure would such a mixture correspond to in our three-dimensional world?

12.5 Compare how differently polarisation enhancement is predicted by Maxwell Garnett and Bruggeman models. Take a two-component mixture where spherical inclusions have permittivity $\epsilon_i/\epsilon_e = -2-\mathrm{j}0.01$ relative to the environment. Plot the effective permittivity as a function of the volume fraction according to both models. Try to include a little dispersion to the inclusion permittivity to be able to conclude something intelligent about how differently Fröhlich frequencies appear in these models.

References

[1] CHANG, A.T.C., and WILHEIT, T.T.: 'Remote sensing of atmospheric water vapor, liquid vater, and wind speed at the ocean surface by passive microwave techniques from the Nimbus 5 satellite', *Radio Science*, 1979, **14**, pp. 793-802

[2] KAATZE, U.: 'Microwave dielectric properties of water', *in* KRASZEWSKI, A. (Ed.): 'Microwave aquametry' (IEEE Press, Piscataway, N.J., 1996), Chap. 2, pp. 37-53

[3] BARAJAS, O., and BUCKMASTER, H.A.: 'Calculation of the temperature dependence of the Debye and relaxation activation parameters from complex permittivity data for light water' *in* KRASZEWSKI, A. (Ed.): 'Microwave aquametry' (IEEE Press, Piscataway, N.J., 1996), Chap. 3, pp. 55-66

[4] GERACE, G.C., and SMITH, E.K.: 'A comparison of cloud models', *IEEE Antennas and Propagation Magazine*, 1990, **32**, pp. 32-38

[5] LIEBE, H.J., MANABE, T., and HUFFORD, G.A.: 'Millimeter-wave attenuation and delay rates due to fog/cloud conditions', *IEEE Trans. Antennas and Propagation*, 1989, **37**, (12), pp. 1617-1623

[6] HIPPEL, A. VON: 'Dielectrics and waves' (Artech House, Boston, 1995)

[7] ONSAGER, L.: 'Electric moments of molecules in liquids', *Journal of the American Chemical Society*, 1936, **58**, (8), pp. 1486-1493

[8] KIRKWOOD, J.G.: 'The dielectric polarization of polar liquids', *Journal of Chemical Physics*, October 1939, **7**, pp. 911-919

[9] KREIBIG, U., and VOLLMER, M.: 'Optical properties of metal clusters', Materials Science Series, **25** (Springer, Berlin, 1995)

[10] MAXWELL GARNETT, J.C.: 'Colours in metal glasses and metal films', *Trans. of the Royal Society,* (London), Vol. CCIII, 1904, pp. 385-420

[11] BOHREN, C.F., and HUFFMAN, D.R.: 'Absorption and scattering of light by small particles' (Wiley, New York, 1983)

[12] NEEVES, A.E., and BIRNBOIM, M.H.: 'Composite structures for the enhancement of nonlinear-optical susceptibility', *J. of the Optical Society of America, B*, 1989, **6**, (4), pp. 787-796

[13] HAMMERSLEY, J.M.: 'Origins of percolation theory', *in* DEUTSCHER, G., ZALLEN, R., and ADLER, J. (Eds): 'Percolation structures and processes', 1983, *Annals of the Israel Physical Society*, **5**, pp. 47-57.

[14] STAUFFER, D., and AHARONY, A.: 'Introduction to percolation theory', Second Edition (Taylor & Francis, London and Philadelphia, 1994)

[15] GRIMMETT, G.: 'Percolation' (Springer, New York, 1989)

[16] ADLER, J.: 'Conductivity exponent from the analysis of series expansions for random resistor networks', *Journal of Physics A (Math. Gen.)*, **18**, pp. 307-314.

[17] SIHVOLA, A., SAASTAMOINEN, S., and HEISKA, K.: 'Mixing rules and percolation', 1994, *Remote Sensing Reviews*, **9**, pp. 39-50

[18] LANDAUER, R.: 'Electrical conductivity in inhomogeneous media', *in* GARLAND, J.C., and TANNER, D.B. (Eds.): 'Electrical transport and optical properties of inhomogeneous media' (American Institute of Physics, Conference Proc. 40, 1978), pp. 2-45

[19] COHEN, R.W., CODY, G.D., COUTTS, M.D., and ABELES, B.: 'Optical properties of granular silver and gold films', 1973, *Physical Review B*, **8**, (8), pp. 3689-3701

[20] GLITTEMAN, J.I., and ABELES, B.: 'Comparison of the effective medium and the Maxwell-Garnett predictions for the dielectric constants of granular metals', 1977, *Physical Review B*, **15**, (6), pp. 3273-3275

[21] SHENG, P.: 'Theory for the dielectric function of granular composite media', 1980, *Physical Review Letters*, **45**, (1), pp. 60-63

Chapter 13

Applications to natural materials

In this chapter, let us study how well the proposed mixing theories agree with the real world. After all, an effective description of matter is ultimately a project that grows from practical needs. As we have noticed in the earlier chapters, in none of the mixing models could the exact effective permittivity be derived with full rigour. Approximations had to be made in the analysis to be able to take into account the effect of the randomness of the structure on the dielectric interaction. Were those assumptions justified?

The approximations earn their legitimation from the success of the mixing models in predicting macroscopic properties of heterogeneous media. Because dielectric data exist for many types of natural and human-made materials, it is time to ask what kind of averages of the components are the dielectric properties of these materials and how do they match against the models.

13.1 Water and ice

Water in its different states is a very important substance in the bio- and geosphere, and water indeed makes a component in many heterogeneous materials in nature. Through its effect on the dielectric response of matter, the content of water in various substances can be studied using electrical measurements, and a field of study like microwave aquametry is an important branch of applied science. To analyse such mixtures, we need to be familiar with the differences between the dielectric properties of water vapour, liquid water, bound water, and ice.

13.1.1 Free and bound water

Dielectric properties of water have been discussed earlier in Chapters 11 and 12. Let us here recall again the frequency dependence of water permittivity:

$$\epsilon_w(\omega) = \epsilon_\infty + \frac{\epsilon_s - \epsilon_\infty}{1 + j\omega\tau} \tag{13.1}$$

with

$$\frac{\epsilon_s}{\epsilon_0} = 190.0 - 0.375T \tag{13.2}$$

$$\frac{\epsilon_\infty}{\epsilon_0} = 4.90 \tag{13.3}$$

$$\tau = \frac{1.99}{T} e^{2140/T} \times 10^{-12} \text{ s} \tag{13.4}$$

where T is the temperature in degrees Kelvin. The temperature dependence of these dispersion parameters is a consequence of the temperature dependence of the viscosity of water. For increasing temperatures the relaxation time decreases, which means a quicker response to the field.

This relation gives a quite broadband description of water dispersion up to several hundred GHz. If we move further to the millimetre wave frequencies, it is suggested that another relaxation function be added to this dielectric function with relaxation time constants that are a fraction of a picosecond [1].

Bound water

The Debye relation (13.1) describes well the behaviour of free water. However, when water is mixed with other materials, its response to the electric field is changed. The creation of orientational and rotational polarisation is no more free but hindered to a larger or lesser degree. Water molecules tend to stick at boundaries to hydrophilic surfaces of other components. The degree of this phenomenon, which is called *adsorption*, affects very much the freedom of water in the mixture and it does not respond in the same manner as in bulk form. This type of water is called bound water, and its permittivity is considerably different from that of free water. On the other hand, bound water is not a clearly definiable phase but rather the degree of binding diminishes as a function of the distance from the hydrophilic surface. Therefore, the permittivity of bound water also varies, and in the microwave modelling studies, very different values have been used. The relative permittivity is sometimes approximated to be lossless and around $\epsilon/\epsilon_0 \approx 3$, or the order of ice permittivity [2, 3]. On the other hand, also values as high as $35 - j15$ have been used [4]. Of course, the dispersion characteristics of the permittivity of water are affected by binding. Instead of following a clean Debye dispersion, the permittivity of water in the boundary layer contains a distribution of relaxation times.

In moist and humid materials, also the amount of water that is in free or bound form is different. In very porous and fractally structured mixtures, there is always a boundary quite close to any of the water molecules in the substance and a high degree of binding takes place. Furthermore, the distinction between free and bound water is not necessarily sharp; rather, binding increases with decreasing distance to the local surface. For more detailed discussion, the reader is referred to the books [5, 6] and references therein.

13.1.2 Ice

Ice consists chemically of the same water molecules as liquid water. Nevertheless, the dielectric behaviours of ice and liquid water differ strongly. Pure ice is a solid and therefore questions can be raised about its isotropy, in contrast with water which as a fluid is completely isotropic. The crystallographic structure of ice is indeed hexagonal in normal atmospheric conditions.[1] Even if there exists anisotropy in monolithic ice crystals, usually in geophysical applications the target of interest is polycrystalline ice in which the anisotropy is averaged away.

Pure ice also displays a Debye-type dielectric behaviour with relaxation phenomena. However, the Debye model parameters are very different from those of water. In particular, the relaxation frequency is around the kilohertz region, about six decades lower than for free water. This makes pure ice fairly dispersionless at microwave frequencies where water permittivity is strongly dependent on the wavelength. Also the static permittivity of ice is higher than that of water, around 100 and even higher, depending on temperature.

At microwave frequencies, we are therefore high above the relaxation regime, and ice is practically lossless.[2] The relative permittivity of pure ice has stabilised down to the high-frequency value and is quite independent of frequency. The literature gives values for the high-frequency relative permittivity that quite often remain between 3.1 and 3.2, although some measurements have been reported that give figures below 3 (see, for example, [8]). A set of resonator measurements between 2.4 and 9.6 GHz for the temperature range $-30° - 0°$ C give the basis for the regression curve [9] for the relative permittivity of pure ice

$$\epsilon_{ice}/\epsilon_0 = 3.1884 + 0.00091\, T \tag{13.5}$$

where T is the temperature in degrees Celsius.

With further increase in frequency, another dispersion regime appears. This takes place around 100 THz. And after that, when the optical window is encountered, the permittivity of ice is again real, the value relative to vacuum being 1.72. However, a slight dispersion can be observed across the visible octave: the refractive index of ice ranges from 1.306 for red light to 1.318 for violet [10].

---

[1] This "ordinary" ice is called ice *Ih*, but other types can be created by freezing water in special temperatures and pressures. To create ice with cubic structure (*Ic*), the solidifying temperature has to be between $-80°$ C and $-130°$ C [7].

[2] The loss tangent is less than 10^{-4} at the frequency of 10 GHz. Various studies show that the imaginary part of the permittivity of pure ice has a minimum at 2 to 4 GHz.

Lake and sea ice

Lake and sea ice differ from pure ice in the respect that the freezing process which gives rise to it occurs for saline water. The resulting ice is not homogeneous but rather formed as a mixture where pure ice forms a matrix in which pockets remain that contain saline water. This liquid is called brine. Lake ice has lower salinity than ocean ice. Also in the case of sea ice, there is a large difference between the properties of first-year ice and multi-year ice. First-year sea ice has higher salinity and is anisotropic due to the elongation of the brine pockets. After several seasons, the brine liquid drains out downwards from the pockets and the pockets may even debrine completely. The increase of entropy causes that the remaining air-bubbles are more rounded than younger pockets. Because the losses in lake and sea ice emanate from the saline brine liquid, the losses of sea ice decrease with age. A horizontally cross-sectional photo of multi-year sea ice from Canada can be seen in Figure 13.1, and this photo really shows that geophysical ice is by no means homogeneous single-crystal lattice.

To model the permittivity of real-life ice, the mixture has to be decribed at least as a three-component mixture. Pure ice forms a background matrix, in which two types of inclusions are embedded: brine pockets and air bubbles. Air inclusions

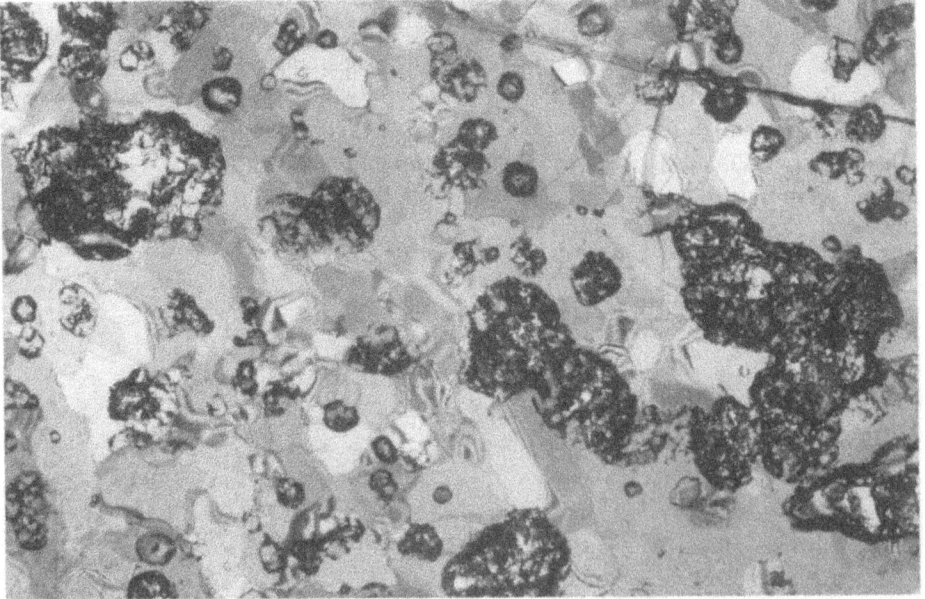

Figure 13.1: *Thin section of multi-year hummock ice. Air bubbles (dark objects) are located around 2 cm below the ice surface. (Reproduced with permission of M. Shokr, Atmospheric Environment Service, Canada.)*

tend to decrease the average permittivity from the pure-ice value but the brine liquid effect is opposite, and in practice, the latter effect is stronger, due to the high dielectric constant of saline water. Sea ice has therefore permittivity values higher than pure ice: values up to $\epsilon_r \approx 4$ have been reported but permittivity is strongly dependent on temperature. This is because when the temperature falls, brine liquid loses more and more of its pure water component, and the brine pockets decrease in size. The electric polarisability hence also decreases, although the remaining brine in the decreased pocket increases in salinity. The dielectric loss tangent can also be quite high for saline ice near melting temperatures, even close to $\tan \delta \approx 1$. Practically all losses are due to the brine inclusions.

The anisotropy of sea ice is caused by the brine pockets that grow elongated because of gravity. And because the losses arise from the brine pockets ice has a considerably larger attenuation for vertically polarised waves than horizontal polarisation. In practise, the uniaxiality of sea ice is not completely vertical with respect to ice surface but it is slightly tilted due to sea water movement during the freezing period.

For a study aiming to predict the dielectric properties of sea ice, see [11]. In remote sensing applications one must not forget the different electromagnetic behaviour of ocean ice compared with low-salinity ice in regions like the Baltic Sea [12].

13.2 Snow

A very essential geophysical material for northern regions of the Earth is snow, and there is no need to stress its relevance in the global hydrological and climatological cycles. From the modelling point of view, snow is a very good example of a dense mixture where clearly discrete component phases can be distinguished. Dry snow is a mixture of air and ice, and in wet snow, liquid water phase forms the third component.

But in the microstructure, snow may exhibit wide variation depending on its history and environmental conditions. This statement can be justified by comparing Figures 13.2 and 13.3 which both depict snow. The surface hoar on lake ice of Figure 13.2 is certainly different from the wet Spring snow in Figure 13.3. One may, still, try to apply mixing rules to both.

Many studies on the dielectric response of snow have been made since the early microwave measurements by Cumming [13]. In the following, let us see how the basic mixing formulas apply to snow.

13.2.1 Dry snow

The dielectric properties of dry snow are fairly well known. Macroscopically, dry snow can be characterised by one parameter, its density ρ_d, and the density renders

Figure 13.2: *Surface hoar on lake ice in Switzerland. The width of the view in the picture is about* 20 cm. *(Reproduced with permission of C. Mätzler, University of Bern.)*

us uniquely the volume fraction of ice (f) in snow:

$$f = \frac{\rho_d}{\rho_{ice}} \tag{13.6}$$

where $\rho_{ice} \approx 0.917 \, \text{g/cm}^3$ is the density of ice. Hence the modelling predictions of the permittivity of dry snow are simple functions of the density and the permittivity of pure ice, which can be taken quite dispersionless across the microwave frequency, and the losses are small. A commonly accepted value for the relative permittivity of ice is 3.19.

Being a two-component mixture with rather low permittivity contrast, dry snow does not offer any surprising behaviour characteristics in its effective permittivity dependence on the density. Neither do the permittivity predictions according to various mixing models deviate very much from one another. Figure 13.4 compares different Maxwell Garnett formulas for dry snow as a function of the density of snow. Seasonal snow very seldom exceeds values of $0.4 \, \text{g/cm}^3$ ($400 \, \text{kg/m}^3$), and hence the relative permittivity of dry snow most often remain below the value 2. Compared are isotropic models where ice particles are located in air, and the inclusions are either spheres, needles, or discs. The orientation distribution is random such that the resulting mixture is isotropic (cf. the results in Section 4.2.4).

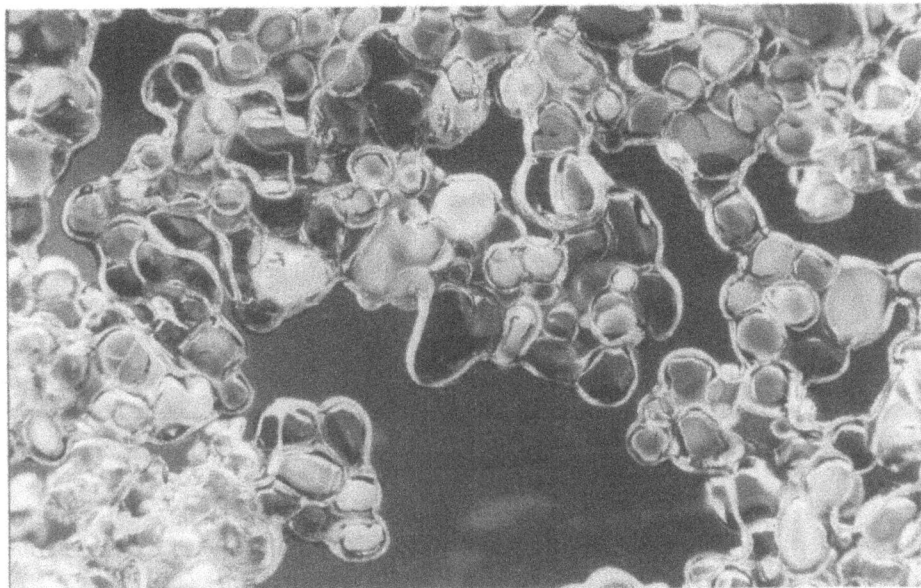

Figure 13.3: *Wet Spring snow in Austria. The grain diameters are around* 1 mm. *The mean grain size is* 0.7 mm. *(Reproduced with permission of A. Denoth, University of Innsbruck.)*

The models are tested against experimental results by Mätzler [14]. One can observe that indeed, the model predictions are very similar, but it is to be noted that the sphere model gives too low values compared to experiments. The disc curve overestimates the effective permittivity, whereas the needle-curve might be the best compromise, at least in this Maxwell Garnett case.

It is certainly to be expected that for low volume fractions, a sphere model cannot be realistic because in snow, the environment is air in which ice elements make the inclusions. The structure has to support itself mechanically, which means that in the limit of very low ice volume fractions, such a structure has to be made of ice sticks, and not spheres. Indeed common sense would speak for the fact that the shape of the ice particles in dry snow has to change with density. Of course also age matters: the destructive metamorphosis which erodes the fine structure of the original crystals causes the structure of snow to evolve towards more and more spherical grains. There is an interesting study [15] where, assuming the ice grains to be ellipsoids, optimal depolarisation factors are estimated from the permittivity behaviour of snow as a function of density. An a priori choice has to be made in [15] whether the grains are randomly oriented prolates or oblates but in both cases, a clear shape change can be observed.

How do the various models apart from Maxwell Garnett explain the behaviour

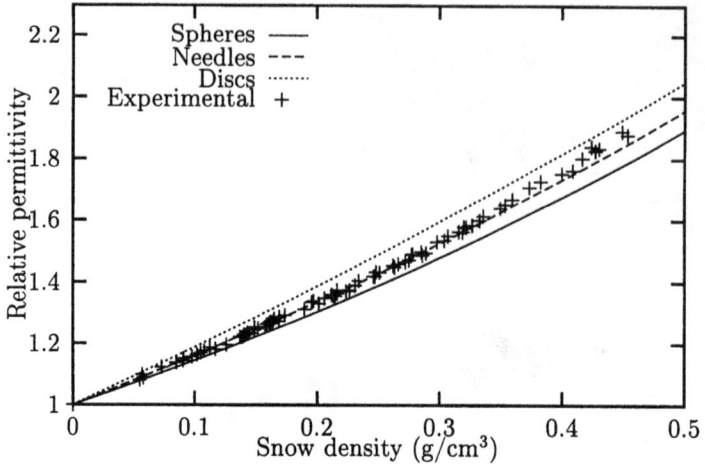

Figure 13.4: *Permittivity of dry snow as a function of its density. Maxwell Garnett mixing rules are shown in which the ice inclusions are randomly oriented spheres, needles, or discs. Also shown are the experimental measurements [14].*

of dry snow? In Figure 13.5 Maxwell Garnett is compared with Bruggeman and Coherent Potential models, where again spherical inclusion geometry is assumed. It seems that Bruggeman and Coherent Potential are indeed more realistic models than Maxwell Garnett. Other authors have found the Polder – van Santen model to be successful for dry snow [14], which is understandable; recall from Section 9.1 that they are the same model. Also towards the higher end of the dry snow density, the Looyenga model has been observed to match the experiments well.

Apart from the mixing models, empirical relations have been presented for the permittivity of dry snow. Mätzler has given the connection [14]

$$\epsilon_{ds}/\epsilon_0 = 1 + 1.5995\rho_d + 1.861\rho_d^3 \tag{13.7}$$

between the relative permittivity and density (in units g/cm^3) of dry snow, limited to values below 0.5 g/cm^3. The relation is fairly close to the Finnish model [16]

$$\epsilon_{ds}/\epsilon_0 = 1 + 1.7\rho_d + 0.7\rho_d^2 \tag{13.8}$$

13.2.2   Wet snow

When temperature increases, snow starts to melt. Wet snow is, from the dielectric point of view, a much more difficult material than dry snow. One reason for this is that the dielectric contrasts are very large in the resulting three-phase mixture of air, ice, and water, but at the same time, the amount of liquid water, which has large permittivity, is low. Small changes in the distribution and small-scale

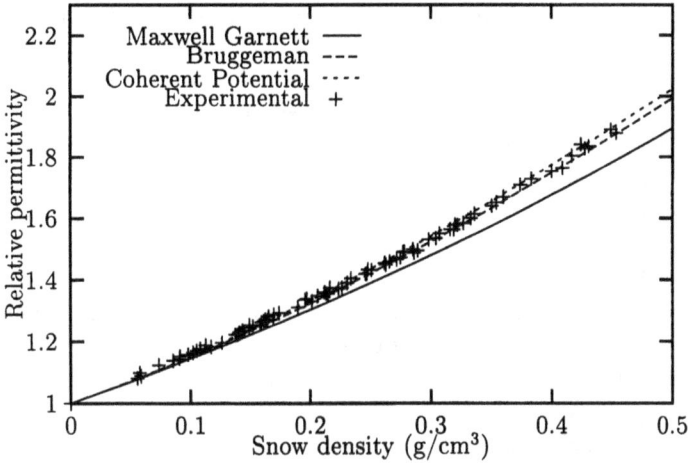

Figure 13.5: *Permittivity of dry snow as a function of its density. Maxwell Garnett, Bruggeman, and Coherent potential mixing rules are shown in which the ice inclusions are spheres. Also shown are the experimental measurements [14].*

structure of the water phase cause relatively large deviations in the permittivity of wet snow. Another reason for the fact that the present models for the dielectric properties are less accurate for wet snow compared to dry snow is certainly the practical difficulty in measuring the liquid water content in snow. Calorimetric methods to determine the wetness require careful treatment of the snow sample in order to eliminate uncontrolled heat exchance and ice phase melting.

Measurement series have been performed to collect dielectric data of wet snow which can be used to derive regression curves for the dielectric properties as functions of density and wetness. The following empirical relation is one example of a model to connect the snow permittivity [16] to its density and wetness:

$$\epsilon_s = \epsilon_{ds} + \Delta\epsilon \tag{13.9}$$

where ϵ_{ds} is the permittivity of the "dry part" of the snow, and the liquid water contribution $\Delta\epsilon$ is related to the complex permittivity of water ϵ_w at the same frequency:

$$\Delta\epsilon = (0.1W + 0.8W^2)\epsilon_w \tag{13.10}$$

Here W is snow wetness, in other words the volume fraction of the liquid water phase. The dry snow contribution ϵ_{ds} is calculated using the dry density ρ_d of the snow which is the difference between the wet snow density ρ_s and wetness: $\rho_d = \rho_s - W$, where the densities have to be given in units g/cm^3. Note that according to this model, the frequency dependence of snow follows water dispersion, and all dielectric losses are caused by the liquid water phase. The model, hence, is not valid at a very broad band of frequencies but limited to the microwave region around 1 GHz.

A more careful look at the distribution of liquid water in wet snow causes difficulties in quantitative description of the structure. A look at Figure 13.3 confirms this. However, in the literature an important observation has been reported regarding the behaviour of the microstucture of the water phase in snow. This is the separation of *pendular* and *funicular* regimes of wet snow [17], and of porous materials in general. The pendular regime, for the lower range of liquid saturations, means that liquid water forms separate bubbles, and the air phase is continuous throughout the medium. When the liquid water content increases, a qualitative change takes place in the microgeometry. In this high-wetness range—the funicular regime—the liquid is continuous through the pore space, and naturally, the water phase can no longer be approximated as separate inclusions.

We have learned from Section 4.2 that the shape of the inclusions in a mixture may have a very strong influence on the macroscopic permittivity. Hence one might expect that the pendular–funicular transition be also seen in the permittivity of wet snow. Studies and measurements of the complex permittivity of wet snow have produced some evidence towards this direction. Ambach and Denoth [18] find the transition wetness to be around 6% in fractional volume of liquid water. Another study, by Hallikainen et al. [19], concludes that the change from pendular to funicular regime would be lower; around 2...3% wetness.

The thermodynamically sensitive balance of liquid water and solid ice phases in wet snow makes it difficult to study the extent to which water is free in the mixture. There is evidence, however, that at the onset of melting in a Spring melt–freeze cycle, liquid water remains in bound form for a certain time, up to more than an hour [3]. This could be inferred by observing the radiometric microwave signatures of the snow cover. The brightness temperature changed much later than the appearance of liquid water in snow, the presence of which could be confirmed by independent tests. The delay in the change of the dielectric response of wet snow gives rise to the conclusion that liquid water may indeed stay in a bound state, and the permittivity of the bound water phase is the same as that of ice.

To collect information about the permittivity characteristics of snow, various dielectric sensors have been designed and developed. Of these, let us mention the resometer [14] and snow fork [20]. For more discussion on the dielectric properties of snow, see also [21].

13.3   Rocks and soil

In geological and geophysical applications, it is important to know the dielectric properties of solid and powdered rocks in various forms. Rocks are aggregates of mineral grains that are bound together by molecular interaction forces. Many measurements are available for the dielectric properties of homogeneous rock materials.

For example, broadband measurements by Campbell and Ulrichs [22] show that the real part of the relative permittivity of these rock materials is around 5...6 between 450 MHz and 35 GHz for various granites. The frequency dependence is very weak, and the loss factor stays around 0.01...0.05. See also the extensive tabulations by Parkhomenko [23] for the dielectric properties of various minerals in the megahertz region and below.

13.3.1 Porous bedrock

In bedrock, however, inhomogeneities exist. Fractures can be found in many scales, from microscopic hair cracks to large-scale heterogeneities. Also, rock is often porous, which means that water can flow through it. The porosity varies, and can be around 20–40 % for shale and sandstone but also as small as less than one percent for certain granites [23]. Water is indeed present in bedrock, and because the permittivity of water is again large compared to the homogeneous rock material, the dielectric properties of water-saturated rocks are strongly dependent on the porosity and the microgeometry of the rock matrix.

Especially at low frequencies, when the conductivity of brine-saturated water dominates the dielectric properties of aqueous mixtures, one can hope to find a simple, approximative relation between the porosity and the conductivity of the medium. Such a relation indeed exists, the so-called *Archie law*, which gives the connection between the conductivities of water σ_w and the mixture σ_{eff}:

$$\frac{\sigma_{\text{eff}}}{\sigma_w} = f^m \tag{13.11}$$

where f is the porosity (volume fraction of the pore space, here assumed to be filled by water). The exponent m is the so-called cementation exponent, and values between 1.3 and 4 have been presented for the cementation exponent for clay-free sedimentary rocks to predict the conduction behaviour, depending upon consilidation and other factors in the rock. Archie's law [24] has been used for decades in borehole logging to estimate the water saturation of rocks from the measured electrical conductivity of the rock matter.

With our experience from Section 12.3 about percolation processes we may immediately conclude that Archie's law is not compatible with the Maxwell Garnett model, nor with the Bruggeman model or Coherent potential model. This is because these models, at least in their basic form, predicted the existence of a percolation threshold f_c, a nonzero transient porosity value at which there would be a sudden change in the macroscopic conductivity. In particular, below the threshold there would be practically no conduction due to the insulating rock matrix, and above the threshold, the conductivity would rather be proportional to $(f - f_c)^m$. Archie's law, for its part, gives us a gradual increase in the effective conductivity starting from zero percolation threshold.[3]

---

[3]The conflict with Archie's law is evident in Equation (4.54) where the effective conductivity is seen to vanish for low frequencies, even if the porosity is nonzero.

However, some mixing models are compatible with Archie's law. One of these is the self-similar model by Sen et al. [25], presented in Section 9.4 in connection with the differential mixing rules. According to it the effective permittivity reads

$$\frac{\epsilon_{\text{eff}} - \epsilon_e}{\epsilon_i - \epsilon_e} = f \left(\frac{\epsilon_{\text{eff}}}{\epsilon_i} \right)^{1/3} \tag{13.12}$$

where the host (ϵ_e) is now the rock matrix, and the inclusions (ϵ_i) stand for the water phase, which has volume fraction f, equal to the porosity.

Let us apply Equation (13.12) to the water-saturated rock matrix. The environment is insulating, and hence ϵ_e is real, while the inclusion is conducting: $\epsilon_i = \epsilon'_i - j\sigma_w/\omega$. For low frequencies, the imaginary part dominates: $\epsilon_i \approx -j\sigma_w/\omega$, and the equation for the effective permittivity reads

$$\left(\frac{\sigma_{\text{eff}}}{\sigma_w} \right)^{2/3} = f \tag{13.13}$$

where also the real part of the mixture permittivity is assumed to be negligible in comparison with its imaginary part: $\epsilon_{\text{eff}} = \epsilon'_{\text{eff}} - j\sigma_{\text{eff}}/\omega \approx -j\sigma_{\text{eff}}/\omega$. This leads to Archie's law: $\sigma_{\text{eff}}/\sigma_w = f^{3/2}$, with the exponent $m = 1.5$.

The Sen et al. model was derived by considering spheres of differential sizes, and hence it is natural that the exponent $1/3$ appears in the model. If the model is extended to ellipsoids with depolarisation factor N, the corresponding model would read [25]:

$$\frac{\epsilon_{\text{eff}} - \epsilon_e}{\epsilon_i - \epsilon_e} = f \left(\frac{\epsilon_{\text{eff}}}{\epsilon_i} \right)^{N} \tag{13.14}$$

and this, in the low-frequency limit, predicts the conductivity behaviour according to $\sigma_{\text{eff}}/\sigma_w = f^{1/(1-N)}$. This would give more freedom in the choice of the cementation exponent because the depolarisation factor for an ellipsoid may have a value in the range from 0 to 1.

However, in the light of present research Archie's law is an oversimplification [26]. For example, in carbonate rocks which are very inhomogeneous and have a complex pore structure, Archie's law has been shown to fail. From the application point of view, this is very significant because of the economical importance of the problem of estimation of the world's hydrocarbon reservoirs.

In rock materials, anisotropy may also play a role. Special directions may even be visible in various stones and sedimented rocks, as can be seen in Figure 13.6. The anisotropy, of course, is also a very scale-dependent phenomenon. As an example of the magnitude of the anisotropy, at low frequencies the dielectric anisotropy[4] may be around 1.5–2 for Gypsum [23].

---

[4]The ratio between two eigenvalues of the permittivity dyadic.

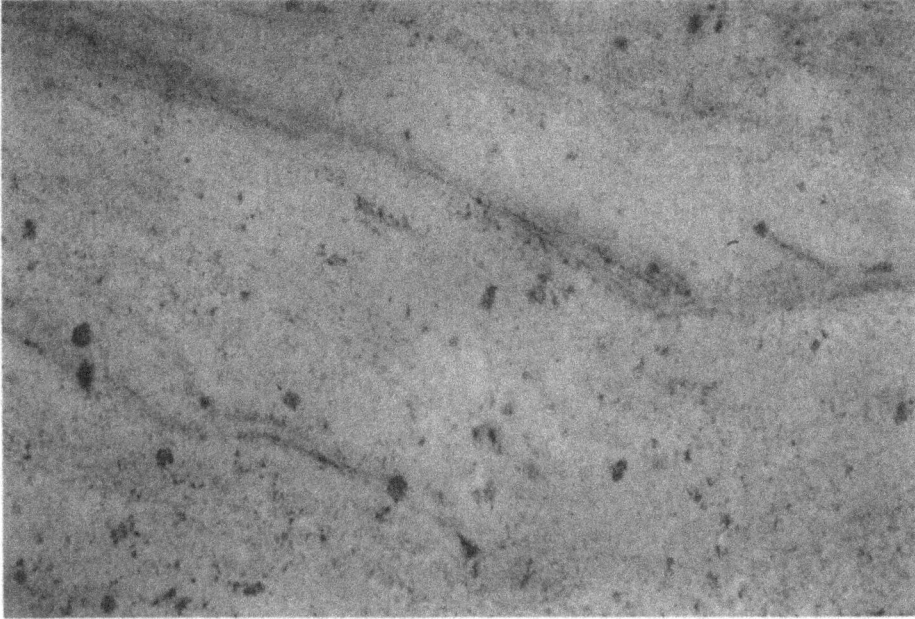

Figure 13.6: *Quartzite, formed from sand layers, from Northern Finland. The broad-size dimension of the cut is around* 3 cm. *Note the garnets (small darker inclusions). (Courtesy of Laboratory of Engineering Geology and Geophysics, Helsinki University of Technology.)*

13.3.2 Soil

Soils of various types cover the surface layer of earth in many places on dry land. Soil is a complex mixture of different solid soil materials, water in free and bound forms, and air. Here again, as in the case of snow, it is helpful to separate the dry soil part and the wet part in the description of the dielectric properties of soil. Not surprisingly, the dry soil permittivity is better understood than its wet part. For example, at a broad range of frequencies, the dielectric losses can be considered arising from the water contribution, although the permittivity of dry soil itself also has an imaginary part.

Dry soil is a two-component mixture, where granular inclusions of rock material occupy a certain volume in air.[5] Depending on the size of the inclusions, soils can be classified into clay, silt, and sand, in increasing coarseness of the particles. But regarding the dielectric properties of dry soil, the texture and grain size do not matter very much. We can approximately relate the permittivity of soil ϵ_s with its density ρ_s.

---

[5]This is of course an enormous simplification for dielectric modelling purposes. Soil is certainly much more than fragmented rock: a complex ecosystem containing various organisms, plants, fungi, microscopic animals, in addition to inorganic matter.

In the literature, common empirical models give the density–permittivity connection as a polynomial. For example, the model used by Russian scientists [27] reads

$$\epsilon_s/\epsilon_0 = (1 + 0.5\rho_s)^2 \qquad (13.15)$$

where the soil density has to be given in g/cm^3. A later study, based on measurements in the range between 1.4 and 18 GHz, by Dobson et al. [4], would give a slightly lower prediction for the permittivity:

$$\epsilon_s/\epsilon_0 = (1.01 + 0.44\rho_s)^2 - 0.062 \qquad (13.16)$$

Although these models are empirical, they can be connected with the Birchak refractive index model (see Section 9.4) where the square roots of the component permittivities are averaged. The density is of course proportional to the volume fraction of the solid soil matter, and in the expression for the permittivity, it becomes raised into the second power.

The Campbell–Ulrichs study on a wide variety of terrestrial rocks [22] also included, in addition to a large set of measurements, classical mixing model predictions for the dielectric properties of powdered rocks. The outcome was that the simple Maxwell Garnett mixing model with spherical inclusions gave a satisfactory agreement with measurements for olivine peridotite powder and olivine basalt. This was, however, not the case for certain other rock types, like aplite granite, where the Bruggeman model seemed to work better.[6] The measurements were conducted at the frequency of 450 MHz.

A recent series of measurements of dry Saharan desert sand [28] gives the permittivity between 245 MHz and 6 GHz. The results show that the real part of the relative permittivity is fairly constant above 1 GHz, and stays around the value $\epsilon_s/\epsilon_0 \approx 2.53$. Connecting this to the sand density which was around $1.45\,g/cm^3$, we may conclude that the measurements give more confirmation to the model of Equation (13.16) than (13.15). The results of [28] also show that the imaginary part of the permittivity of dry sand was decreasing strongly in the microwave region, and a Debye-type spectrum fit to the measurements would suggest a relaxation frequency around 270 MHz.

Although the soil texture does not affect considerably the permittivity of dry material, the situation changes considerably for moist and wet soil. Because the number of particles in a sample of clay is many orders of magnitude larger than in a sample of sand with the same mass, the surface area of the clay sample is also much greater. Consequently the distribution of liquid water is very different in these two cases, and because the water that is adsorbed on surfaces of another phase is more bound than free liquid water, we can expect a soil-type dependence on the permittivity of wet soils.

---

[6]The authors of [22] use the names "Rayleigh mixing formula" instead of Maxwell Garnett, and "Böttcher formula" for the Bruggeman relation.

The extensive study by Hallikainen et al. [29] reveals the behaviour of the wet soil permittivity as a function of frequency, soil type, temperature, and wetness. The results show that indeed, the water phase dominates the magnitude of the permittivity: the real part of the relative permittivity may reach values of 25 for 1 GHz frequency if the volume fraction of water is 0.4. Also the imaginary part may rise to values around 10, which is over two orders of magnitude more than for dry sand.

The results of [29] show also a clear dependence of soil type, especially at the lower end of the frequency range. There the permittivity is higher for coarse-structured soil types, sand and sandy loam, whereas for fine-grained media, silt and clay, the permittivity is lower even if the moisture content is the same. One plausible explanation for this phenomenon is that for the fine-grained soils, the greater specific surface area attaches water to the surfaces to a larger extent than for sand in which case water is more free.[7] The permittivity of bound water is lower than that for free water because of the binding which prevents the water molecules to respond freely to the electric field in the creation of polarisation.

13.4 Rain attenuation

In Chapters 11 and 12, the dielectric losses of water–air mixtures were discussed from the dispersion point of view. There it was pointed out how the Debye-type frequency dependence of water permittivity is translated to the dispersive properties of mixtures like fog, haze, and clouds, and the relaxation frequency moves to higher frequencies.

What can we say about the magnitude of the losses? Certainly fog is very transparent at microwave frequencies but in the case of rain the situation is different. The attenuation of radio waves propagating in rainy troposphere can be calculated from the losses of the effective permittivity of the air–precipitation mixture. For this end, we need to know the volume fraction of water in air.

All raindrops are not similar in rain. Their sizes differ and also depend on how strong the rainfall is. In the literature, the Marshall–Palmer drop size distribution is widely used. According to it, the number of drops of diameter d per unit volume can be calculated from the distribution (see, for example, [30])

$$p(d) = N_0 e^{-bd} \tag{13.17}$$

with $N_0 = 8.0 \times 10^6 \, \text{m}^{-4}$, and the exponent b is related to the rain rate as

$$b = 4100 R_r^{-0.21} \, [1/\text{m}] \tag{13.18}$$

where the rain rate has to be given in units mm/hr. Integration then gives us the dependence between the rain rate and the fractional volume f of water phase in air:

$$f = 8.894 \times 10^{-8} \, R_r^{0.84} \tag{13.19}$$

---

[7]In fact, some investigators argue that practically no part of the water phase in wet sand is bound; all is free.

Now the attenuation A can be calculated from the imaginary part of the effective permittivity of the dilute water–air mixture:

$$A = \frac{8686\pi}{\lambda} \epsilon''_{\text{eff},r} \quad [\text{dB/km}] \tag{13.20}$$

where the wavelength λ has to be given in metres, and the imaginary part of the relative effective permittivity of rain is

$$\epsilon''_{\text{eff},r} = -3f \, \text{Im} \left\{ \frac{\epsilon_w - \epsilon_0}{\epsilon_w + 2\epsilon_0} \right\} \tag{13.21}$$

and here the raindrops are assumed to be spherical.[8]

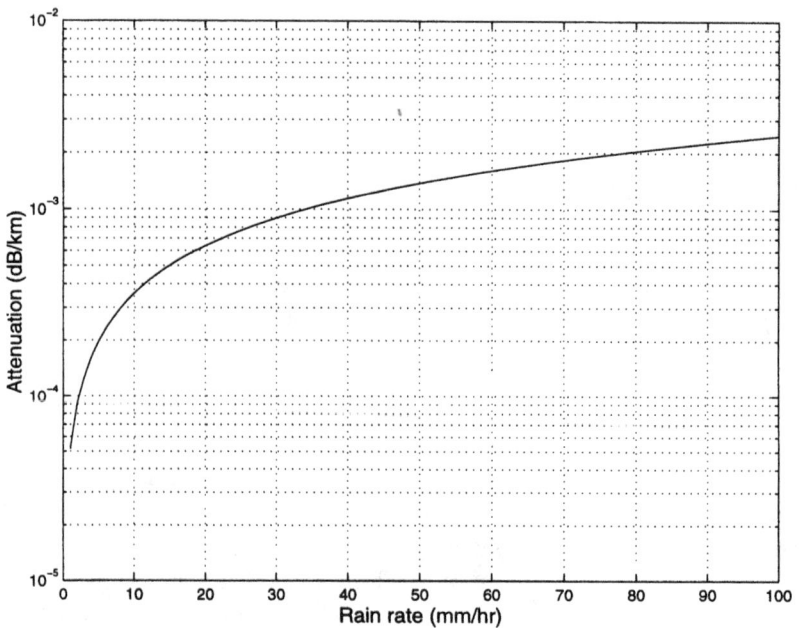

Figure 13.7: *Attenuation of radio waves in the troposphere at 1 GHz as a function of the rain rate. The temperature is 25° C. Only absorption is taken into consideration.*

Figure 13.7 shows the attenuation increasing as a function of rain rate at the frequency of 1 GHz. Of course one has to remember that now only absorption was accounted for in the calculation. As the frequency increases, two mechanisms start to undermine this assumption: first, the scattering contribution increases and, second, the absorption itself cannot be calculated by the quasi-static assumption that the

---

[8] Small raindrops are certainly spherical. But when the size increases, they start to flatten from the bottom side and resemble oblate ellipsoids [31]. For even larger drop sizes, the deviations from the ellipsoidal form can be modelled with the Pruppacher–Pitter expansion for the shape [32].

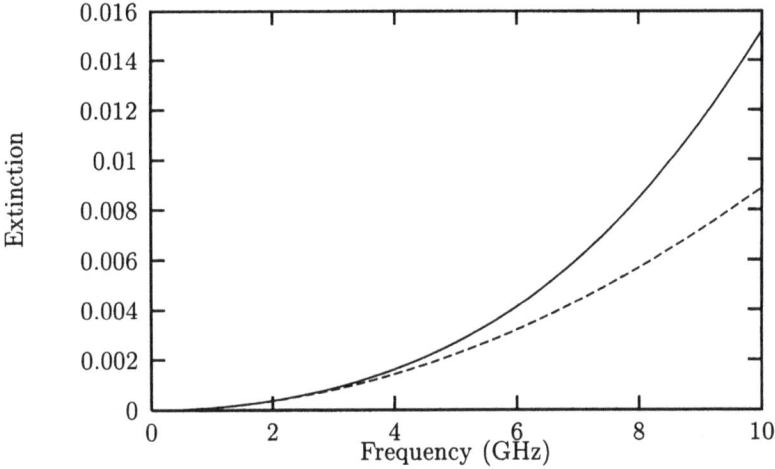

Figure 13.8: *The extinction of a spherical raindrop with 1 mm diameter as a function of frequency up to 10 GHz. The exact Mie solution Q$_{ext}$ (solid line) is shown, and the dashed line gives the approximation where scattering is neglected and the absorption is taken from the imaginary part of the quasi-static Maxwell Garnett formula.*

field inside the raindrop stay uniform throughout the volume. On the other hand, the assumption of a dilute mixture remains valid even for high rain rates, because the volume fraction of inclusions stays below values of the order of 10^{-6}.

To estimate how safe it is to neglect these high-frequency contributions of the 1 GHz calculations of Figure 13.7, one can compare the exact solution of the extinction efficiency Q_{ext} of a spherical raindrop, calculated with the full Mie theory (see Section 10.2), with the absorption according to the quasi-static assumption

$$Q_{abs,qs} = 4x \, \text{Im} \left\{ \frac{\epsilon_w - \epsilon_0}{\epsilon_w + 2\epsilon_0} \right\} \tag{13.22}$$

where, again, $x = k_0 a$ is the size parameter of the sphere.

Figure 13.8 illustrates this comparison. There the extinction is calculated for a raindrop of 1 mm diameter, as a function of frequency. One can observe that for this size, the quasi-static absorption assumption starts to fail when the frequency goes beyond 5 GHz. Of course, in rain there are drops of all sizes, and the higher the rain rate, the greater the average size becomes, as can be seen from the Marshall–Palmer distribution function.

For a survey on scattering from hydrometeors, see [33]. The backscattering problem is connected with a mixing approach in [34] with a mini-review of mixing principles; see also [35].

13.5   Wood, trees, and canopies

Compared to snow, ice, and soils, the dielectric analysis of organic materials causes additional difficulties. Wood, for instance, has a fibrous microstructure, and due to the fibre composition, an inherent anisotropy is present. Furthermore, the problem of the balance between bound and free water is by no means easy in this case of organic structures where water certainly tends to attach to surfaces of the fibres and makes it difficult to model the dielectric properties of wood materials as a mixture.

The cell wall substance in wood consists mostly of cellulose, hemicelluloses, and lignin. The dielectric properties of oven-dry cellulose show a frequency dependence similar to polar molecules, although the polarisation is associated with the displacement of hydroxyl groups ($-OH$), and not that much with orientational polarisation. The real part of the relative permittivity is around 6–7 at low frequencies and decreases to values 3–4 in the gigahertz region [36]. The loss tangent ($\tan \delta = \epsilon''/\epsilon'$) has its maximum around 10 MHz where it can reach values close to 0.1. Lignin does not have as high dielectric susceptibility as cellulose or hemicellulose, and it is also more constant with respect to frequency, having values around 4 for the real part of the relative permittivity.

Altogether, the dry cell substance has dielectric properties determined by the constitutents. The average dry wood cell substance, at a temperature 20–25° C, and for a density 1.53 g/cm³ have the following values: the real part of the relative permittivity decreases from the low-frequency value 6.5 to around 3.3 at microwaves. The loss tangent is at its maximum of 0.67 around 10 MHz, and decreases to 0.043 at 10 GHz.[9]

For moist wood and living trees, the dielectric properties may differ strongly from those of dry cell material. Water in its various forms contributes strongly to the permittivity. Furthermore, the anisotropy of wood is conspicuous. Although the cell wall material itself exhibits slight anisotropy, the fibrous wood structure enhances greatly susceptibility differences in along- and across-fibre directions.

Regardless of whether we treat a trunk, branch, or a needle of a tree, a coordinate system for the natural directions of the material are according to the cylindrical system. This is illustrated in Figure 13.9. There the axial (longitudinal), transversal (azimuthal), and radial directions are marked with L, T, and R, respectively.

The permittivity of the wood matter is then anisotropic with the eigenvectors along these directions with unit vectors $\mathbf{u}_L, \mathbf{u}_T, \mathbf{u}_R$, and the permittivity dyadic reads

$$\overline{\overline{\epsilon}}_{\text{wood}} = \epsilon_L \mathbf{u}_L \mathbf{u}_L + \epsilon_T \mathbf{u}_T \mathbf{u}_T + \epsilon_R \mathbf{u}_R \mathbf{u}_R \qquad (13.23)$$

In practice, the main anisotropy is uniaxial: the longitudinal component of the permittivity is larger than the two other ones: $\epsilon_L > \epsilon_T$ and $\epsilon_L > \epsilon_R$. A slight

---

[9]See Appendix 1 of [36] for extensive tabulations of the dielectric properties as functions of various wood parameters.

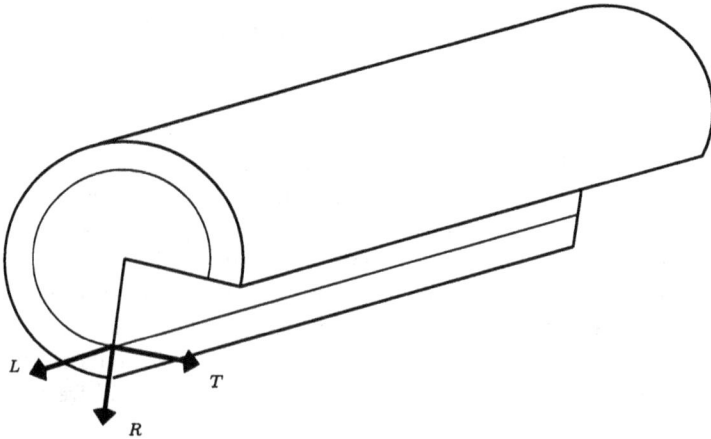

Figure 13.9: *The eigendirections of the tree structure.*

anisotropy may be observed also in the transversal plane, but often the radial and azimuthal components are close to each other in magnitude: $\epsilon_R \approx \epsilon_T$.

But a tree in its dielectric response is not only anisotropic; it is also nonhomogeneous. The core part of a trunk (xylem) has lower permittivity than the surface part below the bark (phloem), and also it has been found that the permittivity also increases with the height from ground. A recent study [37] on the permittivity of conifer trees (fir and spruce) gives interesting results of both the anisotropy and inhomogeneity at microwave frequencies between 1 and 10 GHz.

In [37], the anisotropy ratio ϵ_L/ϵ_T was observed to be roughly 1.5–3. The surface part may have relative permittivity values around $40 - j10$ of the longitudinal component, and the values decrease towards the centre part. New needles may have even higher permittivity than the maximum trunk permittivities. Within this frequency window (1–10 GHz), the real part of the permittivity slightly decreases with increasing frequency, but the imaginary part tends to increase with frequency.

The magnitude of the permittivity is obviously connected with the moisture of the tree, and microwave measurement principles are certainly useful for the determination of water content in trees [6], [38]. The observed anisotropy agrees qualitatively with the predictions of the mixing models of Secion 4.2, where the Maxwell Garnett model for a mixture with aligned ellipsoidal inclusions was derived. For needle-shaped inclusions, the permittivity component is always higher in the needle axis direction compared with the transversal component. Note that this conclusion holds regardless of which one of the components, inclusions or environment, has larger permittivity.

Studies exist also for the dielectric behaviour of agricultural vegetation. For example, leaves and stalks of corn and wheat have been measured at microwaves [2] to determine the dependence of the permittivity on temperature, moisture, and frequency. The measurements could be explained in that study reasonably well by a four-component refractive mixing model.[10] The four components of the model were air, bulk vegetation, and free and bound water. For bulk vegetation, the (lossless) relative permittivity was assumed to be 3. The water phase was assumed to be such that the water volume fraction f_w was divided into bound water part of 0.05, and the rest was free water, with volume fraction $f_w - 0.05$. The permittivity of bound water was assumed that of ice. To give one data point, the four-component model predicts that the relative permittivity of leaves would be around $30 - j10$ for a volumetric moisture of 0.6, a result which agreed rather well with measurements.

For practical use, a very broadband model for the dielectric properties of vegetation has been presented in [39], where a semiempirical formula is given for the complex permittivity of leaves as a function of the moisture and frequency within the range between 1 to 100 GHz.

For the dielectric properties of agricultural products from the point of view of microwave heating, see the review by Nelson [40].

In a larger scale vegetation canopies could also be modelled as effective media but then the scattering properties start to dominate at much lower frequencies than, say, for rain. However, certainly at very low frequencies (LF-region and at longer wavelengths) grain fields and perhaps even (dense) forests could be considered to have a macroscopic permittivity which is a certain average of the various contributions of the bulk vegetation. This effective permittivity, probably, would be again strongly uniaxial with a vertical permittivity as the dominant component.

13.6   Biological tissues

Bioelectromagnetism studies human tissues with most interest in their dielectric properties. From that point of view, living materials have no special status over other media: they are characterised by their complex permittivity. Already in the eighteenth century experiments were conducted on the electrical properties of biological systems when Luigi Galvani discovered the stimulation of dissected and prepared frog legs by electrical sparks (for the history of bioelectromagnetism, see, for example, [41]). The permittivity and conductivity of tissues are functions of their constituents, and hence using electrical measurements—for example bioimpedance analysis—it is possible to gather information about the internal state of the body.

Mammals have a total water content amounting to 65–70 % of their body mass. Therefore in the dielectric description of biological tissues the various relaxation phenomena and water binding play a major role, perhaps even more importantly than what was the case in soils and wood materials. A large amount of dielectric data have been published about the conductivity and permittivity and their dispersion of various tissues, like skeletal muscles, liver, brain matter, blood, bone, etc. [42–45].

---

[10]The Birchak model, where the refractive index of the mixture is a volume average of the component refractive indices.

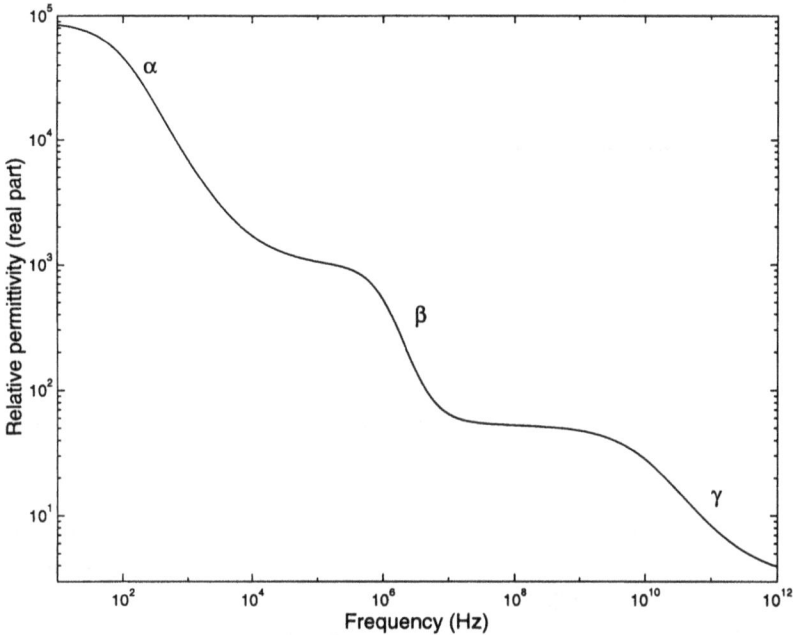

Figure 13.10: *The real part of the relative permittivity of biological tissue as a function of frequency. The various dispersion mechanisms are indicated.*

The dispersion of the biological tissues is a result of several different mechanisms. Figure 13.10 shows the real part of the relative permittivity of a tissue. There different dispersion mechanisms can be distinguished, and in the bioelectromagnetics literature, it is common to label these effects by α, β, and γ-dispersion.

Towards the higher end of the frequency range, we can distinguish the free water relaxation, which is responsible for the γ-dispersion. At lower radio frequencies, however, a broader dispersive range can be observed. This is called the β-dispersion. It is not a perfect Debye spectrum but rather consists of a distribution of relaxation effects with a range of different relaxation times. β-dispersion is caused by the cellular structure of the tissues where the cytoplasm is separated from the extracellular fluid by a membrane. This membrane is a better insulator than the cell fluid and acts as a capacitor to accumulate charge. The relaxation time constant is therefore connected to the electric circuit parameters of the cell system. This interfacial polarisation, briefly discussed in Section 2.1, often carries the name Maxwell–Wagner effect.

In the low-frequency end of the dispersion curve, another dispersion mechanism appears. This α-dispersion makes the real part of the permittivity increase to extremely high values in the audio frequencies. Detailed biochemical phenomena responsible for this peculiar dispersion are still unknown. However, ionic diffusion

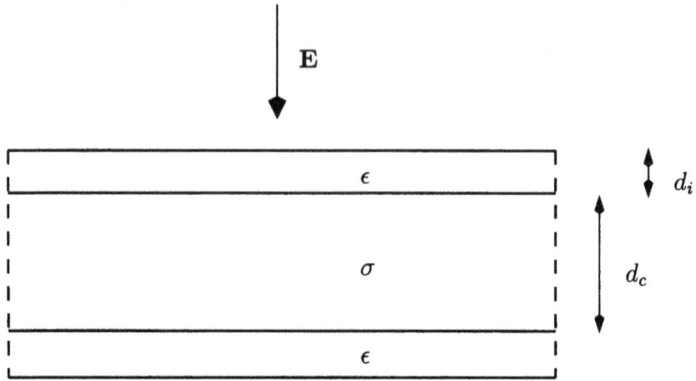

Figure 13.11: *A model for the cell membrane: conducting electrolyte between thin insulating sheets.*

in the electrical double-layers, or the structure of the complex tubular system of the membranes with associated electrolytes have been suggested as possible mechanisms behind the α-dispersion [42].

It is certainly true that a wide range of effective permittivities and conductivities can be designed by laminating different types of insulators and conductors. This makes it possible—by fractal layering of sheets, for example—to manufacture supercapacitors with a great charge-accumulating power. A similar effect with cell membranes as sheets that insulate conducting regions within cells is certainly one of the explanation mechanisms for the high values of the permittivity in the low-frequency end. Why such a supercapacitor effect could take place can be understood from a simple mixing theory.

Consider the situation depicted in Figure 13.11, where conducting electrolyte is confined between two planar layers of insulating material, and where the electric field vector is perpendicular to the layer surfaces. The thickness of the insulating layers d_i is small compared to the thickness of the conducting layer d_c. The permittivity of the insulator is ϵ and the conductivity of the electrolyte σ.

At low frequencies, again, we can define an effective complex permittivity for this system, along with the basic Maxwell Garnett principles. This is like a mixture of aligned discs, and hence the depolarisation factor is $N = 1$ (cf. Equation 4.25). The effective permittivity is

$$\epsilon_{\text{eff}} = \left(\frac{f_i}{\epsilon} + \frac{f_c}{-j\sigma/\omega} \right)^{-1} \tag{13.24}$$

where the volume fractions of the insulator phase (f_i) and the conductor phase (f_c) are

$$f_i = \frac{2d_i}{d_c + 2d_i} \qquad \text{and} \qquad f_c = \frac{d_c}{d_c + 2d_i} \tag{13.25}$$

The real part of the effective permittivity is, then

$$\epsilon'_{\text{eff}} = \frac{\epsilon/f_i}{1 + \left(\frac{\omega}{\omega_0}\right)^2} \tag{13.26}$$

where the "relaxation" frequency is $\omega_0 = (f_i\sigma)/(f_c\epsilon)$. From this relation one can see that at low frequencies (below ω_0), the real part of the effective permittivity can reach very high values if $f_i \ll 1$, in other words if the insulator membrane is thin compared to the thickness of the electrolyte layer.

This phenomenon of interface-caused dispersion of the dielectric permittivity carries the name Maxwell–Wagner effect [46]. Although it would be tempting to declare the strong α-dispersion in biological tissues to be caused by the Maxwell–Wagner effect, this seems not to be the case. More complex and intricate phenomena in the electrolytes in connection with cell surfaces are probably causing the observed low-frequency enhancements of the permittivity [47, 48].

Problems

13.1 Study the temperature dependence of the relaxation frequency $f_r = 1/(2\pi\tau)$ of water at microwave frequencies. Make its Taylor series expansion around $T = 20°$ C. How many gigahertz per degree does the relaxation frequency increase?

13.2 In Section 9.4.2, some related "differential mixing models" were introduced, of which the Sen–Scala–Cohen model was one. On the other hand, in Section 13.3.1, Archie's law (13.11) could be seen to be compatible with the SSC model. Can Archie's law be derived from the other differential mixing models?

13.3 Consider the simple membrane–electrolyte model shown in Figure 13.11 to explain α-dispersion in biological cells. It turned out that the extremely large values for the real part of the effective permittivity were compatible with this model. Very large capacitances can be generated with thin layers. Plot the frequency dependence of the real part $\epsilon'(\omega)$.

Study also the imaginary part of the effective permittivity for the same system. Also plot $\epsilon''_{\text{eff}}(\omega)$.

13.4 For a material with a distribution of relaxation times, a pure Debye spectrum is insufficient to describe the dispersion of permittivity [49]. Various modifications are suggested, among which is the following:

$$\epsilon(\omega) = \epsilon_\infty + \frac{\epsilon_s - \epsilon_\infty}{1 + (j\omega\tau)^{1-\beta}}$$

Here, in addition to the ordinary Debye parameters $\epsilon_s, \epsilon_\infty$, and τ, there is the dimensionless parameter β. The Debye spectrum (13.1) corresponds to the special case $\beta = 0$.

Plot the Cole–Cole diagram (locus of the permittivity in the ϵ', ϵ''-plane, as the frequency varies) for this modified relaxation permittivity. Plot also the curves $\epsilon'(\omega)$ and $\epsilon''(\omega)$. Use the value $\beta = 0.2$.

13.5 Repeat the previous problem for the Cole–Davidson permittivity model:

$$\epsilon(\omega) = \epsilon_\infty + \frac{\epsilon_s - \epsilon_\infty}{(1 + j\omega\tau)^\beta}$$

References

[1] KAATZE, U.: 'Microwave dielectric properties of water', *in* KRASZEWSKI, A. (Ed.): 'Microwave aquametry' (IEEE Press, Piscataway, N.J., 1996), Chap. 2, pp. 37-53

[2] ULABY, F.T., and JEDLICKA, R.P.: 'Microwave dielectric properties of plant materials', *IEEE Transactions on Geoscience and Remote Sensing*, 1984, **22**, (4), pp. 406-415

[3] MÄTZLER, C., STROZZI, T., WEISE, T., FLORICIOIU, D.-M., and ROTT, H.: 'Microwave snowpack studies made in the Austrian Alps during the SIR-C/X-SAR experiment', *International Journal of Remote Sensing*, 1997, **18**, (12), pp. 2505-2530

[4] DOBSON, M.C., ULABY, F.T., HALLIKAINEN, M.T., and EL-RAYES, M.: 'Microwave dielectric behavior of wet soil–Part II: Dielectric mixing models', *IEEE Transactions on Geoscience and Remote Sensing*, 1985, **23**, (1), pp. 35-46

[5] HASTED, J.B.: 'Aqueous dielectrics' (Chapman & Hall, London, 1973)

[6] KRASZEWSKI, A. (Ed.): 'Microwave aquametry' (IEEE Press, Piscataway, N.J., 1996)

[7] HOBBS, P.: 'Ice physics' (Clarendon Press, Oxford, 1974)

[8] ULABY, F.T., MOORE, R.K., and FUNG, A.K: 'Microwave remote sensing – Active and passive', Vol. III (Artech House, Norwood, Mass., 1986)

[9] MÄTZLER, C., and WEGMÜLLER, U.: 'Dielectric properties of fresh-water ice at microwave frequencies', *Journal of Physics D: Applied Physics*, 1987, **20**, pp. 1623-1630

[10] FLETCHER, N.H.: 'The chemical physics of ice' (Cambridge University Press, 1970)

[11] SIHVOLA, A.H., and KONG, J.A.: 'Effective permittivity of dielectric mixtures', *IEEE Transactions on Geoscience and Remote Sensing*, 1988, **26**, (4), pp. 420-429; 'Correction', *ibid.*, 1989, **27**, (1), pp. 101-102

[12] HALLIKAINEN, M.T.: 'Microwave remote sensing of low-salinity sea ice', *in* CARSEY, F.D. (Ed.): 'Microwave remote sensing of sea ice', *Geophysical Monograph* **68**, (American Geophysical Union, Washington, DC, 1992) Chap. 20, pp. 361-373

[13] CUMMING, W.A.: 'The dielectric properties of ice and snow at 3.2 centimeters', *Journal of Applied Physics*, 1952, **23**, (7), pp. 768-773

[14] MÄTZLER, C.: 'Microwave permittivity of dry snow', *IEEE Transactions on Geoscience and Remote Sensing*, 1996, **34**, (2), pp. 573-581

[15] MÄTZLER, C.: 'Improved Born approximation for scattering of radiation in a granular medium', *Journal of Applied Physics*, 1998, **83**, (11), pp. 6111-6117

[16] TIURI, M.E., SIHVOLA, A.H., NYFORS, E.G., and HALLIKAINEN, M.T.: 'The complex dielectric constant of snow at microwave frequencies', *IEEE Journal of Oceanic Engineering*, 1984, **9**, (5), pp. 377-382

[17] COLBECK, S.C.: 'The geometry and permittivity of snow at high frequencies', *Journal of Applied Physics*, 1982, **53**, (6), pp. 4495-4500

[18] AMBACH, W., and DENOTH, A.: 'The dielectric behavior of snow: a study versus liquid water content', Proceedings of *NASA Workshop of Microwave Remote Sensing of Snowpack Properties*, NASA Conf. Publ. **CP-2153**, Fort Collins, Colorado, 1980, pp. 69-81

[19] HALLIKAINEN, M.T., ULABY, F.T., and ABDELRAZIK, M.: 'Dielectric properties of snow in the 3 to 37 GHz range', 1986, *IEEE Transactions on Antennas and Propagation*, **34**, (11), pp. 1329-1340

[20] SIHVOLA, A., and TIURI, M.: 'Snow fork for field determination of the density and wetness profiles of a snow pack', 1986, *IEEE Transactions on Geoscience and Remote Sensing*, **24**, (5), pp. 717-721

[21] MÄTZLER, C.: 'Applications of the interaction of microwaves with the natural snow cover', 1987, *Remote Sensing Reviews*, **2**, p. 259-387

[22] CAMPBELL, M.J., and ULRICHS, J.: 'Electrical properties of rocks and their significance for lunar radar observations', *Journal of Geophysical Research*, November 1969, **74**, (25), pp. 5867-5881

[23] PARKHOMENKO, E.I.: 'Electrical properties of rocks' (Plenum Press, New York, 1967)

[24] ARCHIE, G.E.: 'The electrical resistivity log as an aid in determining some reservoir characteristics', *Trans. American Institute of Mining and Metallurgical Engineers*, 1942, **146**, pp. 54-62

[25] SEN, P.N., SCALA, C., and COHEN, M.H.: 'A self-similar model for sedimentary rocks with application to the dielectric constant of fused glass beads', *Geophysics*, 1981, **46**, (5), pp. 781-795

[26] SEN, P.N.: 'Resistivity of partially saturated carbonate rocks with microporosity', *Geophysics*, March–April 1997, **62**, (2), pp. 415-425

[27] SHUTKO, A.M.: 'Microwave radiometry of lands under natural and artificial moistening', *IEEE Transactions on Geoscience and Remote Sensing*, 1982, **20**, (1), pp. 18-29

[28] MÄTZLER, C.: 'Microwave permittivity of dry sand', *IEEE Transactions on Geoscience and Remote Sensing*, 1998, **36**, (1), pp. 317-319

[29] HALLIKAINEN, M.T., ULABY, F.T., DOBSON, M.C., EL-RAYES, M., and WU, L-K.: 'Microwave dielectric behavior of wet soil–Part I: Empirical models and experimental observations', *IEEE Transactions on Geoscience and Remote Sensing*, 1985, **23**, (1), pp. 25-35

[30] ULABY, F.T., MOORE, R.K., and FUNG, A.K: 'Microwave remote sensing – Active and passive', Vol. I (Artech House, Norwood, Mass., 1986)

[31] DOVIAK, R.J., and ZRNIĆ, D.S.: 'Doppler radar and weather observations' (Academic Press, Orlando, Florida, 1984)

[32] OGUCHI, T.: 'Scattering properties of Pruppacher-and-Pitter form raindrops and cross polarization due to rain: Calculations at 11, 13, 19.3, and 34.8 GHz', January–February 1977, *Radio Science*, **12**, (1), pp. 41-51

[33] OGUCHI, T.: 'Scattering from hydrometeors: A survey', September–October 1981, *Radio Science*, **16**, (5), pp. 691-730

[34] BOHREN, C.F., and BATTAN, L.J.: 'Radar backscattering by inhomogeneous precipitation particles', *Journal of the Atmospheric Sciences*, 1980, **37**, pp. 1821-1827

[35] BOHREN, C.F., and BATTAN, L.J.: 'Radar backscattering of microwaves by spongy ice spheres', *Journal of the Atmospheric Sciences*, 1982, **39**, pp. 2623-2628

[36] TORGOVNIKOV, G.I.: 'Dielectric properties of wood and wood-based materials', Springer Series in Wood Science (Springer-Verlag, Berlin, 1993)

[37] FRANCHOIS, A, PIÑEIRO, Y., and LANG, R.H.: 'Microwave permittivity measurements of two conifers', *IEEE Transactions on Geoscience and Remote Sensing*, 1998, **36**, (5), pp. 1384-1395

[38] NYFORS, E., and VAINIKAINEN, P.: 'Industrial microwave sensors' (Artech House, Norwood, Massachusetts, 1989)

[39] MÄTZLER, C.: 'Microwave (1–100 GHz) dielectric model of leaves', *IEEE Transactions on Geoscience and Remote Sensing*, 1994, **32**, (5), pp. 947-949

[40] NELSON, S.O.: 'Dielectric properties of agricultural products. Measurements and applications', *IEEE Transactions on Electrical Insulation*, 1991, **26**, (5), pp. 845-869

[41] MALMIVUO, J., and PLONSEY, R.: 'Bioelectromagnetism' (Oxford University Press, New York, 1995)

[42] FOSTER, K.R., and SCHWAN, H.P.: 'Dielectric properties of tissues and biological materials: a critical review', *CRC Critical Reviews in Biomedical Engineering*, 1989, **17**, (1), pp. 25-104

[43] GABRIEL, C., GABRIEL, S., and COHOUT, E:: 'The dielectric properties of biological tissues: I. Literature survey', *Physics in Medicine and Biology*, 1996, **41**, (11), pp. 2231-2249

[44] GABRIEL, S., LAU, R.W., and GABRIEL, C.: 'The dielectric properties of biological tissues: II. Measurements in the frequency range 10 Hz to 20 GHz', *Physics in Medicine and Biology*, 1996, **41**, (11), pp. 2251-2269

[45] STUCHLY, M.A., and STUCHLY, S.S.: 'Dielectric properties of biological substances—tabulated', *Journal of Microwave Power*, March 1980, **15**, (1), pp. 19-26

[46] WAGNER, K.W.: 'Erklärung der dielektrischen Nachwirkungsvorgänge auf Grund Maxwellscher Vorstellungen', *Archiv für Elektrotechnik*, 1914, **2**, (9), pp. 371-387

[47] SCHWARZ, G.: 'A theory of the low-frequency dielectric dispersion of colloidal particles in electrolyte solution', *Journal of Physical Chemistry*, 1962, **66**, pp. 2636-2642

[48] BOCKRIS, J. O'M., GILEADI, E., and MÜLLER, K.: 'Dielectric relaxation in the electric double layer', *The Journal of Chemical Physics*, 1966, **44**, (4), pp. 1445-1456

[49] SCAIFE, B.K.P.: 'Principles of dielectrics' (Clarendon Press, Oxford, 1989)

Chapter 14

Concluding remarks

Our path through various attempts to explain and understand dielectric properties of heterogeneous materials brought us finally to the comparison of the models with the properties of real-life materials. In the last chapter on applications, we saw that very often a gap remains between our theory-based mixing models and the true properties of actual existent materials. Engineers are painfully aware of this, and they—instead of using a classical mixing rule—often resort to empirical relations between the nonelectrical and dielectric properties of media of their interest. Such formulas may be weighted averages of the properties of the constituents, for example the refractive model or the Looyenga formula, or in the extreme case such a relation can be a regression curve upon experimental data.

But in the end, such formulas are vulnerable to the charge that they are empirical and lack a deeper theoretical foundation. Nothing in these relations really touches electromagnetically the structure of the medium. Therefore I would like to once again underline the message of the present book: despite their shortcomings and idealisations, electromagnetic mixing formulas are able to convey deep and meaningful information about the effective behaviour of heterogeneous materials.

The pragmatic motivation for the use of mixing rules is the need to predict characteristics of such materials that cannot be treated in their full complexity. For some users, the main "product" mixing rules can offer is their ability to decrease dramatically the number of degrees of freedom of the mixture. In the extreme case, such a user is perhaps only interested in the macroscopic permittivity or conductivity, and the structural details can be left aside totally. Of course, some guarantee is needed that the result can be reliably used in a given application with given external conditions. Such a universal warranty is certainly not easy to provide. However, the quasi-static assumption as an upper-frequency limit is certainly always good to keep in mind when applying mixing rules. Furthermore, similarly to industrial products, a consumer has to face the fact that the quality is subject to a natural variation. The bounds which were discussed in Chapter 8, between which the effective permittivity of true samples have to fall, can be thought of as tolerances to which the output products of a factory's conveyor belt conform.

Another appealing property of the mixing rules is their generality with respect to variations in geometry and matter. In the mathematical appearance of the mixing rules, these effects can be separated to a certain degree. A Maxwell Garnett rule has similar structure for spherical and ellipsoidal inclusions; some factors may only vary. This fact makes it possible for a user to distinguish which particular property of the prediction of the effective permittivity is caused by structural geometry and which perhaps by the behaviour of the dielectric response of the component materials. And if the user is willing to motivate himself into mastering dyadic and six-vector algebra, the mixing rules are readily available for anisotropic and bi-anisotropic mixtures. Then only the symbols for certain quantities have to be reinterpreted in a more general manner. A complex edifice grows naturally from simple building blocks.

A third important character of mixing principles that is worth remembering in this short conclusion is their ability to contain interesting physics hidden behind the deceivingly simple appearance. We analysed examples such as how the dispersive character of homogeneous materials could change drastically when they formed particles in a host matrix. Conducting materials in particulate form may produce an insulator with resonant frequency dispersion. Strong dielectric responses, polarisation enhancements, and nonlinear percolation processes are really existent phenomena in nature, and we saw that indeed, with mixing formulas these effects could be approached and also partly understood.

Despite the generality of the treatment of the present book, in one respect its focus has been narrow. The objective throughout has been on electromagnetic description of matter, and particularly its dielectric characterisation. Dielectric and magnetic properties of matter, albeit extremely important in their description, are only one cross-cut of the full understanding of matter. The electromagnetic picture has to be complemented with, for example, mechanical and thermodynamic descriptions.

However, much has been gained by a comprehensive electromagnetic mixing theory. In addition to having an understanding of the effective dielectric character of materials, the principles of the theory can be very much reused in the effective description of matter regarding its other physical properties. For example, the expression for the effective (electric) conductivity σ_{eff} of a heterogeneous sample as a function of the conductivities of the components is equally valid for the effective thermal conductivity of the same sample. This is a consequence of duality. The same Laplace equation holds for the electric potential as for the thermal temperature potential, and the boundary conditions are determined by the structure of the heterogeneous sample. Also, instead of heat, particles or momentum can be transferred which makes the conductivity treatment applicable for viscous and diffusive media. Likewise, in the literature describing elastic properties of heterogeneous media, Maxwell Garnett type mixing rules can be found, although there one must keep in mind the higher-order tensorial character of the response of solid materials.

Finally, let us return briefly to the role of scales, both in space and time, in mixing theories and their application in engineering. If we use a band-limited electromagnetic signal in our interaction with matter, we are aware of the spatial scale into which we can penetrate in its structure. The smaller details beyond this limit in a way form the domain of mixing theories; these details are collectively contributing, according to the mixing rule, to the effective parameters. But in the light of the very fast pace of present technology into smaller sizes in integrated circuits, and also toward narrow pulses and high frequencies, it may seem that this domain reserved for the mixing rules shrinks at an alarming rate. Is the rain forest in Amazonas a suitable metaphor for the ecology of random media, for which the diversity has been sustainably developed by traditional mixing rules?

Perhaps not. "There's plenty of room at the bottom." These words of Richard Feynman emphasise the depth of scales in which matter is composed and into which it can be artificially moulded. Visions of very small components and micromachines are becoming true with fields of research called nanotechnologies. But from the point of view of effective electromagnetic description, one has to remember that the spatial and temporal scales are connected. The drive in technical applications towards higher frequencies is always scalewise compensated by the miniaturisation of the composite structures. This means that the need for homogenisation principles does not vanish with the emergence of electrical signals with ever-faster time variation. The absolute scale is not important. The building blocks in these always-smaller structures, no matter how tiny, have to be described by effective terms.

Collection of dyadic relations

A dyad **ab** is a pair of vectors **a** and **b**, which are multiplied with a dyadic product. A dyadic is a polynomial of dyads:

$$\overline{\overline{A}} = \sum_{i=1}^{n} \mathbf{a}_i \mathbf{b}_i \tag{A.1}$$

Notationally, vectors are of bold face type, and dyadics are denoted by a double overbar.

The double products are defined by

$$
\begin{align}
(\mathbf{ab}) : (\mathbf{cd}) &= (\mathbf{a} \cdot \mathbf{c})(\mathbf{b} \cdot \mathbf{d}) \tag{A.2} \\
(\mathbf{ab}) {\overset{\times}{\times}} (\mathbf{cd}) &= (\mathbf{a} \times \mathbf{c})(\mathbf{b} \times \mathbf{d}) \tag{A.3}
\end{align}
$$

The unit dyadic $\overline{\overline{I}}$ has the properties $\mathbf{a} \cdot \overline{\overline{I}} = \overline{\overline{I}} \cdot \mathbf{a} = \mathbf{a}$, and $\overline{\overline{A}} \cdot \overline{\overline{I}} = \overline{\overline{I}} \cdot \overline{\overline{A}} = \overline{\overline{A}}$ for any vector **a** and any dyadic $\overline{\overline{A}}$.

The transpose of $\overline{\overline{A}}$ is denoted by $\overline{\overline{A}}^T$, with $\mathbf{a} \cdot \overline{\overline{A}} = \overline{\overline{A}}^T \cdot \mathbf{a}$.

The antisymmetric dyadic $\mathbf{a} \times \overline{\overline{I}} = \overline{\overline{I}} \times \mathbf{a}$ obeys

$$\left(\mathbf{a} \times \overline{\overline{I}} \right)^T = -\mathbf{a} \times \overline{\overline{I}} \qquad \text{and} \qquad \left(\mathbf{a} \times \overline{\overline{I}} \right) \cdot \overline{\overline{A}} = \mathbf{a} \times \overline{\overline{A}} \tag{A.4}$$

Definitions:

$$\overline{\overline{A}}^2 = \overline{\overline{A}} \cdot \overline{\overline{A}} \qquad\qquad \overline{\overline{A}}^{(2)} = \frac{1}{2} \overline{\overline{A}} {\overset{\times}{\times}} \overline{\overline{A}} \tag{A.5}$$

$$\text{tr}\,\overline{\overline{A}} = \overline{\overline{A}} : \overline{\overline{I}} \qquad\qquad \text{(trace)} \tag{A.6}$$

$$\mathrm{spm}\,\overline{\overline{A}} = \frac{1}{2}\overline{\overline{A}} \overset{\times}{\times} \overline{\overline{A}} : \overline{\overline{I}} \qquad \text{(sum of principal minors)} \qquad (A.7)$$

$$\mathrm{spm}\,\overline{\overline{A}} = \frac{1}{2}\left[\left(\mathrm{tr}\,\overline{\overline{A}}\right)^2 - \mathrm{tr}\,\overline{\overline{A}}^2\right] \qquad (A.8)$$

$$\det\overline{\overline{A}} = \frac{1}{6}\overline{\overline{A}} \overset{\times}{\times} \overline{\overline{A}} : \overline{\overline{A}} \qquad \text{(determinant)} \qquad (A.9)$$

$$\mathrm{adj}\,\overline{\overline{A}} = \overline{\overline{A}}^{(2)T} = \frac{1}{2}(\overline{\overline{A}} \overset{\times}{\times} \overline{\overline{A}})^T \qquad \text{(adjoint)} \qquad (A.10)$$

$$\overline{\overline{A}}^{-1} = \frac{\mathrm{adj}\,\overline{\overline{A}}}{\det\overline{\overline{A}}} = \frac{\overline{\overline{A}}^{(2)T}}{\det\overline{\overline{A}}} = 3\frac{(\overline{\overline{A}} \overset{\times}{\times} \overline{\overline{A}})^T}{\overline{\overline{A}} \overset{\times}{\times} \overline{\overline{A}} : \overline{\overline{A}}} \qquad \text{(inverse)} \qquad (A.11)$$

$$\overline{\overline{A}}^3 - \left(\mathrm{tr}\,\overline{\overline{A}}\right)\overline{\overline{A}}^2 + \left(\mathrm{spm}\,\overline{\overline{A}}\right)\overline{\overline{A}} - \left(\det\overline{\overline{A}}\right)\overline{\overline{I}} = 0 \qquad (A.12)$$

Properties

$$\mathrm{adj}\left(\overline{\overline{A}} + \overline{\overline{B}}\right) = \mathrm{adj}\,\overline{\overline{A}} + \mathrm{adj}\,\overline{\overline{B}} + \left(\overline{\overline{A}} \overset{\times}{\times} \overline{\overline{B}}\right)^T \qquad (A.13)$$

$$\begin{aligned} \det\left(\overline{\overline{A}} + \overline{\overline{B}}\right) &= \det\overline{\overline{A}} + \det\overline{\overline{B}} + \mathrm{tr}\left[\left(\mathrm{adj}\,\overline{\overline{A}}\right)\cdot\overline{\overline{B}}\right] + \mathrm{tr}\left[\overline{\overline{A}}\cdot\left(\mathrm{adj}\,\overline{\overline{B}}\right)\right] \\ &= \det\overline{\overline{A}} + \det\overline{\overline{B}} + \frac{1}{2}\overline{\overline{A}} \overset{\times}{\times} \overline{\overline{A}} : \overline{\overline{B}} + \frac{1}{2}\overline{\overline{B}} \overset{\times}{\times} \overline{\overline{B}} : \overline{\overline{A}} \end{aligned} \qquad (A.14)$$

$$\overline{\overline{A}} \overset{\times}{\times} \overline{\overline{I}} = \left(\overline{\overline{A}} : \overline{\overline{I}}\right)\overline{\overline{I}} - \overline{\overline{A}}^T \qquad (A.15)$$

For the three-dimensional unit dyadic:

$$\overline{\overline{I}} \overset{\times}{\times} \overline{\overline{I}} = 2\overline{\overline{I}} \qquad \mathrm{tr}\,\overline{\overline{I}} = \overline{\overline{I}} : \overline{\overline{I}} = 3 \qquad \mathrm{spm}\,\overline{\overline{I}} = 3 \qquad \det\overline{\overline{I}} = 1 \qquad (A.16)$$

For the two-dimensional unit dyadic $\overline{\overline{I}}_t = \overline{\overline{I}} - \mathbf{uu}$:

$$\overline{\overline{I}}_t \overset{\times}{\times} \overline{\overline{I}}_t = 2\mathbf{uu} \qquad \overline{\overline{I}}_t : \overline{\overline{I}}_t = 2 \qquad \mathrm{spm}\,\overline{\overline{I}}_t = 1 \qquad \det\overline{\overline{I}}_t = 0 \qquad (A.17)$$

$$\left(\mathbf{u}\times\overline{\overline{I}}_t\right)^2 = \left(\mathbf{u}\times\overline{\overline{I}}\right)^2 = -\overline{\overline{I}}_t \qquad (A.18)$$

$$\mathbf{uu}:\mathbf{uu} = 1 \qquad \mathbf{uu}:\overline{\overline{I}}_t = 0 \qquad \mathbf{uu}:\mathbf{u}\times\overline{\overline{I}} = 0 \qquad (A.19)$$

$$\mathbf{uu} \overset{\times}{\times} \mathbf{uu} = 0 \qquad \mathbf{uu} \overset{\times}{\times} \overline{\overline{I}}_t = \overline{\overline{I}}_t \qquad \mathbf{uu} \overset{\times}{\times} \left(\mathbf{u}\times\overline{\overline{I}}\right) = \mathbf{u}\times\overline{\overline{I}} \qquad (A.20)$$

$$\left(\mathbf{u}\times\overline{\overline{I}}\right):\left(\mathbf{u}\times\overline{\overline{I}}\right) = 2 \qquad \left(\mathbf{u}\times\overline{\overline{I}}\right):\overline{\overline{I}}_t = 0 \qquad (A.21)$$

$$\left(\mathbf{u}\times\overline{\overline{I}}\right) \overset{\times}{\times} \left(\mathbf{u}\times\overline{\overline{I}}\right) = 2\mathbf{uu} \qquad \left(\mathbf{u}\times\overline{\overline{I}}\right) \overset{\times}{\times} \overline{\overline{I}}_t = 0 \qquad (A.22)$$

Collection of basic mixing rules

In this appendix, some of the important mixing rules and other formulas are repeated. The effective permittivity ϵ_{eff} is for a mixture where the environment has permittivity ϵ_e and the inclusions, which occupy a volume fraction f, are of permittivity ϵ_i.

Maxwell Garnett rule

Isotropic materials, spherical inclusions

$$\epsilon_{\text{eff}} = \epsilon_e + 3f\epsilon_e \frac{\epsilon_i - \epsilon_e}{\epsilon_i + 2\epsilon_e - f(\epsilon_i - \epsilon_e)} = \epsilon_e + \frac{n\alpha}{1 - \dfrac{n\alpha}{3\epsilon_e}} \tag{B.1}$$

Also in the form

$$\frac{\epsilon_{\text{eff}} - \epsilon_e}{\epsilon_{\text{eff}} + 2\epsilon_e} = f \frac{\epsilon_i - \epsilon_e}{\epsilon_i + 2\epsilon_e} = \frac{n\alpha}{3\epsilon_e} \tag{B.2}$$

Polarisability of a dielectric sphere, referred to the environment:

$$\alpha = V(\epsilon_i - \epsilon_e) \frac{3\epsilon_e}{\epsilon_i + 2\epsilon_e} \tag{B.3}$$

d-dimensional spheres

$$\frac{\epsilon_{\text{eff}} - \epsilon_e}{\epsilon_{\text{eff}} + (d - 1)\epsilon_e} = f \frac{\epsilon_i - \epsilon_e}{\epsilon_i + (d - 1)\epsilon_e} \tag{B.4}$$

Multiphase mixture, isotropic spheres

$$\epsilon_{\text{eff}} = \epsilon_e + 3\epsilon_e \frac{\sum_{i=1}^{K} f_i \frac{\epsilon_i - \epsilon_e}{\epsilon_i + 2\epsilon_e}}{1 - \sum_{i=1}^{K} f_i \frac{\epsilon_i - \epsilon_e}{\epsilon_i + 2\epsilon_e}} \tag{B.5}$$

Aligned ellipsoids

$$\epsilon_{\text{eff},x} = \epsilon_e + f\epsilon_e \frac{\epsilon_i - \epsilon_e}{\epsilon_e + (1-f)N_x(\epsilon_i - \epsilon_e)} \tag{B.6}$$

Randomly oriented ellipsoids

$$\epsilon_{\text{eff}} = \epsilon_e + \epsilon_e \frac{\frac{f}{3} \sum_{j=x,y,z} \frac{\epsilon_i - \epsilon_e}{\epsilon_e + N_j(\epsilon_i - \epsilon_e)}}{1 - \frac{f}{3} \sum_{j=x,y,z} \frac{N_j(\epsilon_i - \epsilon_e)}{\epsilon_e + N_j(\epsilon_i - \epsilon_e)}} \tag{B.7}$$

Depolarisation factors of ellipsoids $(N_x + N_y + N_z = 1)$

$$N_x = \frac{a_x a_y a_z}{2} \int_0^\infty \frac{ds}{(s+a_x^2)\sqrt{(s+a_x^2)(s+a_y^2)(s+a_z^2)}} \tag{B.8}$$

Prolate spheroids: $\left(a_x > a_y = a_z, \quad e = \sqrt{1 - a_y^2/a_x^2} \right)$

$$N_x = \frac{1-e^2}{2e^3} \left(\ln \frac{1+e}{1-e} - 2e \right) \tag{B.9}$$

Oblate spheroids: $\left(a_z < a_x = a_y, \quad e = \sqrt{a_x^2/a_z^2 - 1} \right)$

$$N_z = \frac{1+e^2}{e^3}(e - \tan^{-1} e) \tag{B.10}$$

Anisotropic inclusions

Spheres:

$$\overline{\overline{\epsilon}}_{\text{eff}} = \epsilon_e \overline{\overline{I}} + 3\epsilon_e f \left[\overline{\overline{\epsilon}}_i + 2\epsilon_e \overline{\overline{I}} - f(\overline{\overline{\epsilon}}_i - \epsilon_e \overline{\overline{I}}) \right]^{-1} \cdot (\overline{\overline{\epsilon}}_i - \epsilon_e \overline{\overline{I}}) \tag{B.11}$$

Ellipsoids:

$$\overline{\overline{\epsilon}}_{\text{eff}} = \epsilon_e \overline{\overline{I}} + f\epsilon_e \left[\epsilon_e \overline{\overline{I}} + (1-f)\overline{\overline{L}} \cdot (\overline{\overline{\epsilon}}_i - \epsilon_e \overline{\overline{I}}) \right]^{-1} \cdot (\overline{\overline{\epsilon}}_i - \epsilon_e \overline{\overline{I}}) \qquad (B.12)$$

Anisotropic inclusions and environment

$$\overline{\overline{\epsilon}}_{\text{eff}} = \overline{\overline{\epsilon}}_e + f\left(\overline{\overline{\epsilon}}_i - \overline{\overline{\epsilon}}_e\right) \cdot \left[\overline{\overline{\epsilon}}_e + (1-f)\overline{\overline{L}}' \cdot (\overline{\overline{\epsilon}}_i - \overline{\overline{\epsilon}}_e)\right]^{-1} \cdot \overline{\overline{\epsilon}}_e \qquad (B.13)$$

with

$$\overline{\overline{L}}' = \frac{\det \overline{\overline{A}}}{2} \int_0^\infty ds\, \overline{\overline{\epsilon}}_r \cdot \frac{\left(\overline{\overline{A}}^2 + s\overline{\overline{\epsilon}}_r\right)^{-1}}{\sqrt{\det\left(\overline{\overline{A}}^2 + s\overline{\overline{\epsilon}}_r\right)}} \qquad (B.14)$$

Hyperpolarisabilities of a nonlinear sphere

$$\alpha = 3\epsilon_e \frac{\epsilon_i - \epsilon_e}{\epsilon_i + 2\epsilon_e} V \qquad (B.15)$$

$$\beta = \left(\frac{3\epsilon_e}{\epsilon_i + 2\epsilon_e}\right)^3 \epsilon_0 \chi^{[2]} V \qquad (B.16)$$

$$\gamma = \left(\frac{3\epsilon_e}{\epsilon_i + 2\epsilon_e}\right)^4 \epsilon_0 \chi^{[3]} V \qquad (B.17)$$

Other mixing rules

Bruggeman (Polder – van Santen)

Spheres:

$$(1-f)\frac{\epsilon_e - \epsilon_{\text{eff}}}{\epsilon_e + 2\epsilon_{\text{eff}}} + f\frac{\epsilon_i - \epsilon_{\text{eff}}}{\epsilon_i + 2\epsilon_{\text{eff}}} = 0 \qquad (B.18)$$

Randomly oriented ellipsoids:

$$\epsilon_{\text{eff}} = \epsilon_e + \frac{f}{3}(\epsilon_i - \epsilon_e) \sum_{j=x,y,z} \frac{\epsilon_{\text{eff}}}{\epsilon_{\text{eff}} + N_j(\epsilon_i - \epsilon_{\text{eff}})} \qquad (B.19)$$

Coherent potential

Spheres:

$$\epsilon_{\text{eff}} = \epsilon_e + f(\epsilon_i - \epsilon_e) \frac{3\epsilon_{\text{eff}}}{3\epsilon_{\text{eff}} + (1 - f)(\epsilon_i - \epsilon_e)} \tag{B.20}$$

Randomly oriented ellipsoids:

$$\epsilon_{\text{eff}} = \epsilon_e + \frac{f}{3}(\epsilon_i - \epsilon_e) \sum_{j=x,y,z} \frac{(1 + N_j)\epsilon_{\text{eff}} - N_j\epsilon_e}{\epsilon_{\text{eff}} + N_j(\epsilon_i - \epsilon_e)} \tag{B.21}$$

Unified mixing formula (MG ($\nu = 0$); Bruggeman ($\nu = 2$); CP ($\nu = 3$))

$$\frac{\epsilon_{\text{eff}} - \epsilon_e}{\epsilon_{\text{eff}} + 2\epsilon_e + \nu(\epsilon_{\text{eff}} - \epsilon_e)} = f \frac{\epsilon_i - \epsilon_e}{\epsilon_i + 2\epsilon_e + \nu(\epsilon_{\text{eff}} - \epsilon_e)} \tag{B.22}$$

and its form as a second-degree polynomial equation for $\chi_{\text{eff}} = \epsilon_{\text{eff}} - \epsilon_e$

$$\nu \chi_{\text{eff}}^2 + [\epsilon_i + 2\epsilon_e - f(1 + \nu)(\epsilon_i - \epsilon_e)] \chi_{\text{eff}} - 3f(\epsilon_i - \epsilon_e)\epsilon_e = 0 \tag{B.23}$$

Bruggeman nonsymmetric

$$\frac{\epsilon_i - \epsilon_{\text{eff}}}{\epsilon_i - \epsilon_e} = (1 - f) \left(\frac{\epsilon_{\text{eff}}}{\epsilon_e} \right)^{1/3} \tag{B.24}$$

Sen–Scala–Cohen

$$\frac{\epsilon_{\text{eff}} - \epsilon_e}{\epsilon_i - \epsilon_e} = f \left(\frac{\epsilon_{\text{eff}}}{\epsilon_i} \right)^{1/3} \tag{B.25}$$

Power-law (exponential) models

$$\epsilon_{\text{eff}}^\beta = f\epsilon_i^\beta + (1 - f)\epsilon_e^\beta \tag{B.26}$$

High-frequency correction (dilute mixtures)

$$\epsilon_{\text{eff}} = \epsilon_e + 3f\epsilon_e \frac{\epsilon_i - \epsilon_e}{\epsilon_i + 2\epsilon_e} \left[1 - j\frac{2}{3}(k_e a)^3 \frac{\epsilon_i - \epsilon_e}{\epsilon_i + 2\epsilon_e} \right] \tag{B.27}$$

Index

www.ingramcontent.com/pod-product-compliance
Lightning Source LLC
Chambersburg PA
CBHW050523190326
41458CB00005B/1645